Processor and System-on-Chip Simulation

Rainer Leupers · Olivier Temam
Editors

Processor and System-on-Chip Simulation

Editors
Rainer Leupers
RWTH Aachen
Templergraben 55
52056 Aachen
Germany
leupers@iss.rwthaachen.de

Olivier Temam
INRIA Saclay
Batiment N
Parc Club Universite
rue Jean Rostand
91893 Orsay Cedex
France
olivier.temam@inria.fr

ISBN 978-1-4899-9607-7 ISBN 978-1-4419-6175-4 (eBook)
DOI 10.1007/978-1-4419-6175-4
Springer New York Dordrecht Heidelberg London

Printed on acid-free paper.

Springer is part of Springer Science+Business Media (www.springer.com)

To Bettina (Rainer) and To Nathalie, Lisa, and Nina (Olivier).

Contents

Contributors

Daniel Aarno Intel, Stockholm, Sweden, daniel.aarno@intel.com

Michael Adler Intel, Hudson, MA, USA

Jung-Ho Ahn Seoul National University, Seoul, Korea, gajh@snu.ac.kr

Ghiath Al-kadi NXP Semiconductors, Eindhoven, The Netherlands

Matthias Alles Microelectronic Systems Design Research Group, University of Kaiserslautern, Kaiserslautern, Germany, matthias.alles@eit.uni-kl.de

Eduardo Argollo HP Labs, Barcelona, Spain, eduardo.argollo@hp.com

David August Princeton University, Princeton, NJ, USA, august@cs.princeton.edu

Nathan Binkert HP Labs, Palo Alto, CA, USA, binkert@hp.com

Gabriel Black VMware, Palo Alto, CA, USA, gblack@vmware.com

David Black-Schaffer University of Uppsala, Uppsala, Sweden

Torben Brack University of Kaiserslautern, Kaiserslautern, Germany, torben.brack@eit.uni-kl.de

Brad Calder Microsoft, Redmond, WA, USA, bcalder@microsoft.com

Nathan Chong ARM Ltd., Cambridge, UK, nathan.chong@arm.com

Veerle Desmet Ghent University, Gent, Brussels, Belgium, veerle.desmet@elis.ugent.be

Marc Duranton NXP Semiconductors, Eindhoven, The Netherlands, marc.duranton@gmail.com

Lieven Eeckhout Ghent University, Ghent, Belgium, lieven.eeckhout@elis.ugent.be

David Eklöv University of Uppsala, Uppsala, Sweden

Joel Emer Intel/Massachusetts, Hudson, MA, USA, joel.emer@intel.com

Jakob Engblom Wind River, Kista, Sweden; Intel, Stockholm, Sweden, jakob.engblom@windriver.com

Ayose Falcón Intel Labs, Barcelona, Spain, ayose. falcon@intel.com

Babak Falsafi EPFL IC ISIM PARSA, Lausanne, Switzerland, babak.falsafi@epfl.ch

Paolo Faraboschi HP Labs, Barcelona, Spain, paolo.faraboschi@hp.com

Krisztian Flautner ARM Ltd., Cambridge, UK, krisztian.flautner@arm.com

Björn Franke The University of Edinburgh, Edinburgh, Scotland, UK, bfranke@inf.ed.ac.uk

Lei Gao RWTH Aachen University, Aachen, Germany, gao@iss.rwth-aachen.de

Davy Genbrugge Ghent University, Ghent, Belgium, davy.genbrugge@ugent.be

Sylvain Girbal University of Paris, Paris, France, sylvain.girbal@thalesgroup.com

Daniel Gracia Pérez CEA LIST, Saclay, France, daniel.gracia-perez@cea.fr

Surendra Guntur NXP Semiconductors, Eindhoven, The Netherlands

Erik Hagersten Acumem AB, Uppsala, Sweden, eh@it.uu.se

Greg Hamerly Baylor University, Waco, TX, USA

David Heine Tensilica Inc., Santa Clara, CA, USA, dlheine@tensilica.com

Jan Hoogerbrugge NXP Semiconductors, Eindhoven, The Netherlands

Wei Huang IBM Research, Austin, TX, USA, huangwe@us.ibm.com

Tsuyoshi Isshiki Tokyo Institute of Technology, Tokyo, Japan, isshiki@vlsi.ss.titech.ac.jp

Daniel Jones The University of Edinburgh, Edinburgh, Scotland, UK, daniel.jones@inf.ed.ac.uk

Norman P. Jouppi Hewlett-Packard Co., Palo Alto, CA, USA, norm.jouppi@hp.com

Stefan Kraemer RWTH Aachen University, Aachen, Germany, kraemer@iss.rwth-aachen.de

Timo Lehnigk-Emden Microelectronic Systems Design Research Group, University of Kaiserslautern, Kaiserslautern, Germany, timo.lehnigk-emden@eit.uni-kl.de

Rainer Leupers RWTH Aachen University, Aachen, Germany, leupers@iss.rwth-aachen.de

Grant Martin Tensilica Inc., Santa Clara, CA, USA, gmartin@tensilica.com

Naveen Muralimanohar HP Labs, Palo Alto, CA, USA, naveen.muralimanohar@hp.com

Nenad Nedeljkovic Tensilica Inc., Santa Clara, CA, USA, nenad@tensilica.com

Daniel Ortega Intel Labs, Barcelona, Spain, daniel.ortega@intel.com

Emre Özer ARM Ltd., Cambridge, UK, emre.ozer@arm.com

Angshuman Parashar Intel, Hudson, MA, USA

Michael Pellauer Massachusetts Institute of Technology, Cambridge, MA, USA

Erez Perelman Intel, Cupertino, CA, USA

Daniel Powell The University of Edinburgh, Edinburgh, Scotland, UK, d.c.powell@sms.ed.ac.uk

Steve Reinhardt Advanced Micro Devices, Inc., Vancouver, WA, USA, stever@gmail.com, steve.reinhardt@amd.com

Ali Saidi ARM Ltd., Austin, TX, USA, ali.saidi@arm.com

Christoph Schumacher RWTH Aachen University, Aachen, Germany, schumacher@iss.rwth-aachen.de

Timothy Sherwood University of California, Santa Barbara, CA, USA

Kevin Skadron University of Virginia, Charlottesville, VA, USA, kadron@cs.virginia.edu

Kevin Smart Synopsys (Northern Europe) Ltd, Livingston, Scotland, UK, ksmart@synopsys.com

Mircea Stan University of Virginia, Charlottesville, VA, USA, mircea@virginia.edu

Olivier Temam INRIA Saclay, Orsay Cedex, France, olivier.temam@inria.fr

Andrei Terechko Vector Fabrics B.V., Eindhoven, The Netherlands

Nigel Topham The University of Edinburgh, Edinburgh, Scotland, UK, npt@staffmail.ed.ac.uk

Norbert Wehn Microelectronic Systems Design Research Group, University of Kaiserslautern, Kaiserslautern, Germany, norbert.wehn@eit.uni-kl.de

Bengt Werner Intel, Stockholm, Sweden, bengt.werner@intel.com

Chapter 1
Introduction

1.1 Recent Evolutions in System Modeling and Impact on Software Simulation

Architecture simulation by virtual prototypes is a fundamental design and validation tool of system architects. It consists in replicating an architecture in software in order to validate its functional behavior, to evaluate its performance, to explore architecture design options, or to enable early software development and non-intrusive debugging. In this book, we highlight some of the most recent developments in architecture software simulation. This book is motivated by multiple and drastic evolutions in architecture, technology, application domains, and manufacturing, all happening at the same time, and which greatly challenge the development of architecture simulation or virtual platforms.

The first fundamental evolution lies in the application domains. For several decades, the architecture domain was rather neatly partitioned into high-performance computing architectures (from workstations to supercomputers) and embedded architectures (mobile devices, custom devices, etc.). However, embedded architectures have been pulled toward greater functionality and performance (e.g., smartphones, 3G and 4G wireless communication, set-top boxes), while, at the same time, stringent low-power requirements and the appeal of a new broad market has attracted high-performance architecture manufacturers toward embedded techniques and domains. As these two domains converge, they bring together architecture design practices, and especially software simulation practices, which are very different. In the high-performance domain, simulation has been traditionally focused on complex architecture exploration, which privileges slow cycle-accurate modeling, while in the embedded domain, simulation has been focused on hardware/software co-design of system-on-chip, which privileges fast and approximate transaction-level modeling. As a result of the merger of the high-performance and embedded domains, each domain must now borrow from the other in order to tackle changing architectures.

Architecture is where lays the second evolution. For several decades, high-performance architectures were mostly fast and complex single-core architectures, while embedded architectures were mostly simple, slow and low-power, single-core,

R. Leupers, O. Temam (eds.), *Processor and System-on-Chip Simulation*,
DOI 10.1007/978-1-4419-6175-4_1,

multi-core, or system-on-chip (cores and accelerators) architectures. Multi-cores are now mainstream in high-performance architectures (from a few cores in Intel processors to several hundreds in Nvidia GPGPUs), and embedded cores are now becoming increasingly complex (such as the superscalar 1 GHz ARM Cortex A9, with support for cache coherence). As a result, software simulation must now efficiently tackle several tens, possibly several hundreds of IP blocks, corresponding to either homogeneous or heterogeneous cores, and accelerators, a challenge, particularly in terms of simulation speed.

Beyond more complex architectures, software simulation is now used to model entire devices or platforms. The progressive transition from hardware prototyping to software prototyping is the third evolution, which takes place in system design. For a broad range of products involving electronic systems, engineers are shifting away from cumbersome hardware prototyping, which requires to physically assemble a prototype of the final product, to software prototyping where the product is almost entirely simulated in software. Virtual prototyping speeds up the design process and facilitates the exploration of alternative components, including cases where they are not yet available. It is used for domains as diverse as telecommunications infrastructure, onboard car computers, and data centers. The need to tackle full devices and platforms brings new software design and productivity challenges for simulation.

Increasingly large and diverse target systems, coupled with the complexity of underlying multi-core architectures, call for more high-level software simulation approaches, as mentioned before. However, at the same time, technology-related metrics and constraints (power, temperature, area, clock domains, etc.) now have a dominant impact on architecture design and require precise circuit-level modeling techniques. This fourth evolution, and the tension between low-level and high-level modeling needs, is forcing to develop fast modeling techniques capable of capturing key technological trends.

Finally, a few words on terminology. The domain of software simulation is multifaceted. The term software simulation often covers many different notions, and some clarifications are necessary before the reader goes through the different chapters.

1.1.1 Abstraction Level

Software simulation denotes the high-level transactional-level modeling used in virtual prototyping, as well as cycle-level modeling used in micro-architecture exploration, or even the Verilog gate-level description of architectures for FPGA prototyping. In *gate-level* modeling, the individual logic gates and state elements (latches, SRAMs) are described; in cycle-level modeling, all operations are described at the granularity of a clock cycle and often of individual bits; in transaction-level modeling, it is assumed that what happens within a core or accelerator has little influence on the whole system behavior except for the memory transactions (hence the term) or the communications among cores and accelerators.

1.1.2 Simulator, Functional Simulator, Detailed Simulator

Several terms are often indifferently used to denote the same notion. A *functional simulator* only simulates the program instructions, updates processor registers, and performs ALU operations, but it does not provide any performance information. A *performance, detailed or cycle-level, simulator* often complements a functional simulator, and it provides performance information (usually time expressed in number of clock cycles, but also power consumption, and other metrics).

1.1.3 Full-System Simulator

A full-system simulator often covers two notions simultaneously: (1) the modeling of the full *hardware system*, not just the processor but also peripherals, etc., (2) the modeling of the full *software system*, not only a user application but also the full operating system.

1.2 Book Overview

The book is arranged in 19 chapters distributed into four parts, each part corresponding to one of the aforementioned evolutions.

Part I details several recent approaches to system modeling, how they can be used for system exploration, and their impact on simulator construction; these approaches range from single-core exploration to multi-core simulation, virtual device prototyping, and data center modeling. In Chapter 2, researchers from Synopsys present several years of experience in the development of a virtual prototyping platform geared toward embedded devices. In Chapter 3, researchers from Virtutech present a virtual prototyping approach which ranges from parallel systems to embedded devices. In Chapter 4, researchers from HP Labs show how to model full-scale data centers efficiently and reliably. In Chapter 5, we show how to use modular simulators to explore multi-core and system architectures, and the corresponding software design issues are studied in detail. In Chapter 6, we show how modular simulation can even be used to tackle the multiplicity of single-core architecture design options.

Part II covers key existing approaches for fast simulation. Many and fairly different concepts have been proposed to tackle the simulation speed challenge. As of now, it is not yet clear which approach(es) will prevail, and it may well happen that the simulator designer will end up combining several such methods together. In Chapter 7, we present a hardware-assisted simulation method (based on FPGAs) for fast and accurate simulation. Parallel simulation, where the simulator is parallelized and leverages existing multi-core hosts, is presented in Chapter 8. In Chapter 9, a fast simulation method based on dynamic binary translation is presented; it is shown to serve both early software development and architecture exploration. In Chapters 10 and 11, two approaches at sampling are presented. Sampling allows to use a detailed (and slow) simulator for only a fraction of the application execution and still obtain precise performance modeling for the full application; the approaches differ in the way samples are chosen. To a lesser extent, sampling techniques with specific goals and characteristics are also discussed in Chapters 4 and 8. In Chapters 12 and 13, two techniques based on statistical simulation are presented. The technique presented in Chapter 12 by Acumem is focused on modeling cache memories and relies on sparse but low-overhead run-time information collection. The technique presented in Chapter 13 shows that it is possible to precisely characterize the performance behavior of an application using traces of only a fraction of the size of the full application trace.

Part III of this book is devoted to more hardware-related simulation issues arising from current and future chip integration technologies. Chapter 14 focuses on the power, area, and latency modeling of memory subsystems. Chapter 15 emphasizes the need for temperature-aware chip design and discusses the corresponding methods for accurate thermal modeling. Last but not least, Chapter 16 shows how a technology-aware design space exploration methodology that directly takes into account key technology metrics like component power, area, and performance can be applied to quickly arrive at optimized MPSoC design points for given applications.

Finally, Part IV emphasizes practical simulation issues and approaches largely rooted in various embedded system application domains. Chapter 17 provides insights into modeling, simulation, and validation technologies at one of the most successful embedded RISC core vendors. The perspectives from another popular processor IP provider are discussed in Chapter 18. It highlights challenges encountered in customizable processor design and simulation as well as system integration by means of contemporary electronic system level design technologies. Chapter 19 focuses on wireless baseband processing and presents the key simulation problems and solution approaches in this particular domain. It exemplifies how a mix of hardware- and software-based validation can help to combat the huge simulation times required for DSP algorithm optimization and validation. Chapter 20 finally presents another simulation technology to efficiently guide MPSoC design space exploration. The proposed trace-driven approach provides high simulation speed and reasonable accuracy for early software performance estimation at the same time.

Part I
System Simulation and Exploration

Chapter 2
The Life Cycle of a Virtual Platform

Kevin Smart

Abstract A virtual platform is a fully functional software model of a complete system, typically used for software development in the absence of hardware, or prior to hardware being available. In this chapter we describe how virtual platforms are created and deployed, providing a case study on how virtual platforms are used for USB software development and verification at Synopsys.

2.1 Creation

What is a virtual platform? A virtual platform is a fully functional software model of a complete system, typically used for software development in the absence of hardware, or prior to hardware being available. It is complete in that it models not only System-on-Chip (SoC) devices but also board components, user interfaces, and real-world I/O too. To be suitable for productive software development it needs to be fast, booting operating systems in seconds, and accurate enough such that code developed using standard tools on the virtual platform will run unmodified on real hardware. As it is a software program it can provide unsurpassed visibility into internal events, provides a deterministic solution for multicore debugging, and is very easy to deploy (Fig. 2.1).

I have personally contributed to and managed the development of over sixty commercial virtual platforms throughout the past few years at Virtio and Synopsys. In this chapter I would like to provide you with some insight into how they are developed and what they are used for. *Throughout this chapter, I will add further detail, italicized, with respect to the USB case study that is summarized in the final section.*

K. Smart (✉)
Synopsys (Northern Europe) Ltd, Alba Centre, Alba Campus, Livingston EH54 7EG, Scotland
e-mail: ksmart@synopsys.com

R. Leupers, O. Temam (eds.), *Processor and System-on-Chip Simulation*,
DOI 10.1007/978-1-4419-6175-4_2,

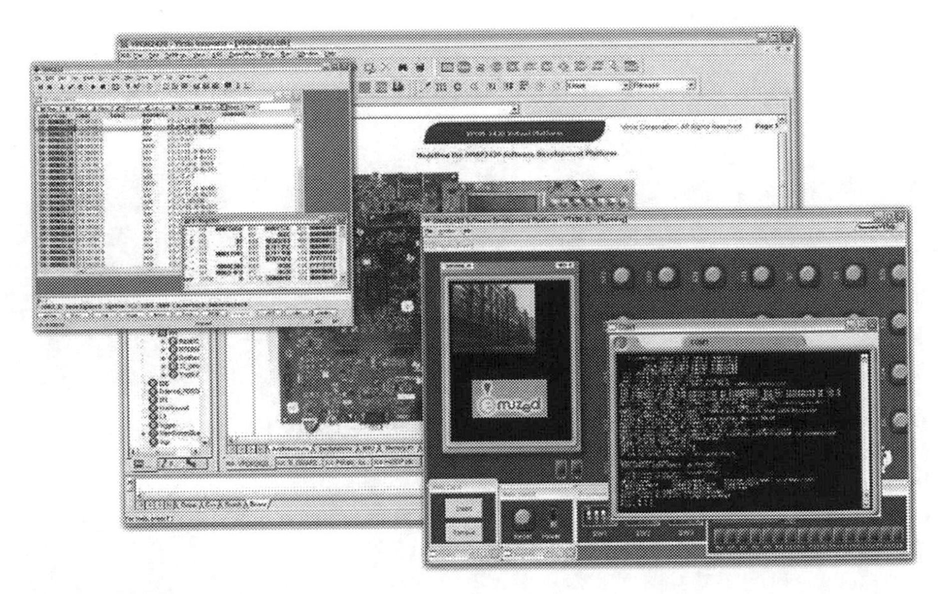

Fig. 2.1 A virtual platform

2.1.1 TLM-2.0. Abstraction and Design Choices

First some further background. A virtual platform is typically constructed of several transaction-level models (TLMs). TLMs are an abstraction, a higher level representation of hardware behavior, focusing on discrete events such as register reads/writes and information passing such as clock frequency notification, rather than bus signals and clock transitions (Fig. 2.2). Through this abstraction, focusing on only the information software can access or cares about, it is possible to meet the performance and software binary compatibility goals, with the additional benefit of reducing model development time. And reduced model development time is important to provide

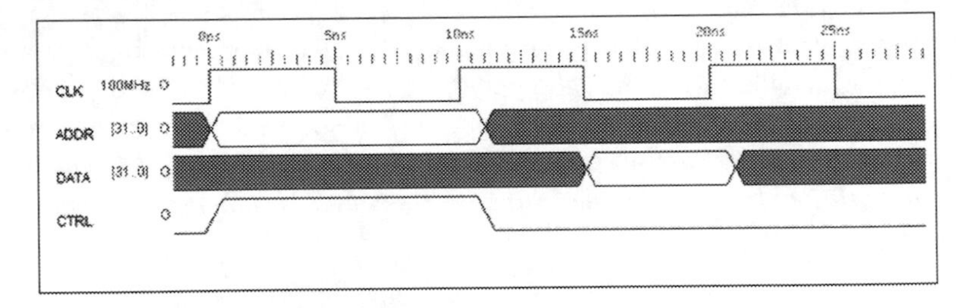


```
clk.setFrequency(100 000 000);
bus.read(addr, &data);
```

Fig. 2.2 Hardware abstraction

software developers with earlier access to the virtual platform and ultimately product schedule improvement.

In June 2008 at DAC the SystemC^{TM} transaction-level modeling standard, TLM-2.0 was ratified [1]. This has enabled easier interoperability of transaction-level models from different vendors by standardizing the interfaces of memory-mapped busses, moving virtual platforms from the early adopters' proprietary solutions to the broad mainstream. Synopsys played a leading role in the development of this standard, donating much of the technology.

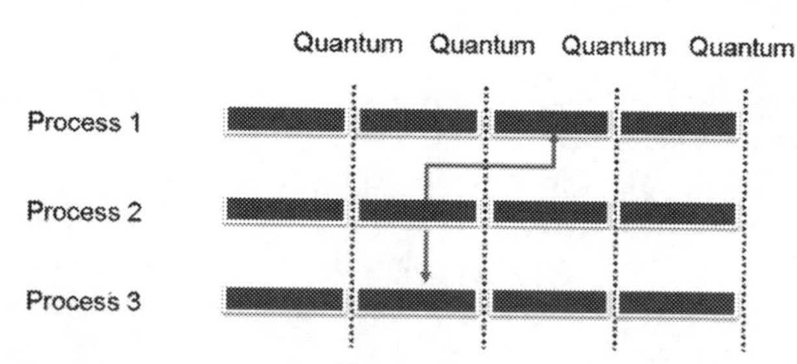

Fig. 2.3 Loosely timed synchronization

TLM-2.0 defines two coding styles: loosely timed or LT and approximately timed or AT. LT is suited to virtual platforms intended for software development and employs techniques such as direct memory interfacing (fast memory access via pointer dereferencing) and temporal decoupling (scheduling each core to run a batch of operations in a quantum period before yielding control, Fig. 2.3) to achieve performance typically in the tens of millions of instructions per second (MIPS). This is suitable for booting complex software such as embedded operating systems in seconds while achieving the goal of software binary compatibility, i.e., software developed on the virtual platform should run unmodified on target hardware. Often instruction set simulators that cache blocks of translated instructions, rather than repeatedly interpreting, are employed to maximize performance.

However, where additional timing accuracy is required, typically for software performance estimation and architectural analysis use cases, the AT style is employed. This supports non-blocking overlapping transactions to more accurately model hardware behavior such as bus contention and prioritization of multiple masters, but at a price, reducing performance by perhaps an order of magnitude, due to increased synchronization within the SystemC kernel, and subsequent increase of model development and validation times, due to greater complexity (Fig. 2.4).

A single model can support both LT and AT modes. There are techniques such as booting software in LT mode and then switching to AT mode at a particular breakpoint, or simulation save and restore: the resumption of execution from a saved state that can alleviate some of the performance issues for software development and validation.

But there is always a performance/accuracy/development time trade off. Virtual platforms typically do not contain many cycle-accurate models of complex components because of the performance impact. And these models can take so long

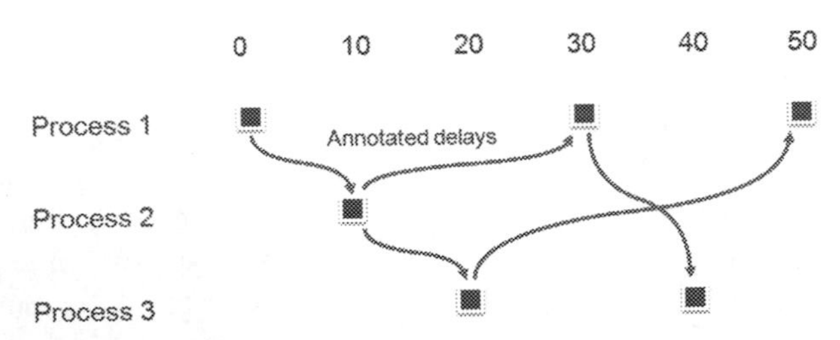

Fig. 2.4 Approximately timed synchronization

to develop and validate that the benefits of early virtual platform availability for pre-silicon software development are lost.

However, if RTL is available there are techniques such as automated RTL to SystemC model conversion, RTL co-simulation, or FPGA co-emulation that can be employed to eliminate the overhead of manually creating a cycle-accurate model, and in the FPGA instance, can result in hardware acceleration. We will look at some applications of these later.

So in terms of abstraction level the choice really boils down to the end users' requirements, concerning performance of the simulation, acceptable accuracy to address their use cases and the desired availability schedule.

In the case of our modeling of USB cores for software development, the loosely timed LT abstraction was selected.

2.1.2 Capability Including Real-World I/O

After the abstraction level of a transaction-level model is chosen, decisions need to be taken on the model's interface requirements. A master component or an initiator, like a processor core, will probably have a TLM-2.0 compatible interface, generating external memory transactions. Similarly a slave peripheral, or target, will have a TLM-2.0 interface to receive transactions.

It is recommended that a component has a clock frequency input so that it can receive its operating frequency, typically in Hertz. This permits the model to alter its behavior if the clock is disabled or the frequency is changed.

And sometimes if power estimation support is required a voltage input and the means to notify calculated power will be added.

At the time of writing there is no standardization of these and other simple sideband signals such as interrupt outputs and DMA requests or more complex connectivity such as I2C or USB. It is hoped that this will change soon, to promote easier interoperability of different vendors' models, reducing the number of interface adapters required.

With the interfaces defined, our attention can focus on how model functionality may be realized. One of the advantages of running on a PC is that its facilities can be leveraged to implement the support of physical interfaces. This "real-world I/O" capability extends to USB host control, serial ports, ethernet, audio input/output, camera, firewire, SIM cards, etc. Other capabilities not always physically supported by a PC such as USB device mode support, MMC/SD/SDIO, HDMI, and SATA are implemented virtually and referred to as "virtual I/O." For instance, an MMC/SD card model receives commands via a transaction interface from its controller, interprets these, and stores data persistently to a file. Tools on the PC can be used to transfer data to and from the card's file system and a button on a cockpit window pressed to virtually insert or remove the card.

The initial development of USB real-world and virtual I/O support was effort intensive, especially on Windows where an appropriate low-level API was not available. We had to acquire specialized Windows USB stack and device driver knowledge to implement a suitable interface, a somewhat rare commodity at the time. In total probably over a staff year was invested in this development. The subsequent implementation of this support on Linux was much more straightforward. Third-party model developers can now benefit from the investment that was made in these interfaces by using the model authoring kits that are available within the DesignWare™ System-Level Library [3].

Major benefits of a virtual platform over hardware include the ease of configuration and analysis capabilities. Most transaction-level models will include configuration parameters to, for instance, enable tracing, alter timing, or even support different hardware revisions. Often the platform configuration and control is scriptable facilitating use within an automated regression test environment.

Users expect virtual platforms to support the software debuggers they are familiar with, but a virtual platform can provide increased hardware insight to aid software debugging, for instance, by tracing interrupt or DMA requests, cache behavior, clock frequencies, pin multiplexing, and power states.

Unfortunately as with sideband signals, there is no current ratified standard for configuration, control, and analysis, but in February 2009, the OSCI configuration control and inspection (CCI) working group was established [2]. This group will define standards for SystemC that complement TLM-2.0 and allow models from different suppliers to be fully exploited in any compatible SystemC environment.

2.1.3 Development Process and Tools

Figure 2.5 describes the model development flow employed by Synopsys.

The top half of the diagram indicates the component model development flow, and the lower portion, platform development and model integration.

The starting point is the specification for the component, "Component Specification" at the top of the diagram. For a loosely timed TLM, a programmer's specification describing signals, memory-map, registers, and their functionality is usually sufficient to create a model. For an approximately timed TLM, additional architectural and timing information may be required to implement greater timing accuracy, but the starting point is still the same, entering register details. For this we use Component Creator, which is part of the Innovator tools suite [3].

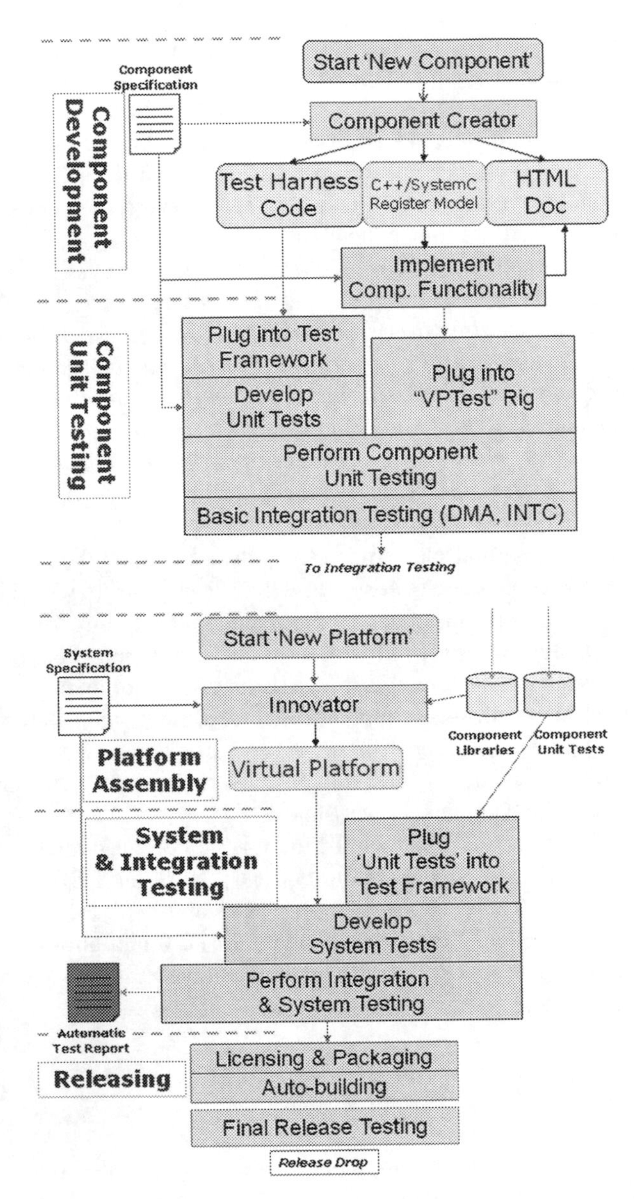

Fig. 2.5 Model development flow

If SPIRIT IP-XACT XML [4] definitions are available we can use these to automate the creation of an initial C++ or SystemC register "stub" model, HTML component documentation, and embedded software-driven verification tests ("Test Harness Code"). A stub model implements register decoding, default register values, and read/write permissions of register bit-fields, but does not implement the functionality behind the registers.

For the USB virtual platform we stubbed out the components not required by the Linux kernel and the applications we were intending to use.

If IP-XACT or other standard register definitions are not available, the model developer must enter them using Component Creator. As this is a manual process, the creation of test harness code should be carried out independently; otherwise any errors in the register implementation would be repeated in verification tests and perhaps not caught until much later.

Automated implementation of model functionality ("Implement Comp. Functionality") is currently the elusive Holy Grail or Nirvana of electronic system-level design. Although there may be some reuse of existing libraries, the task is currently one of traditional manually intensive software development. We have found this to be best suited to software engineers who have some prior electronics or embedded systems knowledge as they typically understand the requirements of the end user and of modeling efficiently at higher levels of abstraction, whereas hardware engineers may initially approach the problem at too low a level, until their mindsets become reconditioned! For a loosely timed model, the task involves reviewing the component's specification, identifying the functionality that will be visible to software, or impact other components, and implementing that functionality in C++ or SystemC.

In many cases the loosely timed model may be generated first and, if required, additional timing detail added to yield an approximately timed model. In some cases, for instance, with a bus interconnect, the loosely timed model may be very abstract, containing no registers, and developed quickly, whereas a more timing accurate model could implement arbitration and prioritization, reflecting contention and other latencies, taking much longer to develop and validate.

In both cases an iterative approach is adopted; the model is integrated into a unit test platform (VPTEST) and embedded software tests written ("Develop Unit Tests") to exercise model functionality as it is added. An approach which we also employ is test-driven development whereby the tests are written first, based on the specifications, and then model functionality implemented, ensuring those tests pass.

With USB controller modeling, test-driven development has proven to be very effective at reducing the number of issues found during integration testing. Here the CppUnit [5] test framework is used to execute developed tests as model functionality is implemented and later as part of regression testing, identifying issues immediately.

Since there may not be full visibility of model functionality, such as signal states, from the software interface, typically BFMs (bus functional models) are employed to perform automated stimulus generation and response capture.

With a model implemented and unit tested, integration testing begins ("Basic Integration Testing"). This testing concentrates on the model's interaction with other components, for instance, interrupt and DMA controllers and displays.

In parallel with individual model development, the main virtual platform is established, with reference to the modeled system's top-level functional specification ("System Specification"), using Synopsys' Innovator ("start 'new platform'"). The platform is constructed from the integration of existing library components from, for instance, the DesignWare System-Level Library and newly developed ones

("Component Libraries"). In both cases, component's unit tests are integrated into the platform's embedded software tests, which specify the platform-specific memory, IRQ, and DMA request mapping ("Plug 'Unit Tests' into Test Framework"). Once integrated, additional platform-specific system tests may be developed, to ensure a group of components within a subsystem, or those that implement a specific software use case, interoperate correctly. An example would be software configuring a camera sensor via I2C, programming a camera controller to DMA frames to memory, and instructing the display controller to display them on an LCD.

With the base platform established it is then added to an automated build machine that builds the latest code and runs the embedded software tests as part of its regression testing.

An incremental approach is applied to the implementation of the virtual platform model content. To provide most benefit to customers the release content and delivery is aligned with customer software development schedules. I will describe this in more detail in the next section.

It is essential to employ good project management and source code configuration practices to ensure that releases are made in a timely fashion and any changes or reprioritization, which will happen especially with pre-silicon development, is tracked and accommodated. Collaborative project management tools such as Microsoft Project Server have proven to be very effective at keeping development staff informed of project changes.

2.2 Deployment

With a virtual platform developed we can now look at its usage.

2.2.1 Pre-RTL and Pre-silicon Software Development

The majority of virtual platforms developed by Synopsys have been used for pre-RTL or pre-silicon software development spanning everything from low-level ROM code through operating systems base porting to end-user applications.

Typically the initial release of the virtual platform is aligned with ROM code requirements. Since this code eventually resides within the SoC it is important to start its development early and ensure that it is correct first time, to avoid costly silicon re-spins. To this end, and to reduce risk, RTL co-simulation and FPGA prototyping may be used for additional verification. With this approach it is feasible to achieve same-day integration of software developed pre-silicon, on the first silicon samples. Here the process is to develop code on the virtual platform, prior to RTL, and once RTL becomes available verify the developed code on an RTL simulator, such as Synopsys' VCS [6], and for better performance FPGA boards. If discrepancies are found between RTL and virtual platform an analysis is performed to determine the root cause. This is useful because it may indicate that the reference specifications are inaccurate or ambiguous; there are issues with the RTL; or the virtual platform is incorrect. Obviously it is highly desirable and more cost-effective to discover errors at source, as early as possible.

Subsequent virtual platform releases add functionality required for OS base porting and multimedia driver development with the model deployed to a larger number of software developers, where features such as virtual and real-world I/O are fully utilized. For instance, Linux software developers can share a common file system via NFS over ethernet, develop USB drivers that communicate with physical devices, test HD video output via HDMI, etc. One other advantage that the virtual platforms have is the ability to provide feedback to software developers on possible programming violations, flagging up warnings if incorrect programming sequences are detected. Another advantage is that they can provide full-system simulation, modeling, for instance, pin multiplexing (the routing of SoC internal signals to external pins), and clock gating; features which tend not to be supported on FPGA prototypes.

Once initial software has been developed on the virtual platform, the combination can be used as an early marketing tool to potential customers, even providing a mockup of an intended device's look and feel via the virtual platform's cockpit window. Furthermore the virtual platform plus base software, such as ported embedded operating systems, can be deployed to the customer's customers: the independent software vendors and such like, to allow them to also jump-start their software development, further accelerating software content availability and product differentiation.

Even when hardware becomes available, the virtual platform still has a role to play. Hardware prototypes may be in short supply, difficult to configure, or unreliable, whereas a software virtual prototype is more easily deployed, available in greater numbers, provides deterministic results, and offers analysis capabilities hardware simply cannot match.

2.2.2 Co-simulation

Co-simulation is where the virtual platform is integrated with an HDL simulator such as Synopsys' VCS, which supports cycle-accurate RTL simulation.

Co-simulation may be used in the following scenarios:

1. *RTL equivalence checking*: to verify correctness of a TLM against existing RTL, to improve user confidence in the accuracy of the TLM
2. *RTL testbench development*: use a TLM to jump-start the testbench infrastructure and test cases development, allowing the verification environment and tests to be developed and debugged much faster
3. *Software verification*: to verify software, which was developed on a TLM, against RTL when it becomes available, to ensure functional correctness

4. *RTL integration*: to reduce TLM development effort where an IP title is particularly complex or full information is not available

Let us examine all four of these use cases.

2.2.2.1 RTL Equivalence Checking

In some development flows the TLM may have been developed independently of the RTL and may not be considered the golden reference because of its higher level of abstraction.

The TLM may be inserted into the RTL testbench and then test cases/scenarios executed and compared against the results from RTL. As the TLM is at a higher level of abstraction, adaptors may be required to convert testbench pin-level signals into transactions compatible with the TLM.

Alternatively if a layered testbench architecture is adopted, such as advocated by the Synopsys Verification Methodology Manual (VMM) [7], where there are scenario, functional, and command layers, transactions can be exchanged at the functional layer. Because of the higher level of abstraction at this layer they are typically easier to map to TLM interfaces.

In a software-driven verification approach, self-checking software is executed on a device under test (DUT), and bus functional models (BFMs) are used to provide stimuli/response to DUT signals, often under the control of a script. Rather than swapping out the DUT for the TLM and writing adaptors to convert pin-level BFM signals to TLM transactions, it is in some cases simpler just to model the testbench infrastructure itself as compatible TLMs. Because the BFMs are themselves abstracted, they can be designed to use the same transaction-level interfaces as the transaction-level models they interface with, implementing script commands operating at the functional rather than signal level.

In all cases, with the TLM instantiated in the testbench environment, the existing tests can be run against the model, discrepancies identified and eliminated on an iterative basis, leading to a more accurate TLM that software developers can have confidence in.

2.2.2.2 RTL Testbench Development

This is very similar to the previous use case where a TLM is interfaced with the functional layer of a VMM layered testbench. The goal this time is to use the TLM to develop tests before the RTL implementation of the design is available. However, due to the TLM's abstraction it may not be possible to test all functionality, for instance, low-level signaling, so additional tests will probably have to be written for the RTL. But still, the TLM will allow the majority of the testbench infrastructure and test cases to be developed much earlier than traditionally possible.

2.2.2.3 Software Verification

In this scenario, software such as a device driver that has been developed for a particular peripheral component, for instance, a USB OTG IP, is tested against the RTL when it becomes available. The majority of the virtual platform remains the same, retaining a CPU subsystem which supports the fast booting and execution of an embedded OS, but the TLM that implements the target component is swapped out for RTL.

Transactors are used to map TLM-2.0 transactions to, for instance, AMBA AHB/APB/AXI bus signals, and also side-band signals such as reset inputs and interrupt request outputs, as required by the RTL. Figure 2.6 shows how RTL of an interrupt controller is interfaced to a SystemC virtual platform. Read and write transactions destined for the interrupt controller's address range are converted into APB pin-level signals by the transactor. In essence, if you refer back to Fig. 2.2, we are now converting from the functional level to and from the signal level. For

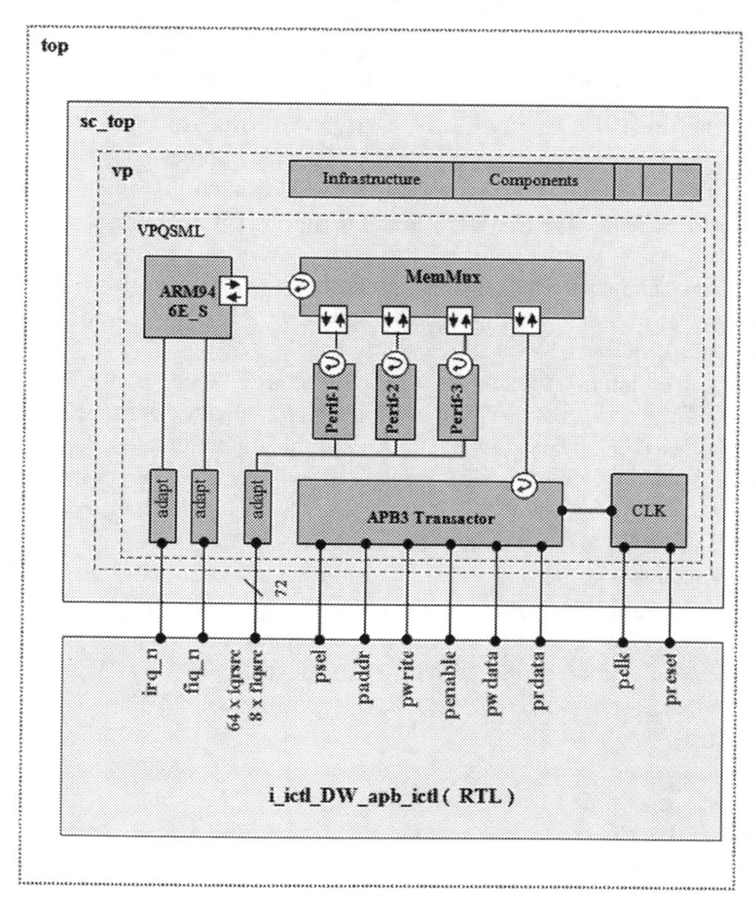

Fig. 2.6 Co-simulation for software verification

complex interfaces, such as AXI, transactors, due to their cycle-accurate nature, can be complex and time consuming to implement and verify, so reuse is extremely desirable.

It may be difficult to realize physical connectivity within the HDL simulator (and this is where the co-emulation approach, described later, has an advantage). Options include using another transactor to map interfaces at the signal level back to transactions, which are then handled by the virtual platform; using bus functional models or verification IP compatible with the interface; or another instance of the IP as RTL, also under software control, to exercise the connection.

Running the software against the RTL may identify issues with the software or RTL/TLM compatibility that require investigation. Testing RTL with software scenarios, even if already verified using an RTL testbench, can increase confidence in the system, due to increased coverage.

This is the approach used by the Synopsys USB software team to verify software created using the TLM, when RTL becomes available.

2.2.2.4 RTL Integration

There may be situations where users of a virtual platform are required to develop driver software for a complex new IP block, such as a graphics controller or video codec, where an existing TLM is not available, and it is not feasible to wait for a model to be developed, due to project time pressure. Or it may be that insufficient specification detail is available for a TLM to be created. If RTL is available, one option is to instantiate it within an HDL simulator. As with the software verification use case, above, transactors are required to translate virtual platform transactions into RTL signals and vice versa.

An alternative solution, which does not require an HDL simulator, could be to use a C/SystemC model that has been generated automatically from the RTL, using third-party tools. The advantage is that it can be integrated directly into the virtual platform and an HDL simulator is not required, but as the C model is essentially at the same abstraction level as RTL (pin-level cycle accurate), transactors are still necessary, and the model may require explicit clocking to operate.

With both approaches virtual platform performance will suffer if the RTL is regularly accessed by software, due to the latencies introduced fulfilling transactions. However, in other cases, where the RTL is accessed infrequently, it may be possible to introduce clock optimization mechanisms, to effectively idle the RTL by stopping its clock source during periods of inactivity, assuming that this does not introduce any unwanted side effects.

Benchmarks have shown SystemC models generated from RTL to be at least 20x slower than abstract hand-coded models. An example of this would be a video codec rendering a 1080p frame in 2 h, compared to 6 min with a TLM. In our USB co-simulation application, described in the case study, software running on the virtual platform with USB OTG TLM was measured to be 30x faster than with the RTL running on an HDL simulator.

The next section describes another option to improve performance, using co-emulation.

2.2.3 Co-emulation

This is where a virtual platform is connected to an emulator or rapid prototyping system, such as Synopsys' FPGA-based CHIPit [8], rather than an HDL simulator. The prototyping system can instantiate synthesizable RTL with the advantages of near-hardware performance and real-time I/O interfacing, albeit with potentially less debug visibility than with an HDL simulator. Typically a transaction-based interface such as the Standard Co-Emulation Modeling Interface (SCE-MI) [9] is employed to connect the two environments together, with normally a processor-based subsystem or SoC model residing on the virtual platform (software side) and a peripheral or component subsystem in FPGA (hardware side). Synthesizable transactors will be required on the hardware side to convert transactions passed across the SCE-MI interface to signals and vice versa.

Although version 2 of SCE-MI offers improved performance over version 1, due to its transaction-oriented, rather than event-oriented nature, the physical interface used, e.g., PCI may offer high bandwidth but with relatively high latencies. This means it is more efficient to use fewer larger size transactions than many smaller size transactions. Such constraints should be considered when partitioning the system for maximum performance. For instance, it may be necessary to introduce local caching on the virtual platform side to amalgamate and increase the average size of transactions passed across the interface. Also if the FPGA instantiated subsystem, such as a display controller, requires frequent access to memory, it may be more efficient to place that memory also on the hardware side, and implement some caching mechanism on the software side, if the software will also access it frequently.

Ironically the latencies introduced by co-emulation integration can assist with software performance optimization. By reducing the number of transactions between virtual platform and RTL we can achieve noticeable performance improvement. It can soon become apparent, by analyzing transaction traces, where redundant register accesses are being performed, an example of which might be the repeated reading of interrupt mask registers in the interrupt handler, when caching the last written value would suffice.

In the USB OTG example shown in Fig. 2.7, we were able to reduce modprobe and mount times over 5x, by reducing the number of SCE-MI transactions or maximizing their size.

One of the drawbacks of this hybrid environment is that, depending upon the instantiated RTL, it may not be possible to keep the time synchronized on both sides, or pause the hardware side at a software debugger breakpoint. This can lead to debug issues due to non-deterministic behavior. The FPGA prototyping environment may also be constrained in terms of how much trace it can capture to assist with debug. However, overall, the co-emulation approach with an FPGA prototyping

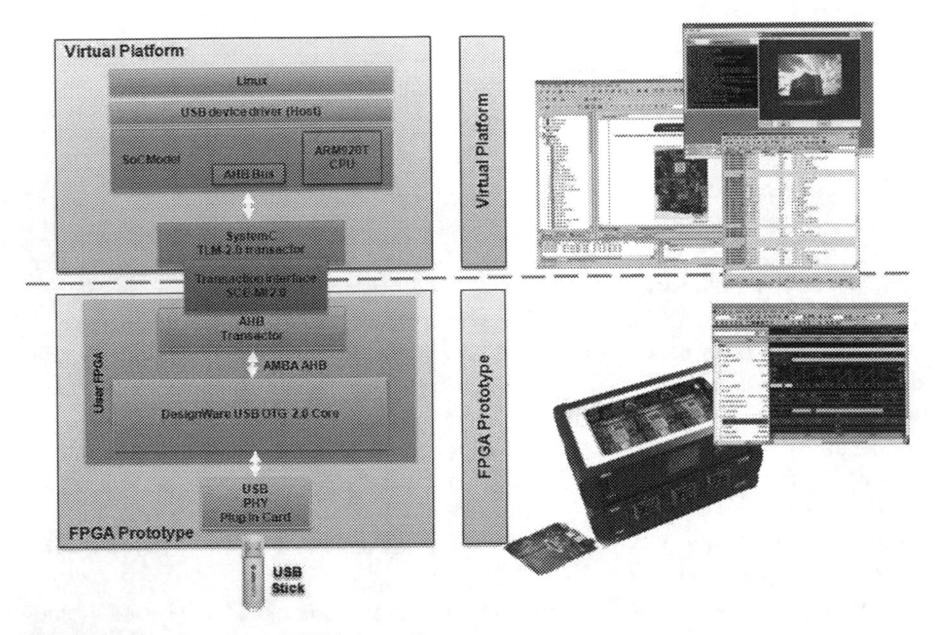

Fig. 2.7 Integration with an FPGA prototype

environment offers better performance for software development than co-simulation with an HDL simulator.

In benchmarks we observed a 12x performance increase with USB OTG RTL instantiated in FPGA rather than an HDL simulator. We would expect to see higher performance gains for more data intensive cores such as codecs.

2.2.4 Regression Testing

As a virtual platform is a software solution it lends itself to easier automation. For instance, scripting support can be used to change the software image loaded or even the version of SoC modeled. You can now imagine the scenario whereby several embedded operating systems are booted on a model of the latest SoC and regression tested on a still-in-circulation older version. Even changing the version of software loaded into flash on hardware could take several minutes, whereas the virtual platform can be updated from files almost instantaneously.

The embedded software may support back-door test interfaces, for instance, via UART serial connections, which are used in hardware testing. These can be supported virtually. Furthermore as the user interface of the virtual platform, including keypad, LCD, and touch screen, is realized as a GUI on the host PC, scripting or traditional software GUI testing tools could be used to automate the user interaction with the platform, extending test coverage beyond what is easily achievable with hardware. So regression testing using virtual platforms has a role to play even when hardware is available.

The same approach can be used in regression testing of RTL, with TLMs swapped out for RTL, as described in the software verification use case (Section 2.2.2.3).

2.3 Evolution

We have seen how virtual platform models are created and deployed. At some point a model may become redundant and no longer supported if the IP it is based on becomes obsolete.

However, the life cycle continues; models from a previous platform may be updated and used as the basis of a derivative or next-generation architecture. If a model does not continue to exist in its present form, some of its generic capability may be extracted to form reusable libraries, employed in new designs.

In the next section we will see the life cycle in action, with reference to USB model development.

2.4 Case Study

2.4.1 SuperSpeed USB 3.0 Linux Driver Development and Verification

First some background. Synopsys provides USB cores and other silicon-verified Intellectual Property as part of its DesignWare® IP offering [10]. Some of these titles are available as SystemC transaction-level models in the DesignWare System-Level Library, after originally being created to enable pre-RTL software development by internal software teams and their partners.

This approach was successfully initially deployed for USB 2.0 OTG software development. At that time the USB team was using ARM920T SoC-based FPGA development boards for Linux driver development. It made sense for us to model that target with sufficient SoC functionality such that Linux images, intended for the hardware board, would run unmodified on the virtual platform.

We created a loosely timed virtual platform based around our existing fast ARM920T instruction set simulator, to ensure that Linux would boot within seconds. Core functionality modeled included interrupt controller, timers, DMA, UARTs, clock and power management, I/O ports, LCD/touch screen, LAN controller and memories, including NAND flash. The functionality that was not required for our use case was stubbed, accelerating development; basic register models were created for I2C, SD, SPI, I2S, and similarly for the SoC USB host and device controllers, since we would be replacing the SoC's USB capability with our own DesignWare USBOTG model.

In all cases the available SoC user's manual and FPGA developer's board manual were sufficiently detailed to create the required models. The standard established development process, as described earlier (Fig. 2.5), was followed, creating initial models, embedded test code and component documentation using Component Creator, adding functionality and unit testing, integrating into the ARM920T base platform using Innovator and system testing.

Finally with the required functionality implemented, the boot loader, Linux kernel and file system were loaded into the platform via its configuration script, the platform debugged, and then handed over to the USB software team as a packaged installer.

The LAN controller model leveraged our real-world I/O capability allowing NFS network mounted file system support for convenience. And the USBOTG model, the target for driver software development, supported real-world and virtual I/O capability, allowing the virtual platform to conveniently control physical USB devices, such as a memory stick, or appear as a device plugged into the PC (Fig. 2.8).

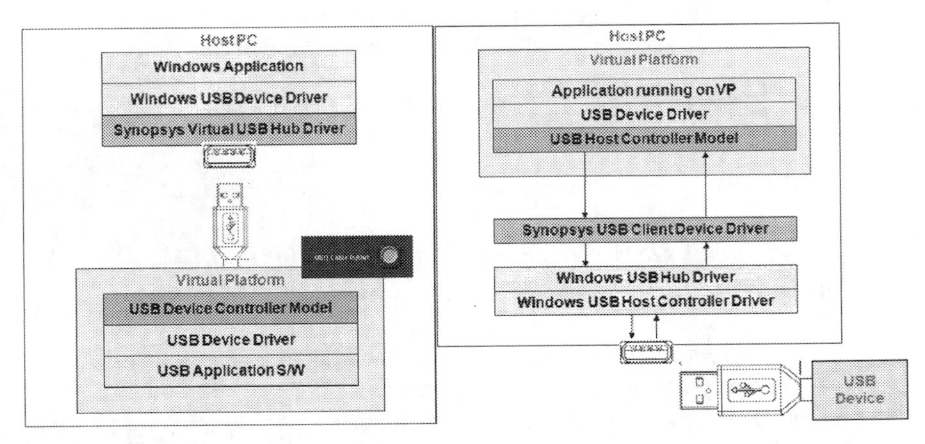

Fig. 2.8 USB virtual and real-world I/O

This capability has enabled the USB software team to develop and test their driver updates for several IP revisions prior to the RTL becoming available, accelerating software support for new USB 2.0 standards such as Link Power Management (LPM) and High-Speed Inter-Chip (HSIC).

In addition we created a co-simulation variant of the platform, on a Linux host, with VCS, to allow the developed drivers to be tested against the RTL, when it became available. This consisted of essentially two platforms, each running embedded Linux, one operating in host mode, the other in device mode, and interfacing to the USBOTG RTL via transactors. The RTL instances were connected back-to-back at their PHY interfaces for data exchange. Pre-RTL developed drivers and applications could then be tested against the RTL, increasing confidence.

The other option available to the USB team is to instantiate the RTL on a CHIPit rapid prototyping system, retaining the rest of the system as a virtual platform (see Fig. 2.7 above). The advantage over VCS co-simulation is that the USB PHY interface can be physically tested and the system offers higher performance for extended testing, while still retaining the debug and visibility benefits of the virtual platform.

A similar approach was deployed with USB 3.0 IP software development. We initially created a USB 3.0 Device model, also featuring virtual I/O support, and instantiated it in the ARM920T-based SoC platform. This was used by the software team for the creation of the initial driver, with the first version of the driver available within 4 weeks. The USB team, however, decided to move from standalone ARM based development boards to plug-in PCI express-based FPGA cards. This unfortunately meant that the developed drivers required modification to support both the AMBA and the PCI interfaces. This was further exasperated when validating the USB 3.0 host model. As this model was compliant with Intel's extensible host controller interface (xHCI) specification, the Intel-developed Linux driver would be used, rather than Synopsys developing one, and this driver required a later kernel version, not easily ported to the ARM920T-based platform. It was finally concluded that it would be more efficient to develop and use an x86-based virtual platform rather than maintaining multiple driver and kernel versions.

Since PCs supporting USB 3.0 were not generally available at the time of the initial USB 3.0 modeling, the virtual and real-world I/O support was only capable of supporting USB 3.0 devices in USB 2.0 mode. To circumvent this we created a back-to-back platform similar to the co-simulation arrangement, which connected the USB 3.0 host xHCI model to the USB 3.0 device model via a transaction interface. This enabled initial testing of USB 3.0 functionality beyond that offered by the I/O support.

In general the software teams using virtual platforms have managed to implement and test new functionality several weeks or months in advance of FPGA prototype availability. Virtual platforms are found to be more reliable than hardware boards, easier to use and deploy to internal and external software developers. Once hardware is available the software bring-up time is significantly reduced, due to the majority of issues having been addressed on the virtual platform, with perhaps only hardware-related timing issues left to resolve.

2.5 Conclusion

In this chapter we have defined what a virtual platform is and how it can be created and deployed, ending with a practical example of usage and benefits to USB software teams. We detailed how Synopsys develops virtual platforms and how they can be used for software development and verification, significantly reducing time to market.

It has been exciting, over the past decade, witnessing virtual platforms cross the chasm from early pioneers' adoption to mainstream deployment, where significant time and cost benefits are being realized. The future will probably bring increased automation and accelerated development at all levels of abstraction. One thing is certain; there still remain significant challenges ahead to realize the full potential of electronic system-level design!

Acknowledgments To my team, colleagues, and ex-colleagues at Virtio and Synopsys who made this possible.

References

1. OSCI TLM-2.0 Language Reference Manual, http://www.systemc.org/downloads/standards/. Accessed 13 Feb (2010).
2. Configuration, Control & Inspection Working Group (CCIWG), http://www.systemc.org/apps/group_public/workgroup.php?wg_abbrev=cciwg. Accessed 13 Feb (2010).
3. Synopsys Virtual Platforms, http://www.synopsys.com/Tools/SLD/VirtualPrototyping. Accessed 13 Feb (2010).
4. The SPIRIT Consortium IP-XACT Releases, http://www.spiritconsortium.org/tech/docs. Accessed 13 Feb (2010).
5. CppUnit C++ Unit Testing Framework, http://sourceforge.net/apps/mediawiki/cppunit/index.php?title=Main_Page. Accessed 13 Feb (2010).
6. Synopsys Verification, http://www.synopsys.com/Tools/Verification. Accessed 13 Feb (2010).
7. Verification Methodology Manual for SystemVerilog, http://vmm-sv.org/. Accessed 13 Feb (2010).
8. Synopsys FPGA Implementation, http://www.synopsys.com/Tools/Implementation/FPGAImplementation. Accessed 13 Feb (2010).
9. Standard Co-Emulation Modeling Interface (SCE-MI) Reference Manual, Version 2.0, http://www.eda.org/itc/scemi200.pdf. Accessed 13 Feb (2010).
10. Synopsys DesignWare Interface and Standards IP, http://www.synopsys.com/IP/InterfaceIP. Accessed 13 Feb (2010).

Chapter 3
Full-System Simulation from Embedded to High-Performance Systems

Jakob Engblom, Daniel Aarno, and Bengt Werner

Abstract This chapter describes use cases for and benefits of full-system simulation, based on more than a decade of commercial use of the Simics simulator. Simics has been used to simulate a wide range of systems, from simple single-processor embedded boards to multiprocessor servers and heterogeneous telecom clusters, leading to an emphasis on scalability and flexibility. The most important features and implementation techniques for a high-performance full-system simulator will be described and the techniques to achieve high simulation performance will be discussed in detail. As the ability to efficiently model systems is critical for a full-system simulator, tools and best practices for creating such models will be described. It will be shown how full-system simulation plays a significant role in the development of complex electronic systems, from system definition through development to deployment.

3.1 Introduction

Over the past decade, *full-system simulation* (FSS), also known as virtual platform technology, has proven to be a very useful tool for the development of computer-based systems of all kinds, including small deeply embedded systems, telecom infrastructure, servers, and high-performance computing solutions. FSS indicates that the simulation of the hardware is complete enough to run the real target software stack and fast enough to be useful for software developers. In a full-system simulator, there are models of processors, memories, peripheral devices, buses, networks, and other interconnects; and the software cannot tell the difference

J. Engblom (✉)
Wind River, Finlandsgatan 52, SE-164 93 Kista, Sweden; Intel, Drottningholmsvägen 22, SE-112 42 Stockholm, Sweden
e-mail: jakob.engblom@windriver.com

R. Leupers, O. Temam (eds.), *Processor and System-on-Chip Simulation*,
DOI 10.1007/978-1-4419-6175-4_3, © Springer Science+Business Media, LLC 2010

from a physical system. The systems being simulated are called *target systems* and the computer used to run the simulation is referred to as the *host*.

Virtutech Simics [1] is a full-system simulator which has been in commercial use since 1998. This chapter provides an overview of how real-world usage has shaped the Simics product. The main users of Simics are software and systems developers, and their main problem is how to develop complex systems involving both hardware and software.

The target systems simulated with Simics range from single-processor aerospace boards to large shared-memory multiprocessor servers and rack-based telecom and datacom systems containing thousands of processors across hundreds of boards. The systems are heterogeneous, containing processors with different word lengths, endiannesses, and clock frequencies. For example, there can be 64-bit Power Architecture processors running control software, with 8-bit microcontrollers managing a rack backplane, mixed with data-plane boards containing dozens of 32-bit VLIW DSPs [2]. The systems are typically built from standard commercial chips along with some custom FPGAs or ASICs.

Most target systems are networked. There can be networks of distinct systems and networks internal to a system (such as VME or Ethernet-based rack backplanes). Multiple networks and multiple levels of networks are common.

The target system is dynamic. During simulation (just like when using the physical machines modeled), users can add and remove boards, bring new processors online, reconfigure network topologies, introduce faults, or plug and unplug hot-pluggable hardware. The software will perceive these events like it would on physical hardware, allowing users to test and develop all aspects of the software stack, including automatic configuration, load balancing, fault detection, and recovery.

Simulation runs can cover many hours or days of target time and involve multiple loads of software and reboots of all or part of the system. Even a simple task such as booting Linux and loading a small test program on an eight-processor SoC can take over 30 billion instructions (this was measured on an early version of Freescale's P4080). Profiling and instrumentation runs can take tens of billions of instructions [3, 4].

Full-system simulation is used in several ways during the product life cycle. It helps to define systems, by providing an executable model of the hardware interface and hardware setup. FSS supports hardware and software architecture work, and it validates that the hardware can be efficiently used from the software stack [4]. FSS is used to develop system software, including debug and test. The software development schedule can be decoupled from the availability of hardware when using FSS and it improves software development productivity by providing a better environment than hardware. When systems are in deployment, FSS is used instead of physical systems for tasks like sales demonstrations, end-user training, and target system configuration.

Section 3.2 describes the particular features that we have found useful for software and systems developers. Achieving high simulation speed is critical for a full-system simulator and is addressed in more detail in Section 3.3. Finally, Section 3.4 discusses how to build target system models.

3.2 Important Features of a Full-System Simulator

The feature set of Simics has been developed and adjusted for almost 20 years in order to meet the needs of system developers (the first code in what was to become Simics was written in 1991). This section discusses the most important features of Simics and why they were important enough to be incorporated in the product.

3.2.1 Modular Design and Binary Distribution

Simics is modular; each device model, processor, or other Simics model or feature is shipped in its own self-contained dynamically loaded object file (as shown at the bottom of Fig. 3.1). This fine-grained structure makes it possible to supply the exact set of models and features needed for any specific user. The object file and its associated command files are referred to as a Simics *module*.

Simics models can be distributed as binary-only modules, with no need to supply source code to the users. Binary distribution simplifies the installation for end users, as they do not have to compile any code or set up build environments. It also offers protection of intellectual property when different companies exchange models.

Simics modularity enables short rebuild times for large systems, as only the modules which are actually changed have to be recompiled. The rest of the simulation is unaffected, and each Simics module can be updated and upgraded independently.

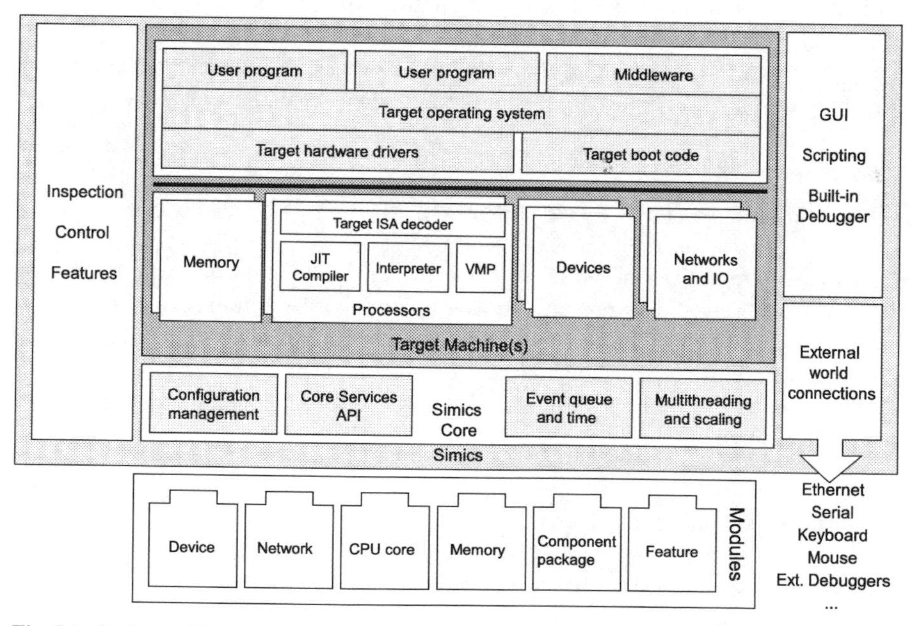

Fig. 3.1 Simics architecture

The modularization of Simics is achieved by the same mechanisms used to implement transaction-level modeling (see Section 3.3.2). A Simics model exposes an arbitrary set of *interfaces* to other models in other modules, and objects can call any model interface in any module. Interfaces are used between device models to model hardware communication paths and to implement other simulator functionality and information flows, such as getting the current cycle count of a processor or finding the address of a variable from the debug module. Unlike SystemC, multiple objects may call the same interface and bindings are not made at compile time. Some interfaces are unidirectional, but bidirectional interfaces are common (for example, a network and a network device sending packets to and from the device). Bidirectional interfaces are simply implemented as two complementary interfaces, one in each direction.

Interfaces are implemented in C, as `structs` containing a set of function pointers. Each interface has a globally unique name within the current simulation. Each class of Simics objects registers the interfaces that it exports with the Simics kernel (when the module containing the class is loaded into the simulation). To call an interface in an object, the simulation kernel is queried for a pointer to the interface of a given name, for some object in the simulation. Since interfaces are implemented in C, the pointer to the receiving object is the first parameter in all interface calls. Essentially, this implements run-time binding between simulation modules, in a way similar to what the Java and Microsoft .net virtual machines allow.

Simics uses the C-level ABI and host operating system dynamic loading facilities. The C++ ABI varies between compiler versions and compiler vendors and is thus not usable in the interface between modules, even though C++ can be used internally in modules. Virtutech provides bindings to write Simics modules using DML (see Section 3.4.1), Python, C, C++, and SystemC, but users can actually use any language they like as long as they can link to C code. For example, a complete JVM has been integrated into Simics, running modules written in Java [3].

3.2.2 Simulation State Checkpointing

Simics *checkpointing* is the ability to save the complete state of a simulation to disk and later bring the saved state back and continue the simulation without any logical interruption from the perspective of the modeled hardware and especially the target software. Checkpoints contain the state of both the hardware and the software, which is implicit in the hardware state. In our experience, checkpointing needs to support the following operations:

- Storing the simulation state to disk.
- Restoring the simulation state on the same host into the same simulation binary.
- Restoring on a different host machine, possibly belonging to another user or organization, where different can mean a machine with a different word length, endianness, operating system, and installed software base.

- Restoring into an updated version of the same simulation model, for example, with bug fixes or expanded functionality.
- Restoring into a completely different simulation model that uses the same architectural state. For example, a detailed clock-cycle-driven model.

Checkpointing can be used to support work-flow optimization, such as a "nightly boot" setup where target system configurations are booted as part of a nightly build and checkpoints saved. During the workday, software developers simply pick up checkpoints of the relevant target states, with no need to boot the target machines themselves.

Another important use of checkpointing is to package bugs and communicate them between testing and engineering, between companies, and across the world. With a deterministic simulator and checkpoint attached to the bug report, reproducing a bug is trivial.

Checkpointing, as it is known today, first appeared in full-system simulators in the mid-1990s [5]. The primary use case at that time was to change the level of abstraction, from a fast functional model to a detailed model, using the fast model to position a workload at an interesting point. This technique is still in use today, supporting both computer architecture research and detailed software performance analysis [1, 6–8].

The key implementation mechanism for checkpointing in Simics is the *attribute* system. Each object in a simulation has a set of attributes. Each attribute has a name that identifies it and a type and two primary interface functions: `get()` to read the value and `set()` to change the value. The type of an attribute is built from primitive types such as integer, floating point, and string, combined into lists or lists of lists. An attribute can accept several different types as the input to `set()`. Attributes also provide user-readable documentation about their purpose. The complete state of the system is captured in its attributes.

3.2.3 Dynamic Configuration and Reconfiguration

A Simics simulation can be reconfigured and extended at any point during a simulation. New modules can be loaded and new hardware models and Simics extensions added. All connections can be changed from scripts and the Simics command line at will. This dynamic nature of a system is necessary to support system-level development work and to support the dynamic nature of the target systems discussed in the introduction.

Configuration flexibility also goes beyond what is physically possible. For example, it is possible to start from a checkpoint of a booted and configured system, add custom hardware models, and load device drivers for the new hardware on the target system [4].

Reconfiguration can also be used to implement fault injection and to test the target software response to faults in the hardware. Typical examples include suddenly removing boards in a rack, stopping a processor, or replacing a device with a faulty variant generating corrupted instead of proper data. Like in checkpointing, the attribute system is the key implementation mechanism for dynamic configuration in Simics.

3.2.4 Repeatability and Reversibility

Simics has been designed from the bottom-up to be a *repeatable*, deterministic simulator, with the same exact simulation semantics regardless of the host. As long as asynchronous input to the simulator is being recorded, any simulation run can be repeated precisely on any host at any time. Note that determinism does not mean that the simulation always runs the same target software in the same way. If the timing of any input or any part of the initial state changes, the simulation will execute differently.

Determinism does not prevent a user from exploring variations in target software behavior [4, 6]. However, the user remains in control and can repeat any simulation run where the variations triggered some interesting behavior.

Based on repeatability, Simics also implements *reverse execution* and reverse debugging, where the user can go back into the history of the system execution. Reverse execution was incorporated in Simics 3.0, launched in March of 2005, which makes it the first usable reverse execution implementation. This is a powerful tool for software debugging, especially for intermittent and timing-dependent bugs which are difficult to reproduce on hardware using classic iterative debugging. Note that Simics reverse execution applies to a complete target systems, possibly containing many processors, boards, and multiple operating system instances. Network traffic, hardware accesses, and everything else going on in the system are reversed, not just a single user-level process as is targeted by most current reverse execution approaches such as gdb 7.0.

Reverse execution is implemented by a combination of checkpointing and deterministic re-simulation. To go back in time, Simics restores the system state from a checkpoint and then re-executes the simulation forward to the desired point in time. The checkpoints used for reverse execution are stored in host memory and not on disk, in order to provide usable performance. Since memory and disk images only need to store the changes between two checkpoints, checkpoints can be kept fairly small. There are also algorithms that intelligently select when to checkpoint and to cull checkpoints in order to keep the memory usage within limits. Note that each checkpoint saves the entire instant state of all objects in the simulation, like a normal Simics checkpoint.

Implementing reverse execution requires more than checkpointing and determinism, however. To achieve useful performance, Simics does not create a new simulation from scratch each time the simulation is reversed. Rather, Simics models are written to accept a reset of their state at any point in time, within the same simulation object instance. Normally, this is simple to achieve for user-provided device models, especially for models written using DML (see Section 3.4.1).

A key enabler for determinism, checkpointing, and reverse execution is that Simics simulation models do not normally use resources outside the simulator – notice that the target machine is internal to Simics in Fig. 3.1. Hardware models live in a completely virtual world and do not open files on the host or drive user interaction directly. All external interaction is handled by the Simics kernel and special infrastructure modules for text consoles, graphical consoles, and network links. Such models are normally provided by Virtutech, handling all the tricky corner cases that are caused by reversibility and providing a sensible user experience. In the case that the outside world needs to be connected to a model, Simics provides a recorder mechanism that can reliably replay asynchronous input under reverse execution.

Reverse debugging and checkpointing have proven to be addictive. Once a user has used them, they never want to be without them again!

3.2.5 Scripting

Simics can be scripted using the Python programming language, as well as using the Simics built-in command-line interpreter (CLI). The CLI offers an accessible way to perform common tasks, while Python offers the ability to write programs to accomplish arbitrarily complex tasks. Scripts can hook into events in the simulation, such as printouts on serial consoles, network packets being sent, breakpoints being hit, interrupts firing in the target hardware, and task switches done by target operating systems.

Scripts in Simics can contain multiple threads, so-called script branches, which allow simple scripting of parallel flows in the simulation, such as code running on different machines in a network. There are barriers and FIFOs available to coordinate the script branches.

Scripting is always available via the Simics command line, and users can work interactively with Simics and later collect sequences of commands into script files, similar to what is typically done with Unix shells.

3.2.6 Visibility

A simulator offers visibility into all parts of the target system. Compared to physical hardware, nothing is hidden inside opaque devices or on the internal buses of SoCs. As virtual time can be stopped, it allows arbitrary complex queries to be performed without affecting the simulated system. Simics offers several ways to access, and manipulate, the state of a target system. There is access to the static state of registers and memory, and Simics device models expose the state of all their programming registers using Simics attributes – see Fig. 3.2 for an example of how this information is presented.

In addition to access to the static state of a system, Simics provides callback hooks to access the flow of transactions in a system. User modules implement

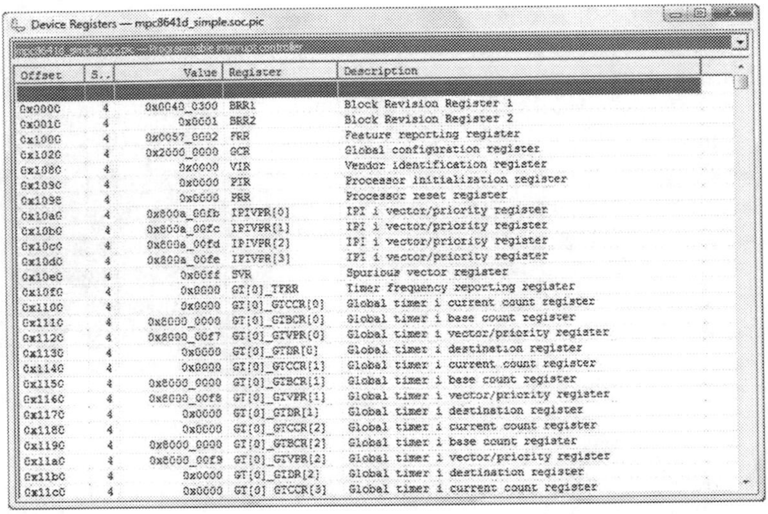

Device Registers — mpc8641d_simple.soc.pic

Offset	S..	Value	Register	Description
0x0000	4	0x0040_0300	BRR1	Block Revision Register 1
0x0010	4	0x0001	BRR2	Block Revision Register 2
0x1000	4	0x0057_0002	FRR	Feature reporting register
0x1020	4	0x2000_0000	GCR	Global configuration register
0x1080	4	0x0000	VIR	Vendor identification register
0x1090	4	0x0000	PIR	Processor initialization register
0x1098	4	0x0000	PRR	Processor reset register
0x10a0	4	0x800a_00fb	IPIVPR[0]	IPI i vector/priority register
0x10b0	4	0x800a_00fc	IPIVPR[1]	IPI i vector/priority register
0x10c0	4	0x800a_00fd	IPIVPR[2]	IPI i vector/priority register
0x10d0	4	0x800a_00fe	IPIVPR[3]	IPI i vector/priority register
0x10e0	4	0x00ff	SVR	Spurious vector register
0x10f0	4	0x0000	GT[0]_TFRR	Timer frequency reporting register
0x1100	4	0x0000	GT[0]_GTCCR[0]	Global timer i current count register
0x1110	4	0x8000_0000	GT[0]_GTBCR[0]	Global timer i base count register
0x1120	4	0x8000_00f7	GT[0]_GTVPR[0]	Global timer i vector/priority register
0x1130	4	0x0000	GT[0]_GTDR[0]	Global timer i destination register
0x1140	4	0x0000	GT[0]_GTCCR[1]	Global timer i current count register
0x1150	4	0x8000_0000	GT[0]_GTBCR[1]	Global timer i base count register
0x1160	4	0x8000_00f8	GT[0]_GTVPR[1]	Global timer i vector/priority register
0x1170	4	0x0000	GT[0]_GTDR[1]	Global timer i destination register
0x1180	4	0x0000	GT[0]_GTCCR[2]	Global timer i current count register
0x1190	4	0x8000_0000	GT[0]_GTBCR[2]	Global timer i base count register
0x11a0	4	0x8000_00f9	GT[0]_GTVPR[2]	Global timer i vector/priority register
0x11b0	4	0x0000	GT[0]_GTDR[2]	Global timer i destination register
0x11c0	4	0x0000	GT[0]_GTCCR[3]	Global timer i current count register

Fig. 3.2 Simics device register viewer

callback functions and register them to get called when transactions occur. For example, memory bus transactions can be intercepted and network packet transmissions monitored. The *hap* mechanism offers notifications when events trigger inside of modules, such as processors getting exceptions and changing between kernel mode and user mode. Haps also provide hooks for changes in the simulation environment, such as the simulation being started or stopped or new simulation objects created. The user can also define new haps, for example, to react to interesting events in a device.

One problem that has to be solved is to know when to trace events in the target system. In a system with multiple processors and many hardware devices, running multitasking operating systems, there is so much going on that tracing everything will overwhelm the user and fill up disk space at a rapid pace. The solution to this problem is to determine points in the software load where tracing should be started and stopped. These points can then be identified in a number of ways. One simple solution is to use *magic instructions*. A magic instruction is an instruction which is a no-op on the physical hardware, but which the simulator considers special. When such an instruction is executed, user modules get a callback. In most cases, using magic instructions does require modifying and recompiling the source code of the target program.

Magic instructions have also been used to prototype special instructions or hardware accelerators, by reading out and modifying the processor state in the magic instruction callback.

As discussed in Section 3.2.8, you can also use debug features to detect interesting points in the code, using breakpoints and hooks for task switches to follow a particular program's execution.

3.2.7 Extensibility

Simics was designed from the start as an open and general simulation platform, where any user can extend the simulator in arbitrary ways. This does not only include modeling new hardware devices or processors, but also add new simulator features and instrumentation modules. The Simics API provides users with full access to the system state and the configuration of the simulation. Using the Simics API and the interfaces presented by simulation models, Simics functionality can be arbitrarily extended without source code access to any part of the simulator.

The Simics API has been used to implement some custom features that really exploit the visibility offered by the simulator. For example, Wright et al. describe an implementation of an introspection system that traces the execution of a Sun HotSpot JVM while running under Solaris on a Simics UltraSPARC target. This framework unintrusively inspects target memory and interprets its contents using Java programs running in a JVM embedded into a Simics module, reusing instrumentation code that is also used on live HotSpot runs [3].

The most common extension to Simics are custom hardware models, from processors and device models, to boards and racks and networks. This is covered in Section 3.4.

3.2.8 Software Abstraction

Like physical hardware, Simics simulations work at the level of machine instructions, processor registers, and memory locations. However, many users are interested in programs, source code, and OS-level processes. To bridge this gap, Simics provides features that map from the machine state to software abstractions.

Symbolic debugging provides a mapping from locations in memory to functions and variables in the executing program. Debugging is used interactively by a user, as well as from scripts and user extension modules to automatically trace, profile, and debug target code. Figure 3.3 shows an example of this in action.

Operating System awareness allows Simics to investigate the state of the target system and resolve the current set of executing threads and processes. OS awareness modules need to be customized for each operating system and involve knowing the layout of operating system task structures in memory and how they are linked together. The detailed offsets in the structures can usually be determined using heuristics at run time. At the time of writing, Simics has OS support for Linux, VxWorks, OSE, and QNX. OS awareness provides scripts and extension hooks to detect when programs and processes are executed, terminated, or switched in and out.

Using these mechanisms, scripts can for example be created to detect deadlock by inspecting a concurrent software load as it is running by detecting accesses to shared variables. To give another example, a commercial OS vendor wrote a module which intercepted all writes to an operating system log function to create a log trace

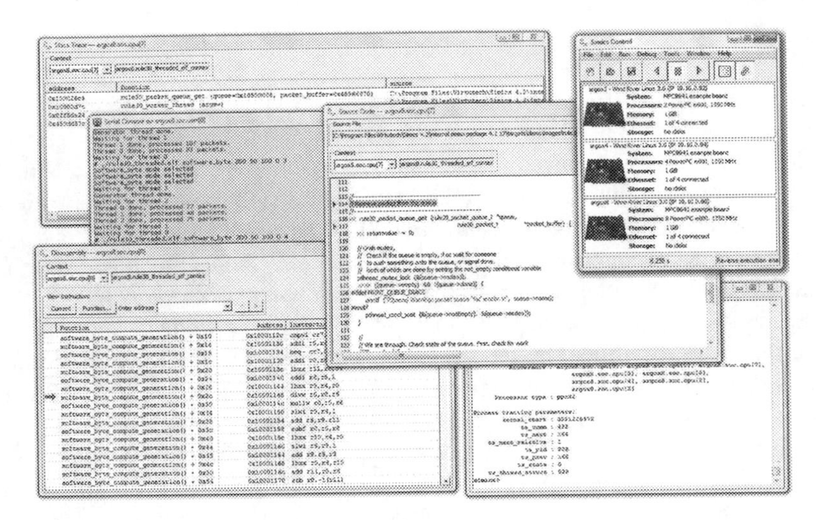

Fig. 3.3 Simics visibility and software abstractions

in the simulator, independent of whether the target system ever got to a state where the trace could be read. Albertsson describes a Simics-based system that automatically creates the information necessary to debug across all layers of the software stack [9].

3.2.9 Connectivity

A simulator used for system development and software development is not really a stand-alone tool. To get maximum value from a simulator, users need to integrate it with other tools. One way to do this is to provide a *real network* facility which connects simulated machines to the physical network. As illustrated in Fig. 3.6, this has been used to drive traffic into virtual networks from physical test equipment, as well as to provide Internet access to the target machines. It can also be used to connect a virtual platform to a piece of physical hardware to perform hardware-in-the-loop testing or to connect a debug agent running on a virtual machine to a debugger running on the same host.

To help write external tools that interact with simulated targets, Simics also provides a *time server* feature that lets external tools synchronize with the time of the virtual world. This makes it possible to set test timeouts in terms of virtual time rather than wall clock time, providing more precision and reliability.

Simics also provides interfaces to connect directly to various debuggers such as GDB, including reverse execution as implemented in GDB version 7, Wind River Workbench, and Freescale CodeWarrior.

3.3 Achieving High Simulation Speed

Achieving high simulation speed is critical to enable users to run the workloads they need. The simulator performance is expected to be comparable to hardware speed even if both the simulated system and its workload are very large. Key aspects that need to be addressed when building a fast simulator are as follows: deciding on the right level of timing abstraction, using a transaction-based modeling methodology of hardware devices, creating fast processor models, using temporal decoupling, and multi-threading the simulation. Simics is optimized for running software as quickly as possible with a simple machine timing model, while allowing detailed timing to be added when needed.

3.3.1 Simics Timing Abstraction

The native Simics level of timing abstraction is *software timing* (*ST*). In ST, the large-scale timing of the system is modeled: time progresses as instructions are being executed. Normally, all memory accesses complete in zero time. Devices can post timed events to model software-visible event timing like timer expiry or delays to completion interrupts. This level of timing abstraction is used by software-oriented simulation solutions like Simics, Qemu, IBM CECsim [10], IBM Mambo [8], and SimOS [5].

On the hardware-oriented side of simulation, the SystemC TLM-2.0 library [11] defines two other levels of abstraction, *loosely timed* (*LT*) and *approximately timed* (*AT*). Unlike ST, LT allows all calls to a device model to be blocking in the SystemC sense, which means that the called function might call `wait()`, thus requiring the simulation kernel to deal with threading. This does carry a performance cost. It also allows for timing annotations on operations, which make writing really fast processor simulators more complicated as time management has to take timing annotations into account. A threaded execution model also makes checkpointing much more difficult, as state is stored on the execution stack [12, 13].

AT is designed to model hardware buses with effects such as arbitration, contention, and split requests and response phases. This allows for detailed simulation of hardware performance, but carries a high cost in simulation speed, since much more function calls and code needs to execute for each data unit transferred and simulation time executed. In particular, to correctly account for contention to shared resources, AT cannot really use temporal decoupling (see Section 3.3.3).

The usefulness of *ST* is proven by the success of several pre-silicon software ports using the ST level of abstraction. IBM's CECsim system has been used to speed the development of new zSeries servers [10, 14]. Simics was used to port Windows, Linux, and NetBSD to the AMD 64-bit x86 architecture in 2001. In 2008, Freescale announced the QorIQ P4080 processor, and the Simics model of the chip was used to port operating systems like Linux, VxWorks, OSE, and Integrity to the chip long before silicon was available. Once physical hardware arrived, operating systems were up and running within a matter of days.

In addition to supporting the ST level of abstraction, Simics supports adding timing details through timing models attached to memory maps, leading to a timing model roughly similar to SystemC TLM-2.0 LT.

Simics can also be used in a *hybrid* mode where detailed clock-cycle (CC)-level models are mixed with ST-level models. The ST models are used to boot operating systems and position workloads, and the simulation is switched over to CC models using checkpoints. In addition, parts of the target system can be simulated using ST models even in CC mode, avoiding the need to create CC models for some hardware. Essentially, the resulting combined solution lets users zoom in on performance details when and where they need to without compromising on the ability to run large workloads [2, 7].

A secondary benefit of modeling at the ST level is that models are easier to reuse. For example, a standard x86 PC contains an i8259 interrupt controller. Over time, the hardware implementation of the i8259 has changed from being a discrete chip to a part of a south bridge and then onto an integrated processor SoC. Regardless of these implementation changes, the exact same Simics model has been used all along.

3.3.2 Pervasive Transaction-Level Modeling

Simics applies transaction-level modeling (TLM) throughout a target system. In TLM, each interaction with a device, for example, a write to or read from the registers of the devices by a processor, is a single simulation step: the device is presented with a request, computes the reply, and returns it in a single function call. Essentially, communication is reduced to function calls between simulation models without involving the simulation kernel. Examples of transactions commonly found in Simics models are memory operations, network packet transmissions, interrupts, and DMA block transfers. The goal is to always perform as much work as possible in each simulation step.

Device models in Simics are designed as mainly passive reactive units. There is no concept of a thread in a Simics device model and the use of events is kept to a minimum. Transactions hitting a device model are instead executed synchronously: the device model computes the result of the transaction, updates its stored state, and then returns. This enhances locality of execution and avoids the cost of maintaining an execution queue and local stack context for each model. It also removes the complexities of parallel programming and locking from device modeling. An important property of this modeling style is that devices store their state in an explicit heap-allocated data structure, which simplifies both multi-threading of the simulation and the implementation of checkpointing (see Section 3.2.2).

As part of the TLM infrastructure, Simics uses a memory map representation of the bus hierarchy of a system. The model contains the processor-visible memory map for each processor in the system and directly routes memory transactions to the right target. If desired, a bus hierarchy can be modeled by chaining extra memory maps and attaching timing models to them.

Networks in Simics are abstracted to complete packets sent between machines. There is no model of the detailed bit-level protocols and encodings used in a physical network. The size of a packet depends on the nature of the network and how often packets are transmitted. In Ethernet, packets can be up to 64 kB, while serial lines usually transmit single bytes at a time, due to their low data rates. Network simulations usually have no knowledge of the meaning of the packets, apart from the parts related to hardware addressing schemes like the Ethernet MAC address. The software on the simulated target systems interacts with a model of a network interface device, programming it just like a physical network interface.

3.3.3 Temporal Decoupling

Temporal decoupling is a standard technique for improving the performance of a simulation containing multiple concurrent units. Rather than switching between the units at each step of the simulation, such as a clock cycle, each unit is allowed to run for a certain amount of virtual time, its *time quantum*, before switching to the next unit. Temporal decoupling has been in use for at least 40 years in computer simulation [15] and is a crucial part of all fast simulators today [11, 16].

In Simics, temporal decoupling is performed between "clocks," typically, a processor executing code, even though it can be other objects in rare circumstances, and each object in the simulation gets its time from one of the clocks. This means that events between an object and its clock are precisely timed, while events from one clock's set of objects to another clock's set will see the effects of temporal decoupling.

Experience shows that using a fairly large time quantum is critical to achieving really high simulation speeds. Figure 3.4 shows some measurements of the effect of temporal decoupling. The line "Single P4080" shows the performance of a multi-threaded compute-intense program running on all eight cores of a Freescale QorIQ P4080 SoC. Performance increases by a factor of 100 from a time quantum of 10 cycles to a time quantum of 100k cycles, after which it stabilizes. From 1000 to 10,000 cycles, performance is doubled.

As far as the software is concerned, temporal decoupling usually does not affect functionality. It does have a small effect on how the target executes, as communication between cores will sometimes get a larger delay than on physical hardware – but the effect is normally not adverse. In our experience, the optimal time quantum for running a shared-memory multiprocessor system tends to be 10k cycles, and between machines connected with networks it can be 100k cycles to 1M cycles. Note that it is possible to selectively reduce the time quantum if needed. For example,

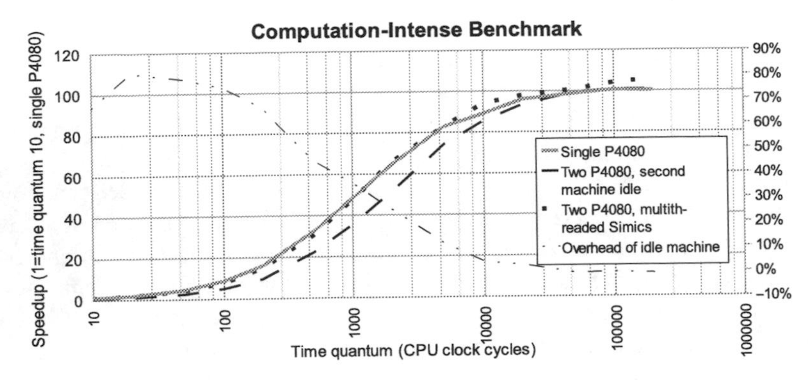

Fig. 3.4 Temporal decoupling and simulation performance

we have seen Simics users position a workload at a suspected fine-grained race condition, switch to a shorter time quantum, and run tests without the potential interference of a long time quantum.

When using a simulator to study detailed performance characteristics of a target system, in particular for caches, shared memory, and other shared resources, temporal decoupling should not be used. It will give a false picture of the contention, as each processor, or other active component, will see itself as alone in the system for its entire time slice. A good rule of thumb is that the time quantum needs to be less than the shortest time in which contention can occur. Most computer architecture researchers tend to set the time quantum to a single or a few cycles, as that does provide the most realistic simulation results.

3.3.4 Fast Processor Simulation

Working at a software-timing level of abstraction with TLM and temporal decoupling, several techniques can be applied to speed up the execution of code in the processors of the target system.

Just-in-time compilation of the target instruction set to the host instruction set is a standard technique used in all fast full-system simulators and virtual platforms (see, for example, Chapters 2, 4, 8, and 17 in this book). As the target processors execute target code, the simulator collects execution statistics and compiles target code that is executed repeatedly to native code. At best, this lets a simulator execute more than one target instruction per host clock cycle (we have measured speeds of 5000 target MIPS for a PowerPC target running on a 3000 MHz x86 laptop, on a simple integer code benchmark). In more typical cases, speeds of hundreds of MIPS can be achieved on workloads that involve operating system activity and device accesses.

Host virtualization extensions can also be used to provide native speed for targets that have the same architecture as the host. Simics uses this on x86 hosts with x86 targets to achieve a simulation speed close to that offered by pure virtualization solutions like VMWare, Xen, and KVM.

Hypersimulation makes the processor skip through idle time in a single simulation step, rather than go through it cycle by cycle. For example, if an x86 processor executes the `HALT` instruction it will not do anything until the next interrupt. Since Simics knows when the interrupt will happen, time is advanced immediately. This is possible because of the isolation of models from the outside world, as shown in Fig. 3.1). Hypersimulation can have simulations run at hundreds of billions of target cycles per second. Hypersimulation can also be applied to any known code pattern indicating idleness or intentional delays. For example, the Power Architecture `bndz 0` instruction, which simply loops until the special `COUNT` register is zero, can be fast-forwarded by Simics.

Hypersimulation is particularly beneficial when executing simulations containing many processors and machines. In practice, most parallel workloads will tend to only load some subset of the system processors at any one time. With hypersimulation, the simulator will spend its time simulating only the active parts of the system, with the inactive parts consuming a very small amount of resources. Figure 3.4 shows an example of the effect of hypersimulation. A second, idle, P4080 was added and the same computation benchmark was ran again. The line "Two P4080, second machine idle" shows the result. Even though twice as many processor cores are being simulated, the simulation is almost as fast as with just a single P4080, provided that a sufficiently large time quantum is used. The line "Overhead of idle machine" shows how the overhead of the idle machine depends on the time quantum length.

3.3.5 Multi-threading the Simulation

With multi-core processors being the norm in server and desktop computers, it is necessary to use multiple host processor cores to speed up the simulation. Simics has supported multi-threading within a simulation process since Simics 4.0 (launched in the Spring of 2008). Before that, Simics used teams of separate Simics processes to utilize multiple clusters of host machines to simulate networks [1]. In all cases, Simics maintains determinism, repeatability, and fully defined simulation semantics.

In practice, multi-threading is applied to networks of target machines, as the network links offer convenient points to separate a simulation into parallel components. It is significantly harder to get any benefit from parallelizing the simulation of tightly connected models like processors sharing memory, as the ratio of computation to synchronization is not as favorable.

Fully defined simulation semantics for parallel simulations is implemented by setting precise propagation delays for data and decoupling the sending and receiving of transactions. For example, if machine A sends an Ethernet packet to machine B at time t, machine B will see it at some point $t{+}d$, where d is the defined propagation delay for that network. The simulation infrastructure makes sure that all threads are synchronized at least as often as d, achieving determinism and repeatability regardless of the actual execution order of the host threads used (or the number of host threads used).

The benefit of multi-threading and distributed simulation heavily depends on the particular latency settings between the target machines and the load balance on the target systems. The progress of the simulation overall is bounded by the target machine that takes the most time to simulate (usually the most loaded target machine). Figure 3.4 shows the efficiency of multi-threading for a small two-machine network. Note how the line "Two P4080, multi-threaded Simics" closely tracks the line "Single P4080," indicating that multi-threading effectively removes any overhead from adding a second machine. Simics users have used clusters containing tens of host processors to speed up the simulation of large target systems with good effect.

3.4 System Modeling

Getting a target system model in place is key to getting the benefits from simulation. In our experience, this is best achieved by an iterative process. For a new system, the starting point tends to be a small core system sufficient to boot an operating system or start some other initial software load. Over time, models are added to cover more of the target system and more of its use cases. The modeling process tends to follow the boot order of the target system.

For modeling, a target system can be partitioned into processors, interconnects (in particular, networks), devices, and memories. Processors, interconnects, and memories tend to be highly reusable across target systems. The same holds for some common devices like timers and interrupt controllers. However, almost every new system contains some new and unique device models. Thus, device modeling tends to be the main effort when modeling a new target system.

The devices, processors, memories, and interconnects are then combined into a hierarchical structure that is called *components* in Simics terminology. As illustrated in Fig. 3.5, Simics components describe processor clusters, SoCs, memories, boards, racks, and other aggregates in a system. They are implemented as Python code, affording users flexibility in describing recurring and dynamic structures. Components often have parameters like the number of processor cores, sizes of memories, and clock frequencies. At the top level, systems are created by Simics CLI scripts that call component creation functions and connect the components.

Simics models are generally created from the same documentation used by software developers, as that provides the best description of *what* a piece of hardware does. Models are also best created by software developers, not hardware developers, as hardware developers in general tend to add too many details to models, making them slower as well as less reusable. Separating the task of creating a model from

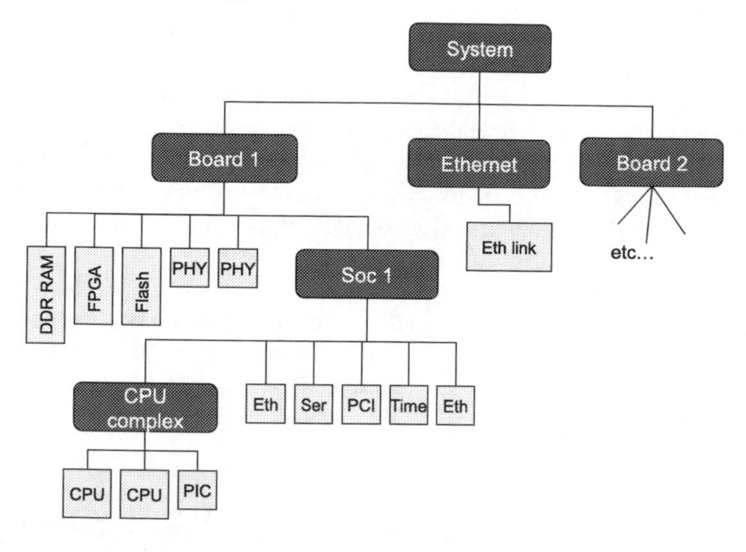

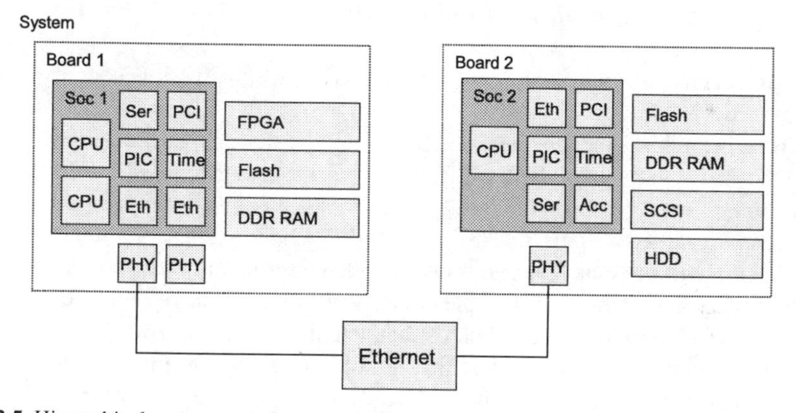

Fig. 3.5 Hierarchical components for an example system

the hardware design process via documentation also has clear benefits to the system development process and the final system quality. It forces documentation to be created and subjects the documentation to a detailed review as a result of building executable models from the documentation.

3.4.1 Device Modeling

To speed up device modeling, Virtutech has created the DML tool, which essentially wraps snippets of C code inside a language designed to make it easy to express device register maps and other common hardware constructs. DML takes the view that *modeling is programming* and tries to make the code required to describe a device model as short as possible.

Key features supported by DML include expressing register maps, bit fields in registers, bit-endianness, byte-endianness, multiple register banks, simulator kernel calls such as event posting, and the connections between cooperating device models. Templates can be used to reuse arbitrarily complex constructions. DML cuts down on repetitive coding and makes device model code easier to write and maintain.

DML can also be used to wrap existing C and C++ models of algorithms and core hardware functionality into a register map and Simics model, enabling their use inside of a Simics virtual platform. DML separates declarations and definitions, allowing reuse of artifacts from the hardware design by automatically converting IP-XACT-like register descriptions to DML declarations.

3.4.2 Stubbing Simulation Components

One way to increase simulation performance and reduce the work needed to build a system model is to replace parts of the target system with *stubs* or *behavioral models*. These behavioral models can be ASICs, boards, devices, entire sub-networks, or anything else that can be isolated from the rest of the system via a well-defined interface.

In complex target systems, consisting of multiple boards, it is common to replace some boards with models that just reflect the behavior of the board at its interface to the rest of the system. For example, when working on control plane software in a telecom system, the data-plane boards that actually process traffic can be stubbed out and replaced by simple models that simply acknowledge their existence and tell the control plane that all is fine. If you consider that each data-plane board can contain tens of processors, this can increase the speed of simulation by an order of magnitude, as well as reducing the need to actually model the processors and their associated devices. Note that there are cases where data-plane boards are modeled in more detail in order to test and validate the combined system software stack. The appropriate configuration depends on the use case for a particular simulation run.

Another example of simplifying a simulation to gain speed is in networking, where you can replace a set of machines used to generate traffic toward a simulated server with a traffic generator. In the most general case, Simics network simulations are interfaced to complete network simulations for mobile phone networks or IP networks. Figure 3.6 shows several of the options available with Simics, including simple traffic generation, rest-of-network simulations, and virtual network test equipment.

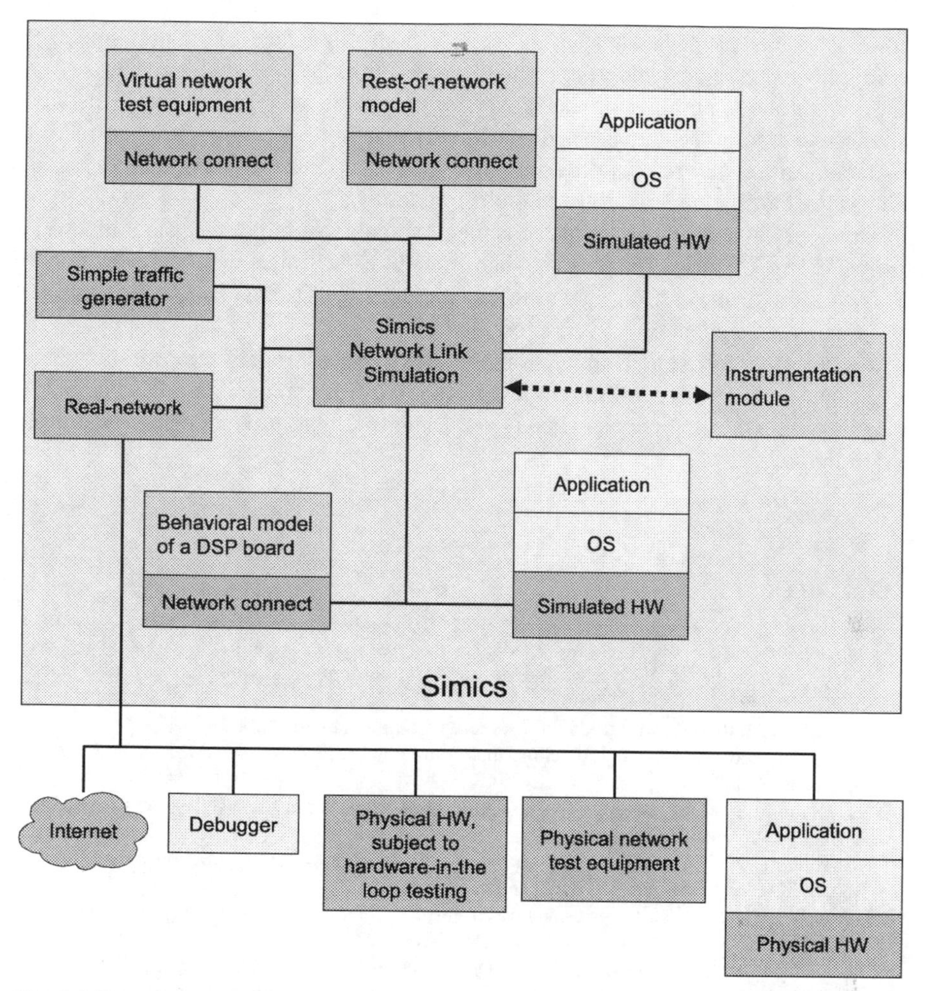

Fig. 3.6 Network simulation with Simics

3.5 Summary

This chapter has described the different use cases and benefits of full-system simulation, as well as the technical challenges that must be overcome to realize the benefits and apply simulation to the use cases. We have addressed the most important fundamental technologies for full-system simulation, including high simulation speed, modularity, binary distribution, dynamic configuration, checkpointing, determinism, and reverse execution. Features such as scripting, OS awareness, state inspection, and connections to the world outside of the simulation have also been addressed. Overall, Simics cannot be seen as a set of independent technologies, but rather as an interconnected system that derives its full power from many mutually reinforcing technologies and design choices.

Techniques for high-performance system simulation were discussed in detail. We discussed basic techniques such as using transaction-based modeling, correctly abstracting time, and temporal decoupling, as well as more advanced techniques such as just-in-time compilation and multi-threading.

An important aspect of a simulator is the practicality of building system models. Tools and best practices for creating such models with Simics were described and ideas for how to make such modeling more efficient, in terms of both simulation speed and development time, were presented.

It is clear that full-system simulation plays a significant role in the definition and development of systems at different levels, ranging from single board OS bring up to large system development and debug and test, through to product demonstrations and training.

References

1. Magnusson, P., Christensson, M., Eskilson, J., Forsgren, D., Hallberg, G., Hogberg, J., Larsson, F., Moestedt, A., Werner, B.: Simics: A full system simulation platform. *Computer* **35**(2), 50–58 (2002).
2. Embedded Multicore, An Introduction. Freescale Semiconductor (2009).
3. Introspection of a java virtual machine under simulation. Tech. Rep. SMLI TR-2006-159, Sun Labs (2006).
4. Engblom, J.: System architecture specification and exploration using a full-system simulator. Virtutech Whitepaper (2009).
5. Rosenblum, M., Varadarajan, M.: Simos: A fast operating system simulation environment. Technical Report CSL-TR-94-631, Stanford University (1994).
6. Alameldeen, A., Wood, D.: Ipc considered harmful for multiprocessor workloads. Micro, *IEEE* **26**(4), 8–17 (2006). DOI 10.1109/MM.2006.73.
7. Engblom, J.: Why virtual platforms need cycle-accurate models. SCDSource (2008).
8. Shafi, H., Bohrer, P.J., Phelan, J., Rusu, C.A., Peterson, J.L.: Design and validation of a performance and power simulator for powerpc systems. *IBM J Res Dev* **47**: 641–651 (2003).
9. Albertsson, L.: "Holistic debugging – enabling instruction set simulation for software quality assurance." In: *Modeling, Analysis, and Simulation of Computer and Telecommunication Systems* (MASCOTS) Monterrey, California, USA, September 11–14 (2006).
10. Stetter, M., von Buttlar, J., Chan, P.T., Decker, D., Elfering, H., Giquindo, P.M., Hess, T., Koerner, S., Kohler, A., Lindner, H., Petri, K., Zee, M.: IBM eserver z990 improvements in firmware simulation. *IBM J Res Dev* **48**: 583–594 (2004).
11. Aynsley, J.: OSCI TLM-2.0 Language Reference Manual. Open SystemC Initiative (OSCI) (2009).
12. Engblom, J., Monton, M., Burton, M.: Checkpoint and restore for systemc models. In: *Forum on Specification and Design Languages* (FDL) France, September 22–24 (2009).
13. Kraemer, S., Leupers, R., Petras, D., Philipp, T.: A checkpoint/restore framework for systemc-based virtual platforms. In: *International Symposium on System-on-Chip* (SoC) pp. 161–167, Tampere, Finland, October 5–7 (2009).
14. Koerner, S., Kohler, A., Babinsky, J., Pape, H., Eickhoff, F., Kriese, S., Elfering, H.: Ibm system z10 firmware simulation. *IBM J Res Dev* **53** (2009).

15. Fuchi, K., Tanaka, H., Manago, Y., Yuba, T.: A program simulator by partial interpretation. In: *SOSP '69: Proceedings of the Second Symposium on Operating Systems Principles*, Princeton, NJ, USA, October 20–22, pp. 97–104 (1969).
16. Cornet, J., Maraninchi, F., Maillet-Contoz, L.: "A method for the efficient development of timed and untimed transaction-level models of systems-on-chip." In: *DATE '08: Proceedings of the Conference on Design, Automation and Test in Europe*, München, Germany, March 10–14, pp. 9–14 (2008). DOI http://doi.acm.org/10.1145/1403375.1403381.

Chapter 4
Toward the Datacenter: Scaling Simulation Up and Out

Eduardo Argollo, Ayose Falcón, Paolo Faraboschi, and Daniel Ortega

It was the best of times; it was the worst of times, Charles Dickens, A Tale of Two Cities

Abstract The computing industry is changing rapidly, pushing strongly to consolidation into large "cloud computing" datacenters. New power, availability, and cost constraints require installations that are better optimized for their intended use. The problem of right-sizing large datacenters requires tools that can characterize both the target workloads and the hardware architecture space. Together with the resurgence of variety in industry standard CPUs, driven by very ambitious multi-core roadmaps, this is making the existing modeling techniques obsolete. In this chapter we revisit the basic computer architecture simulation concepts toward enabling fast and reliable datacenter simulation. Speed, full system, and modularity are the fundamental characteristics of a datacenter-level simulator. Dynamically trading off speed/accuracy, running an unmodified software stack, and leveraging existing "component" simulators are some of the key aspects that should drive next generation simulator's design. As a case study, we introduce the COTSon simulation infrastructure, a scalable full-system simulator developed by HP Labs and AMD, targeting fast and accurate evaluation of current and future computing systems.

4.1 Computing Is Changing

There is general consensus that the computing industry is changing rapidly and both technical and social reasons are at the foundations of this change.

The transition to multi-core, hardware hitting the power wall, and a new emphasis on data-centric computing of massive amounts of information are some of the key disrupting trends. Growth in disk and memory capacity, solid-state storage, and massive increases in network bandwidth complete the picture. Pundits point out that

P. Faraboschi (✉)
HP Labs, Cami de Can Graells, 1-21, Sant Cugat del Valles, 08174 Barcelona, Spain
e-mail: paolo.faraboschi@hp.com

R. Leupers, O. Temam (eds.), *Processor and System-on-Chip Simulation*,
DOI 10.1007/978-1-4419-6175-4_4,

much of the science behind all this comes from long ago. Indeed, shared memory parallel processing and distributed computing are at least four decades old. However, it is now that these technologies are becoming ubiquitous and the IT industry is forced to put these ideas into practice, learning what works and what does not.

Social trends are also contributing to shape the computing world. Enterprises are consolidating their IT infrastructure in warehouse-size datacenters; sometimes their own, sometimes in the cloud. This is happening across the board: end users no longer want to manage shrink-wrap software, SMEs concentrate their resources on core business, and large enterprises farm out their peak workloads at minimum capital expenditures. On the cloud providers' side, cost-saving opportunities and economy of scale are enabling new services and business models.

In this chapter we describe the impact of these trends to the world of modeling tools. We believe that there are important opportunities and challenges for simulation-based approaches. Better decision support data in sizing datacenter-level workloads and their IT infrastructure can provide an important and quantifiable differentiator. However, the problem of right-sizing large datacenters requires tools that can characterize both the target workloads and the hardware architecture space. The reappearance of variety in industry-standard CPUs driven by very ambitious multi-core roadmaps, together with new workloads and metrics, is rapidly making the existing modeling techniques obsolete and here we point to some of the important directions that will differentiate the next generation of modeling tools.

4.1.1 The Resurgence of Hardware Diversification

Analysts [1] predict that over half of all datacenters will be redesigned and relocated in the next 5 years to meet new power, growth, and availability constraints. Long gone is the linear increase of single-thread performance of the uniprocessor era [3] and as we enter the many-core and system-level integration era, hosts of new processors variants are reaching the marketplace. The *variability* is today much higher than in the past, even in industry-standard x86 processors. For example, integrated memory controllers, multiple memory channels, graphics, networking, and accelerators are some of the diversifying parameters.

Choosing the right CPUs for datacenter deployment has become a highly complex task that needs to take into account a variety of metrics, such as performance, cost, power, supply longevity, upgradability, and memory capacity. Cost and performance differences even within the same ISA family are dramatic. In March 2009 [14], we could find a single-core 64-bit x86 at $18 (AMD Athlon XP 1500) and a quad-core at $2,500 (Intel Xeon MP QC X7350). Given the performance, throughput, computation vs. communication balance, cost, and power targets, one may clearly be more appropriate than the other for a given workload. The huge cost and performance difference opens up large opportunities for optimization even within the same ISA family. All of this are coupled with an increasing chipset and board complexity that make these systems extremely difficult to design, even to the extent of picking and choosing among the many available variants optimized for a given use.

And, this is just the beginning. If we look at the foreseeable Moore's law progression, it is easy to predict hundred of computing cores, homogeneous or heterogeneous, integrated in a single component. Additionally, many of these components are gradually absorbing system-level functions such as network interfaces, graphics, and device controllers. Accelerators for anything and everything are also being proposed both for general purpose and for embedded computing.

Finally, the aggressive penetration of non-volatile memory technologies and optical interconnects appearing at the horizon are radically disrupting the memory, networking, and storage hierarchy. This adds other levels of variability, which depend heavily on application characteristics and cannot easily be predicted without a detailed analysis.

4.1.2 Emerging Workloads

The datacenter applications are also evolving and introducing a variability that was unseen in the past. Cloud computing, search and data mining, business intelligence and analytics, user-generated services and mash-ups, sensor network data streams are just a few examples of new applications that are gradually moving to the cloud datacenter. Together with the evolution of traditional enterprise applications, such as databases and business software, these new workloads highly increase the complexity of understanding, monitoring, and optimizing what is running.

Many of these tasks cannot be characterized as being bound to just one resource, such as the CPU or memory. They exhibit a wildly different behavior under varying configurations of storage, memory, CPU, and networking capabilities. On top of this, many applications are expected to run inside virtual environments, ranging from full virtual machines to simple hypervisors. Virtualization clearly offers important benefits such as migration, isolation, consolidation, and hardware independence for application developers and datacenter management. At the same time, virtualization makes it harder to reason about guaranteed performance and the quality of service (QoS) of the virtualized applications and introduces yet another degree of freedom and complexity that requires careful modeling.

Traditionally, workloads have been characterized through representative benchmarks. In the uniprocessor world, comparing alternatives was relatively simple, albeit sometimes misleading. For multi-core and highly integrated systems, the complexity of correlating benchmark performance to user-relevant metrics is quickly exploding. For example, one known issue is the limited scalability of shared-memory performance due to coherency traffic overhead. Understanding where that moving boundary is for each combination of processor cores, interconnect and memory is an unsolved challenge. When looking at workload consolidation in a datacenter, additional complexities arise from the difficulty of predicting the QoS impact of sharing resources. Linear methods that look at resources in isolation miss interference and conflicts at the shared resources. This is where simulation can come to the rescue.

4.1.3 New Metrics

Measuring traditional performance metrics has become much harder. In uniprocessor systems, CPU-bound applications could be characterized by their instructions per cycles (IPC) rate. In multiprocessor systems, only the wall-clock time of an application is what truly defines performance. In transaction-based throughput-oriented processing, the number of transactions per unit of time is what matters. In all scenarios, measuring power consumption is also fundamental for a correct energy management of the whole datacenter, to guide job allocation and dynamic cooling.

If we look at cloud computing, the metrics are once again very different. User experience driven by service-level agreements (SLAs) is what matters, together with the total cost of ownership (TCO) of the computing infrastructure. Being able to *right-size* the infrastructure to provide the best performance and TCO is what can give cloud providers a competitive hedge, allowing them to offer better services at a lower cost. As complex services migrate to the cloud, new opportunities open up, and at the same time the pressure in better characterizing the important metrics of existing and future systems grows. The first consequence on the analysis tools is that a much larger timescale and portion of system execution have to be modeled and characterized.

4.2 Simulation for the Datacenter

Navigating the space of hardware choices at a datacenter scale requires collecting *decision support* data at various levels and that is where simulation can help. This simulation style significantly differs from what is used in the computer-aided design (CAD) world. CAD simulators must reflect the system under design as accurately as possible and cannot afford to take any shortcut, such as improving speed through sampling. *Simulation for decision support* only needs to produce approximate measurements and speed and coverage are equally important as accuracy.

Historically, four different abstraction levels have been used to model future systems: extrapolating real-system measurements, analytical models, simulation, and prototyping. Prototyping has lost much of its appeal for anyone but VLSI designers because of the skyrocketing costs of designing and building integrated circuits, so we do not cover it here. Extrapolating results is the most straightforward method of evaluating some future directions. Unfortunately, in moments of great changes like today, past trends are insufficient to correctly predict the behavior of next-generation parts. Analytical models are very powerful, but they can only model the behavior they know up front. We see an enormous opportunity for simulators, but several obstacles stand in the way, and much of the common wisdom about how to build simulators needs to revisited.

4.2.1 Cycle Accuracy

Many architecture and microarchitecture simulators strive for a cycle-detailed model of the underlying hardware, but are willing to tolerate several orders of magnitude slowdown and a complex software development. With the exception of microprocessor companies, cycle accuracy is rarely validated against real hardware. In academic research, or where the goal is modeling future hardware, the focus typically shifts to validating isolated differential improvements of the proposed techniques.

We believe that cycle-detailed simulators are not always needed and in many cases do not represent the best solution, once you take into account their engineering cost and lack of simulation speed. Many foreseeable uses of simulation for large-scale system modeling favor simpler and less accurate models, as long as they entail faster speed or simpler configurability. This is especially true in the early exploration stages, or for research in fields other than microarchitecture. Absolute accuracy can be substituted with relative accuracy between simulations of different models, which is sufficient for uses that primarily want to discover coarse-grain trends.

4.2.2 Full-System Simulation

Many simulators only capture user-level application execution (one application at a time) and approximate the OS by modeling some aspects of the system calls. Access to devices or other operating system functionalities is wildly simplified. Historically, there have been valid reasons for this, such as engineering simplicity and the small impact of system code in the uniprocessor era. However, as we move to large numbers of cores with shared structures and complex system interactions, accounting for the real impact of system calls, OS scheduling, and accessing devices, memory, or the network has become a necessity

While engineering complexity considerations are important for a development from scratch, virtualization and fast emulation technologies are commonly available today and can be leveraged to build powerful simulators. Adopting a system-level approach is also the only way to deal with closed source, pre-built, and legacy applications, which is mandatory to target datacenter-scale consolidated workloads.

Finally, another advantage of a fast full-system simulator is the practical platform usability. At functional speed (10x slower than native) users can interact with the guest: log on, install software, or compile applications as if running on real hardware.

4.2.3 Profiled-Based Sampling

Sampling is the single largest contributor to simulation speed and coverage for architecture simulation. As it was discussed in (Calder et al., 2010, this volume) and (Falsafi, 2010, this volume), many different research studies have developed very effective sampling techniques.

Unfortunately, some sampling proposals rely on previous offline characterization of the application, and are not adequate for full-system modeling. When dealing with large-scale, full-system, multiple-threads parallel systems, it becomes very difficult to prove that the zones selected by the sampler are representative of the system.

Complex systems change their functional behavior depending on timing. Operating systems use a fixed-time quantum to schedule processes and threads. Threads in a multithreaded application exhibit different interleaving patterns depending on the performance of each of the threads, which in some cases may produce different functional behavior. Networking protocols, such as TCP, change their behavior in presence of congestion or different observed latencies. Messaging libraries change their policies and algorithms depending on the network performance. These are just some examples of the many scenarios that show the fundamental limitations of selecting good representative samples with an offline characterization.

Finally, while statistical sampling has been well studied for single-threaded applications, it is still in its infancy when it comes to modeling multithreaded and multiprocessor workloads. The area of sampling code with locks and how to identify the "interesting" locks worth modeling in details is still an open research problem.

4.2.4 Simulation Speed

An important performance metric is the so-called self-relative slowdown, i.e., the slowdown a program incurs when executed in the simulator vs. what it would take on the target. Looking at the landscape of simulation approaches, from direct execution to cycle-level interpretation, there are at least *five orders of magnitude* of difference in performance (Topham et al., 2010, this volume). A typical cycle-detailed simulator runs in the order of few hundreds of kIPS (thousands of simulated instructions per second), which corresponds to a self-relative slowdown of over 10,000x (i.e., 60s of simulated time in 160 h, about 1 week). On the other extreme, a fast emulator using dynamic binary translation can reach a few 100 s of MIPS, translating to a self-relative slowdown of around 10x (i.e., 60 s of simulation in 10 min).

When targeting datacenter-level complex workloads and hardware, the scale at which important phenomena occur can easily be in the range of minutes. Hence, simulation speed needs to be much closer to the 10x slowdown of dynamic binary translation than the 10,000x of cycle-by-cycle simulators. To reach this target, a simulator must be designed from the beginning with speed in mind. By incorporating acceleration techniques such as sampling upfront, a simulator can ensure that speed and accuracy can be traded off as based on user requirements.

Even the fastest execution-based simulation speed may not be sufficient to cover the datacenter scale of thousands of machines, unless we can afford a second datacenter to run the distributed simulator itself. For this reason, several research groups have proposed the use of analytical models using several different techniques, ranging from simple linear models all the way to neural networks and machine learning [10, 13]. By leveraging this work, a modeling infrastructure could entail a stack of models at increasing accuracy and decreasing speed to be dynamically selected based on target requirements and scale. While we believe this is important work, we do not cover it in this chapter and we primarily focus on execution-based simulation.

4.3 Large-Scale Simulation Use Cases

A full-system scale-out simulator can be very important for *microarchitecture research*, which is traditionally driven by user-level cycle-detailed simulators. Using realistic workloads and looking at the impact of accessing devices open up a whole new set of experiments. Studying the impact of non-volatile memory, 3D-stacked memory, disaggregated memory across multiple nodes or heterogeneous accelerators are all examples that require a full-system simulator.

A second important use is for *architecture exploration*. The evaluation of several design alternatives is a key step in the early stages of every new product design. Even where most components are industry standard, companies are continuously looking for ways to add value and differentiate their products. A typical what-if scenario involves little customization to a datacenter simulator but can highlight the interesting trends that help decide the fate of a particular idea.

Pre-sale engineers are also potential users of a simulation tool, for *right-sizing deployments*. While raw use of simulation is too detailed, simulation can be used to run an off-line parametric sweep of the design space and collect characterization data that can then be used through a simplified (spreadsheet-like) tool. A large deployment usually involves bids from multiple vendors that need to match a particular level of performance and other metrics. It is the responsibility of the pre-sale engineers to “right-size” the IT infrastructure starting from imprecise information, often about components that do not exist yet. In the past, fewer hardware choices and the limited range of well-characterized applications made the problem somewhat tractable.

Today’s rapidly changing workloads and large hardware variety make the task of mapping customers’ requirements on future hardware much more difficult. The problem is to tread the fine line between overly optimistic predictions (risking of being unable to deliver the promised performance), and overly pessimistic predictions (risking of losing the bid). In this world, better modeling translates to lower risks, increased margins, and more customer value loyalty.

One additional important aspect of bidding for large deals is *building prototypes*, especially in high-performance computing. The problem with prototypes is that they are very expensive and become quickly obsolete. So, by using simulation technology to *virtualize* the hardware under test, we can extend the useful life of a prototype, thereby making better use of the capital expenditure.

Software Architects could also use a datacenter simulator. As systems become highly parallel, applications need to be re-thought to *explore parallelism opportunities*, and analytical understanding of performance is also increasingly more difficult. A simulator can help understand scalability issues and can be efficiently used to balance trade-offs that cross architecture, system and programming boundaries. A broader evaluation of alternatives provides guidelines to the application and system architects, reduces risk, and helps make better use of the underlying IT infrastructure. Resilience is another challenge of programming at large scale that efficient simulation tools can help. By using a simulated test bench, we can inject faults to stress test the failure-resistant properties of the entire software system.

4.4 Case Study: The COTSon Simulation Infrastructure

COTSon is a simulator framework jointly developed by HP Labs and AMD [2]. Its goal is to provide fast and accurate evaluation of current and future computing systems, covering the full software stack and complete hardware models. It targets cluster-level systems composed of hundreds of commodity multi-core nodes and their associated devices connected through a standard communication network. COTSon adopts a *functional-directed* philosophy, where fast functional emulators and timing models cooperate to improve the simulation accuracy at a speed sufficient to simulate the full stack of applications, middleware and OSs.

COTSon relies on concepts of reuse, robust interfaces, and a pervasive accuracy vs. speed philosophy. We base functional emulation on established, fast and validated tools that support commodity operating systems and complex applications. Through a robust interface between the functional and timing domain, we leverage other existing work for individual components, such as disks or networks. We abandon the idea of always-on cycle-based simulation in favor of statistical sampling approaches that continuously trade accuracy and speed based on user requirements.

4.4.1 Functional-Directed Simulation

Simulation can be decomposed into two complementary tasks: *functional* and *timing*.

Functional simulation emulates the behavior of the target system, including common devices such as disks, video, or network interfaces and supports running an OS and the full application stack above it. An emulator is normally only concerned with functional correctness, so the notion of time is imprecise and often just a representation of the wall-clock time of the host. Some emulators, such as SimOS [16] or QEMU [5], have evolved into virtual machines that are fast enough to approach native execution.

Timing simulation is used to assess the performance of a system. It models the operation latency of devices simulated by the functional simulator and assures that events generated by these devices are simulated in a correct time ordering. Timing simulations are approximations to their real counterparts, and the concept of accuracy of a timing simulation is needed to measure the fidelity of these simulators with respect to existing systems. Absolute accuracy is not always strictly necessary and in many cases it is not even desired, due to its high engineering cost. In many situations, substituting absolute with relative accuracy between different timing simulations is enough for users to discover trends for the proposed techniques.

A defining characteristic of simulators is the control relationship between their functional and timing components [12]. In *timing-directed* simulation (also called *execution-driven*), the timing model is responsible for driving the functional simulation. The execution-driven approach allows for higher simulation accuracy, since the timing can impact the executed path. For example, the functional simulator fetches and simulates instructions from the wrong path after a branch has been mispredicted by the timing simulation. When the branch simulation determines a misprediction, it redirects functional simulation on its correct path. Instructions from the wrong path pollute caches and internal structures, as real hardware would do.

On the other end of the spectrum, *functional-first* (also called *trace-driven*) simulators let the functional simulation produce an open-loop trace of the executed instructions that can later be replayed by a timing simulator. Some trace-driven simulators pass them directly to the timing simulator for immediate consumption. A trace-driven approach can only replay what was previously simulated. So, for example, it cannot play the wrong execution path off a branch misprediction, since the instructions trace only contains the correct execution paths. To correct for this, timing models normally implement mechanisms to account for the mispredicted execution of instructions, but in a less-accurate way.

Execution-driven simulators normally employ a tightly coupled interface, with the timing model controlling the functional execution cycle by cycle. Conversely, trace-driven simulators tend to use instrumentation libraries such as Atom [17] or Pin [11], which can run natively in the host machine. Middle ground approaches also exist, for example, Mauer et al. [12] propose a *timing-first* approach where the timing simulator runs ahead and uses the functional simulator to check (and possibly correct) execution state periodically. This approach clearly favors accuracy vs. speed and was shown to be appropriate for moderately sized multiprocessors and simple applications.

We claim that the speed, scalability and need to support complex benchmarks require a new approach, which we call *functional-directed*, that combines the speed of a fast emulator with the accuracy of an architectural simulator. Speed requirements mandate that functional simulation should be in the driver's seat and, with sufficient speed, we can capture larger applications and higher core counts. The functional-directed approach is the foundation of the COTSon platform (Fig. 4.1)

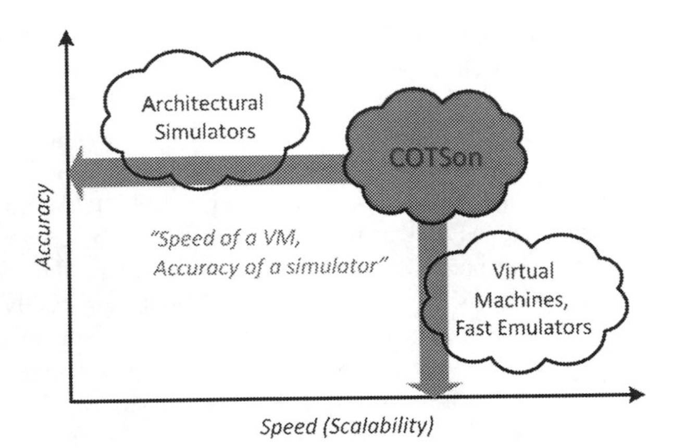

Fig. 4.1 COTSon positioning in the accuracy vs. speed space

that can address the need of many different kinds of users. Network research, usually employing captured or analytically generated traces, may generate better traces or see the impact of their protocols and implementations under real applications load. Researchers in storage, OS, microarchitecture, and cache hierarchies may also benefit from the holistic approach that enables the analysis and optimizations of the whole system while being exercised by full application workloads.

4.4.2 Accuracy vs. Speed Trade-offs

As we previously discussed, when dealing with large-scale modeling, simulation speed is by far one of the most important aspects. Although independent experiments may be run in parallel for independent configurations, sequential simulation speed still fundamentally limits the coverage of each experiment.

COTSon is designed with simulation speed as a top priority and takes advantage of the underlying virtual machine techniques in its functional simulation, e.g., just-in-time compiling and code caching. The functional simulation is handled by the AMD's SimNow simulator which has a typical slowdown of 10x with respect to native execution (i.e, a simulation speed of hundreds of MIPS). Other virtual machines such as VMware [15] or QEMU [5] have smaller slowdowns of around 1.25x, but a lower functional fidelity and limited range of supported system devices make them less interesting for a full-system simulator.

Speed and accuracy are inversely related, and they cannot be optimized at the same time; for example, sampling techniques trade instantaneous fidelity with a coarse grain approximation at a macroscopic level. For datacenter-scale workloads, the goal is to approximate with high confidence the total computing time of a particular application without having to model its specific detailed behavior at every instant. To this effect, COTSon exposes knobs to change the accuracy vs. speed trade-offs, and the CPU timing interfaces have built-in speed and sampling hooks.

This enables skipping uninteresting parts of the code (such as initial loading) by simulating them at lower accuracy. It also enables a fast initial benchmark characterization followed by *zoomed* detailed simulations. All techniques in COTSon follow this philosophy, allowing the user to select, both statically and dynamically, the desired trade-off.

4.4.3 Timing Feedback

Traditional trace-driven timing simulation lacks a communication path from the timing to the functional simulation, and the functional simulation is independent of the timing simulation. In unithreaded microarchitecture research, this is normally not a severe limiting factor. Unfortunately, more complex systems do change their functional behavior depending on their performance as we previously discussed. Having *timing feedback* – a communication path from the timing to the functional simulator – is crucial for studying these kinds of situations.

COTSon combines a sample-based trace-driven approach with timing feedback in the following way. The functional simulator runs for a given time interval and produces a stream of events (similar to a trace) which the respective CPU timing models process. At the end of the interval, each CPU model processes the trace and computes an IPC value. The functional simulator is then instructed to run the following interval at full speed (i.e., not generating events) with that IPC. By selecting different interval heuristics, the user can turn the accuracy vs. speed knob either way and clearly this approach works very well with sampling, enabling the user to select just those intervals which are considered representative.

4.4.4 COTSon's Software Architecture

COTSon decouples functional and timing simulation through a clear interface so that, for example, we can reuse existing functional emulators. Figure 4.2 shows an overview of the software architecture. The emulator for each node is AMD's SimNow [4], a fast and configurable platform simulator for AMD's family of processors that uses state-of-the-art dynamic compilation and caching techniques. The decoupled software architecture is highly modular and enables users to select different timing models, depending on the experiment. Programming new timing models for CPUs, network interfaces or disks is straightforward, as well as defining new sampling strategies. This versatility is what makes COTSon a simulation *platform*.

Timing simulation is implemented through a double communication layer which allows any device to export functional events and receive timing information. All events are directed to their timing model, selected by the user. Each timing model may describe which events it is interested in via a dynamic subscription mechanism.

Synchronous devices are the devices that immediately respond with timing information for each received event, including disks and NICs. One example of

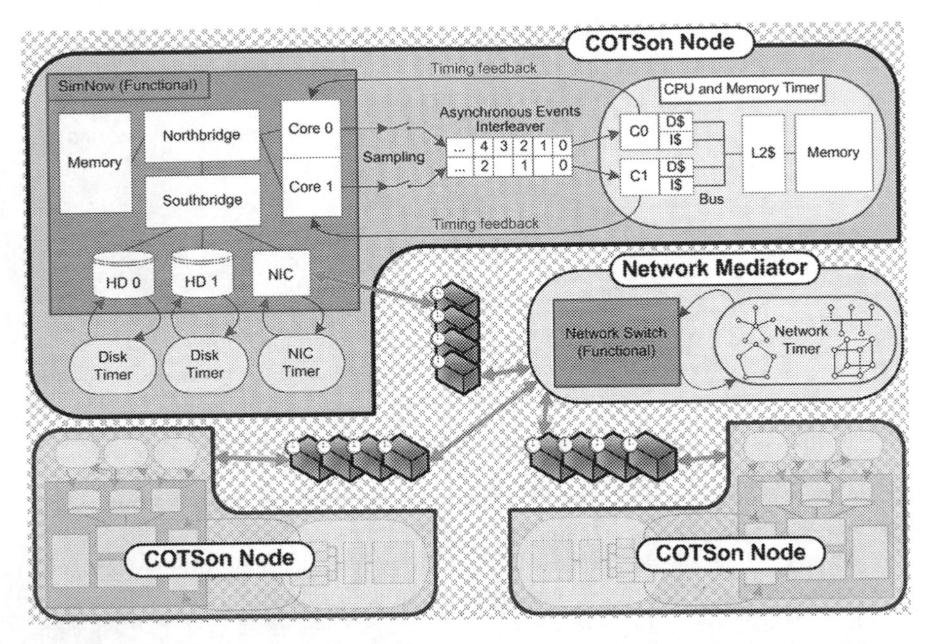

Fig. 4.2 COTSon's software architecture

synchronous communication is the simulation of a disk read. The IDE device in the functional emulator is responsible for handling the read operation, finding out the requested data and making it available to the functional simulator. The functional simulator sends the read event with all the pertinent information to COTSon which delivers it to a detailed disk timing model (such as *disksim* [7]). The latency computed by the timing model is then used by the emulator to schedule the functional interrupt that signals the completion of the read to the OS.

Synchronous simulation is unfortunately not viable for high-frequency events. If each CPU instruction had to communicate individually with a timing model, the simulation would grind to a halt. Virtual machines benefit extraordinarily from caching the translations of the code they are simulating, and staying inside the code cache for as long as possible. A synchronous approach implies leaving the code cache and paying a context switch at every instruction. For this reason we introduced the concept of asynchronous devices.

Asynchronous devices decouple the generation of events and the effect of timing information. Instead of receiving a call per event, the emulator produces tokens describing dynamic events into a buffer. The buffer is periodically parsed by COTSon and delivered to the appropriate timing modules. At specific moments, COTSon asks the timing modules for aggregate timing information (as IPC) and feeds it back to each of the functional cores. The IPC is used by the emulator to schedule the progress of instructions in each core for the next execution interval. Whenever a new IPC value is produced, the scenario in which applications and OS are being simulated changes and because the simulated system time evolves based on the output from the timing modules we achieve the coarse grain approximation goal needed to model large-scale applications.

An additional complexity comes from the many situations in which the information from the CPU timing modules has to be filtered and adapted before being passed to the functional simulator. An example of this occurs with samples where a core is mostly idle: the small number of instructions in that sample may not be enough to get an accurate estimate of the IPC and feeding back the resulting IPC significantly reduces simulation accuracy. The COTSon timing feedback interface allows for correcting the functional IPC through a *prediction* phase that uses mathematical models (such as ARMA, auto-regressive moving-average [6]) borrowed from the field of time-series forecasting.

4.4.5 Multi-core Instruction Interleaving

Functional emulators normally simulate multi-core architectures by sequentially interleaving the execution of the different cores. Each core is allowed to run independently for some maximum amount of time, called multiprocessor synchronization quantum, which can be programmed. At the end of the synchronization quantum, all the cores have reached the same point in simulated time and simulation can progress to the next quantum.

As we described above, the simulation of each core generates a series of events that are stored into asynchronous queues (one per core). This guarantees determinism, but of course diverges from the sequence occurring in a real parallel execution. In order to mimic an execution order that better approximates parallel hardware, COTSon interleaves the entries of the individual queues based on how the CPU timing models consider appropriate for their abstraction. The interleaving happens before sending the instructions (and their memory accesses) to the timing models and differs from what the emulator has previously executed. However, this difference only impacts the perceived performance of the application, which is then handled through the coarse-grain timing feedback.

Unfortunately, some synchronization patterns, such as active waits through spin locks, cannot be well captured by simple interleaving. For example, the functional simulator may decide to spin five iterations before a lock is acquired, while the timing simulator may determine that it should have iterated ten times because of additional latencies caused by cache misses or branch mispredictions. While this discrepancy can impact the accuracy of the simulation, we can reduce the error at the expense of a lower simulation speed by shortening the multiprocessor synchronization quantum. Another possibility consists of tagging all fine-grain high contention synchronization mechanisms in the application code being analyzed. When these tags arrive at the *interleaver*, COTSon can adjust the synchronization quantum to simulate the pertinent instructions based on the timing information. Tagging synchronization primitives requires modifying the guest OS (or application library), but offers the highest-possible simulation accuracy.

4.4.6 Dynamic Sampling

The goal of any sampling mechanism [18, ref-II-09-CALDER] is to identify and simulate only the representative parts of an execution. In COTSon, when a sampler is invoked, it is responsible for deciding *what to do next* and *for how long*. The sampler instructs the emulator to enter one of four distinct phases: *functional*, *simple warming*, *detailed warming* and *simulation*. In the *functional* phase, asynchronous devices do not produce any kind of events, and run at full speed. In the *simulation* phase the emulator is instructed to produce events and sends them to the timing models. In order to remove the non-sampling bias from the simulation, most samplers require that the timing models be warmed up. COTSon understands two different warming phases: *simple warming* is intended for warming the high-hysteresis elements, such as caches and branch target buffers; *detailed warming* is intended for warming up both high-hysteresis elements and also low-hysteresis ones, such as reorder buffers and renaming tables. Normally the sampler inserts one or several warming phases before switching to simulation. The sampler may be controlled from inside the functional system via the use of special backdoor connections to the simulator. For example, users may annotate their applications to send finer-grain phase selection information to the sampler.

In chapter [ref-II-09-CALDER] the fundamental benefits of sampling are amply discussed. However, traditional sampling techniques like SMARTS work at their best when the speed difference between sampled and non-sampled simulations is relatively small. For the environment in which COTSon operates where the functional emulator can over 1,000x faster than the timing simulation, a different approach is required. For example, Amdahl's law tells us that to get half the maximum simulation speed we sample with a 1:1000 ratio, which is a much more aggressive schedule than what statistical theory tells us. To reach this level of sampling, we have to use application-specific knowledge, and that is where *dynamic sampling* techniques come into play.

We observed that program phases are highly correlated to some of the high-level metrics that are normally collected during functional emulation, such as code cache misses and exceptions [9]. Intuitively, this makes sense if we consider that a new program phase is likely to be triggered by new pieces of code being executed, or by new data pages being traversed [ref-II-09-CALDER]. By monitoring these high-level metrics, a dynamic sampler can automatically detect whether it should remain in a functional phase (when the metrics are constant), or switch to a new detailed simulation phase (when variations in metrics are observed). Figure 4.3 shows results of dynamic sampling compared to SimPoint (with and without taking into account offline profiling time) and SMARTS. By adjusting the sensitivity threshold of the COTSon dynamic sampler we achieve different performance-accuracy data points.

4.4.7 Scale-Out Simulation

The foundation of any scale-out (clustered) architecture is its messaging infrastructure, i.e., the networking layer. COTSon supports NIC devices in the computing

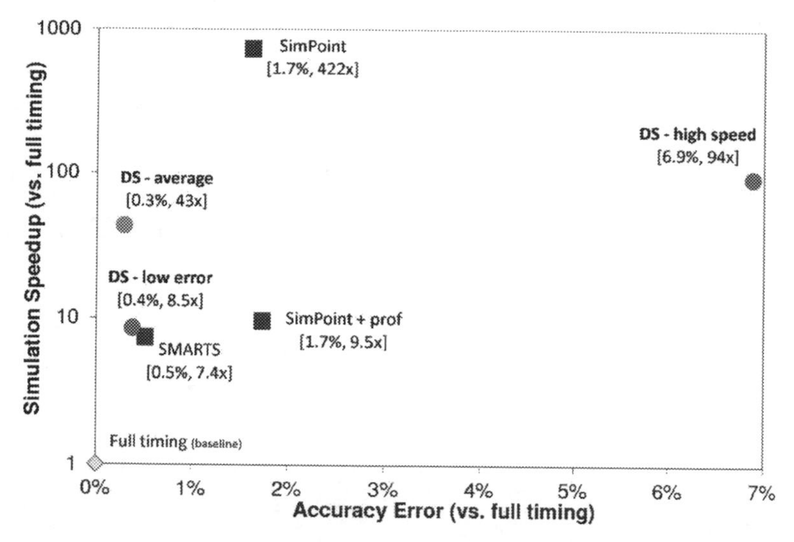

Fig. 4.3 Effects of dynamic sampling vs. fixed and profiled-based sampling

platforms as the key communication primitives that enable nodes to talk to one another. The timing for a NIC device in each node is very simple: it merely determines how long a particular network packet takes to be processed by the hardware. When a particular application needs to communicate using the network, it executes some code that eventually reaches the NIC. This produces a NIC event that reaches the NIC synchronous timing model. The response time from the timing model is then used by the functional simulator to schedule the emission of the packet into the world external to the node. The packets from all nodes get sent to an external entity, called the *network switch model* (or *mediator*). Among the functionalities of the switch model are those that route packets between two nodes, or allow the packet to reach the external world through address translation. It also offers a Dynamic Host Configuration Protocol (DHCP) for the virtual machines and the redirection of ports from the host machine into the guest simulated system.

To simulate a whole cluster, the control structure of COTSon instantiates several node simulator instances, potentially distributed (i.e., parallelized) across different host computers. Each of them is a stand-alone application which communicates with the rest via the network switch model. To simulate network latencies, the switch model relies on network timing models, which determine the total latency of each packet based on its characteristics and the network topology. The latency is then assigned to each packet and is used by the destination node simulator to schedule the arrival of the packet.

The switch model is also responsible for the time synchronization of all the node instances. Without synchronization, each of the simulated nodes would see time advance at a different rate, something akin to having skewed system clocks working at different frequencies in each node. This does not prevent most cluster applications from completing successfully, but it does prevent the simulator to make accurate timing measurements. Our switch model controls the maximum skew dynamically so that overall accuracy can be traded for simulation speed. Depending on the density of network packets in an interval, the maximum skew can increase or decrease, within a constrained range [9]. Figure 4.4 shows accuracy/speed profiles for two distributed applications, the NAS parallel suite and the NAMD molecular dynamics program, under five different policies that manage the maximum skew.

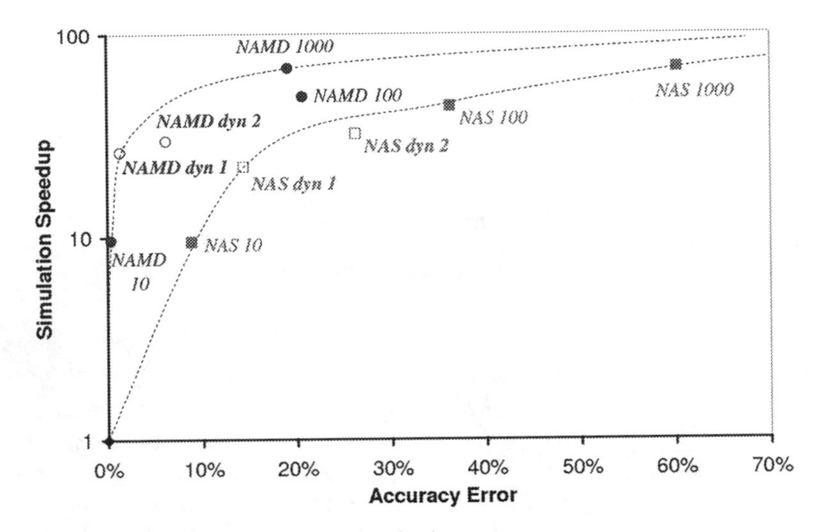

Fig. 4.4 Effects of adaptive quantum synchronization

4.5 Conclusions

The computing industry is changing rapidly and in very dramatic ways. Consolidation is turning cloud datacenters into the "new computer" that hardware, system, and software architects have to optimize. Unfortunately, the sheer number of variations that need to be explored makes the design and analysis of scale-out systems intractable with traditional tools and modeling techniques.

In this chapter, we have discussed some of the directions in which we believe computer simulation should evolve to cover the upcoming wave of new hardware, workloads, and metrics that scale out to the datacenter level. We base our analysis on what we learned in developing and using COTSon, a scalable full-system simulator developed by HP Labs in collaboration with AMD.

Leveraging fast emulation/virtualization techniques and designing for speed, full system, and modularity are the fundamental characteristics that we showed are necessary to build a scalable simulator. Being able to dynamically trade-off speed and accuracy, running unmodified applications and their entire software stack and leveraging existing "component" simulators are the key motivations that should drive the simulator design philosophy.

References

1. AFCOM's Data Center Institute, Five Bold Predictions for the Data Center Industry that will Change Your Future. March (2006).
2. Argollo, E., Falcón, A., Faraboschi, P., Monchiero, M., Ortega, D.: COTSon: Infrastructure for full system simulation. *SIGOPS Oper Syst Rev* **43**(1), 52–61, (2009).
3. Asanovic, K., Bodik, R., Christopher Catanzaro, B., Gebis, J.J., Husbands, P., Keutzer, K., Patterson, D.A., Plishker, W.L., Shalf, J., Williams, S.W., Yelick, K.A.: The landscape of parallel computing research: A view from Berkeley. In: *Technical Report UCB/EECS-2006-183, EECS Department, University of California, Berkeley*, December (2006).
4. Bedicheck, R.: SimNow: Fast platform simulation purely in software. In: *Hot Chips 16*, August (2004).
5. Bellard, F.: QEMU, a fast and portable dynamic translator. In: *USENIX 2005 Annual Technical Conference, FREENIX Track*, Anaheim, CA, pp. 41–46, April (2005).
6. Box, G., Jenkins, G.M., Reinsel, G.C.: *Time Series Analysis: Forecasting and Control*, 3rd ed. Prentice-Hall, Upper Saddle River, NJ (1994).
7. Bucy, J.S., Schindler, J., Schlosser, S.W., Ganger, G.R., Contributors.: The Disksim simulation environment version 4.0 reference manual. In: *Carnegie Mellon University Parallel Data Lab Technical Report CMU-PDL-08-101*, May (2008).
8. Falcón, A., Faraboschi, P., Ortega, D.: Combining simulation and virtualization through dynamic sampling. In: *Proceedings of the IEEE International Symposium on Performance Analysis of Systems & Software (ISPASS)*, San Jose, CA, April 25–27, (2007).
9. Falcón, A., Faraboschi, P., Ortega, D.: An adaptive synchronization technique for parallel simulation of networked clusters. In: *Proceedings of the IEEE International Symposium on Performance Analysis of Systems & Software (ISPASS)*, Austin, TX, April 20–22, (2008).
10. Karkhanis, T.S., Smith, J.E.: A first-order superscalar processor model. In: *Proceedings of the 31st Annual International Symposium on Computer Architecture*, München, Germany, June 19–23, (2004).
11. Luk, C.-K., Cohn, R., Muth, R., Patil, H., Klauser, A., Lowney, G., Wallace, S., Reddi, V.J., Hazelwood, K.: Pin: Building customized program analysis tools with dynamic instrumentation. In: *Proceedings of the ACM Conference on Programming Language Design and Implementation (PLDI)*, Chicago, IL, June 12–15, (2005).
12. Mauer, C.J., Hill, M.D., Wood, D.A.: Full-system timing-first simulation. In: *SIGMETRICS '02: Proceedings of the 2002 ACM SIGMETRICS International Conference on Measurement and Modeling of Computer Systems*, Marina Del Rey, CA, June 15–19, (2002).
13. Ould-Ahmed-Vall, E., Woodlee, J., Yount, C., Doshi, K.A., Abraham, S. Using model trees for computer architecture performance analysis of software applications. In: *Proceedings of the IEEE International Symposium on Performance Analysis of Systems & Software (ISPASS)*, San Jose, CA, April 25–27, (2007).
14. Pricewatch.com data March (2009).
15. Rosenblum, M.: VMware's virtual platform: A virtual machine monitor for commodity PCs. In: *Hot Chips 11*, August (1999).
16. Rosenblum, M., Herrod, S.A., Witchel, E., Gupta, A.: Complete computer system simulation: The SimOS approach. *IEEE Parallel Distrib Technol* **3**(4), 34–43, (1995).
17. Srivastava, A., Eustace, A.: ATOM—a system for building customized program analysis tools. In: *Proceedings of the ACM Conference on Programming Language Design and Implementation (PLDI)*, Orlando, FL, June 20–24, (1994).
18. Yi, J.J., Kodakara, S.V., Sendag, R., Lilja, D.J., Hawkins, D.M.: Characterizing and comparing prevailing simulation techniques. In: *Proceedings of the 11th International Conference on High Performance Computer Architecture*, pp. 266–277, San Francisco, CA, February 12–16, (2005).

Chapter 5
Modular ISA-Independent Full-System Simulation

Gabriel Black, Nathan Binkert, Steven K. Reinhardt, and Ali Saidi

Abstract Research-oriented simulators require flexibility and configurability in addition to good performance and reasonably accurate timing models. The M5 simulator addresses these needs by developing abstractions that promote modularity and configurability without sacrificing performance. This design allows M5 to provide a combination of capabilities—including multisystem simulation, full-system simulation, and ISA independence—that is unique among open-source research simulation environments. This chapter provides an overview of the key abstractions in M5.

5.1 Introduction

Computer architecture research places demands on simulators that are different from those of product development. Researchers care about modeling performance, but wish to explore large design spaces, often including unconventional designs. As a result, they require reasonably accurate but highly flexible performance models and are willing to trade some reduction in accuracy for increased flexibility and higher simulation speeds. In contrast, the "virtual platforms" discussed in Chapter 1, such as Virtutech Simics (see Chapter 2) and AMD's SimNow™[1], are typically used for pre-silicon software development and thus focus on high-speed comprehensive functional modeling rather than performance modeling. At the other extreme, microprocessor product designers use very accurate performance models to evaluate detailed design trade-offs, but these simulators are relatively inflexible, being closely tied to specific product designs, and are inherently slow due to their level of detail.

M5, originally developed by the authors and others while at the University of Michigan, addresses the needs of computer architecture researchers by providing

S.K. Reinhardt (✉)
Advanced Micro Devices, Inc., Bellevue, WA USA
e-mail: steve.reinhardt@amd.com

R. Leupers, O. Temam (eds.), *Processor and System-on-Chip Simulation*,
DOI 10.1007/978-1-4419-6175-4_5,

a flexible, modular, open-source simulation environment. M5 grew out of a desire to research novel system architectures for high-speed TCP/IP network I/O. This goal dictated several requirements. First, because most of a system's networking code—protocol stacks and device drivers—is in the operating system kernel, we needed the ability to simulate the execution of OS kernel code, known as *full-system simulation*. Second, we needed an environment that could faithfully model network interface controllers and other I/O devices, particularly their interaction with the simulated system's memory hierarchy. Third, we required the ability to model multiple networked systems in a deterministic fashion to capture both ends of a network connection. No publicly available tool provided all these features, motivating us to develop a new system to meet our needs.

Although network-oriented simulation drove M5's unique functional requirements, we continued to use M5 for more conventional microarchitecture and memory-system research. These disparate uses forced M5 to span a variety of options: modeling full systems and stand-alone applications via system-call emulation, single and multiple systems, uniprocessors and multiprocessors, simple and detailed timing models, etc. Managing this scope of alternatives required us to develop structures and interfaces that allowed users to flexibly compose a number of different component models into a variety of configurations.

We strove to define these interfaces in a way that would not compromise M5's performance or accuracy. While M5's performance is highly dependent on the workload and system configuration being simulated, running M5 on a 3.2 GHz system achieves 3–6 MIPS for basic functional simulation, 1–3 MIPS for a simple in-order core model with atomic-mode cache hierarchy, and 200–400 KIPS for a detailed out-of-order CPU model and timing-mode cache hierarchy. We have also demonstrated M5's accuracy by comparing its results against real systems running networking benchmarks and found that M5 predicted achieved network bandwidth within 15% in most cases [7].

M5's modularity and flexibility, combined with its unrestrictive open-source license, have made it a popular tool for architecture research at a number of institutions in academia and in industry. In this chapter, we focus on the key abstractions in M5 that contribute to its modular design:

- The *SimObject class* serves as a foundation for pervasive object orientation.
- *Port objects*, used to interface SimObjects within the memory system, allow flexible and interchangeable composition of CPUs, caches, I/O devices, and buses.
- *Python integration* allows flexible configuration and control of simulation from a scripting language, without giving up the performance of C++.
- The *ISA definition system* encapsulates the specifics of various instruction-set architectures (ISAs), including instruction decoding and semantics, virtual address translation, and other platform and OS dependencies. This subsystem enables M5 to support multiple ISAs across multiple CPU models, even for full-system simulation. M5 currently supports the Alpha, ARM, MIPS, POWER, SPARC, and x86 ISAs, with some level of full-system support for Alpha, ARM, SPARC, and x86.

5.2 Object Orientation

Object-oriented design is a key contributor to M5's modularity. The `SimObject` C++ and Python base classes, from which all major component models derive, provide uniform interfaces to users and convenient code reuse to developers. The ability to construct configurations from independent, composable objects leads naturally to advanced capabilities such as multicore and multisystem modeling.

5.2.1 SimObjects

All major simulation components in M5 are considered SimObjects and share common behaviors for configuration, initialization, statistics, and serialization (checkpointing). SimObjects include models of concrete hardware components such as processor cores, caches, interconnect elements, and devices, as well as more abstract entities such as a workload and its associated process context for system-call emulation.

Every SimObject is represented by two classes, one in Python and one in C++. These classes derive from the `SimObject` base classes present in each language. The Python class definition specifies the SimObject's parameters and is used in M5's script-based configuration. The common Python base class provides uniform mechanisms for instantiation, naming, and setting parameter values. The C++ class encompasses the SimObject's state and remaining behavior, including the performance-critical simulation model. The integration of the Python and C++ manifestations of a SimObject is covered in more detail in Section 5.4.

The configuration process assigns a unique name to each SimObject instance based on its position in the configuration hierarchy (e.g., `server.cpu0.icache`). These names are used in statistics, serialization, and debugging, in which numerous instances of the same SimObject type must be distinguished.

The C++ `SimObject` base class provides uniform interfaces for common SimObject operations in the form of virtual functions. These functions must be defined by a SimObject to initialize itself, register its statistics, and serialize and unserialize its state as necessary. The base class also maintains a list of all instantiated SimObjects. Operations such as serializing simulation state consist primarily of walking this list and calling `serialize()` on each SimObject.

Unlike many other simulation systems, M5 does not prescribe any specific mechanisms for SimObjects to interact with virtual time. Each SimObject must schedule events on the global event queue as needed to support its level of timing detail. This flexibility allows different models to support different levels of detail and to exploit model-specific knowledge to minimize the number of scheduled events (an important factor in determining simulator performance). For example, a CPU object typically schedules an event for each of its internal clock cycles, but may decide not to schedule events when it is idle or stalled. Most other SimObjects are more passive, being invoked indirectly when they receive messages from other components, then scheduling events only as needed to model delays in processing those messages.

M5's virtual time is maintained in integer ticks, independent of the clock rate of any particular component. Each tick typically represents 1 ps, though an alternate time unit can be specified. M5's Python configuration infrastructure allows SimObject parameters to be expressed in a variety of units such as nanoseconds or gigahertz, which are automatically converted into ticks.

5.2.2 Multicore Systems

Given the encapsulation properties of object-oriented design, configuring a basic multicore system in M5 is as simple as instantiating multiple instances of a processor core SimObject, then connecting these cores with main memory and devices via a bus object. Introducing caches into the configuration requires a cache coherence protocol (described in Section 5.3.4).

Because each core can be instantiated independently, heterogeneous multicore systems (e.g., combinations of in-order and out-of-order cores) are also easily constructed. The ISA is a compile-time parameter for each CPU model and the current build system supports just a single global ISA setting, so mixed ISA systems are not presently feasible. However, only relatively modest changes would be needed to enable multiple ISAs in the same simulation if desired.

5.2.3 Multisystem Simulation

As already mentioned, the ability to model multiple systems communicating via TCP/IP was a key motivation in the design of M5. As with multicore simulation, this capability is a natural consequence of encapsulation; the user merely instantiates multiple independent sets of system components (cores, memories, devices, and interconnect). A pair of systems (more specifically, the network interface devices on those systems) can be connected by a common SimObject modeling an Ethernet link.

Heterogeneity is particularly important in multisystem modeling, because a typical experiment will want to model a "system under test" in detail, while one or more other load-generating systems need only be modeled in a relatively lightweight functional mode. Additional simplifications can be achieved by exploiting the flexibility of simulation. For example, a single client machine running at a simulated 20 GHz clock rate with a one-cycle main memory latency can generate as much load on a system under test as a large number of more realistic systems. Keeping all these systems in a common timebase is very important to emulate TCP behavior properly while maintaining deterministic simulations.

5.3 Memory Modeling

The goal of the memory system in M5 is to provide a flexible interface for interconnecting memory components. These components, including interconnect, caches, bridges, DRAM, programmed I/O devices, and DMA-capable I/O devices, all use the same generic interface. This allows arbitrary objects to be connected, subject to coherence requirements, without designing for each possible combination.

Unlike some simulators, M5 provides a unified functional and timing view of the memory system (accounting for both time and data). This unified view allows for the accurate modeling of timing-dependent operations as well as making it more difficult for accesses to "cheat," since an incorrectly timed access may also mean incorrect data.

5.3.1 Ports and MemObjects

The flexibility of the M5 memory hierarchy is based on the *ports* interface and the *MemObjects* that provide them. All objects in the memory system inherit from the `MemObject` base class, which extends the `SimObject` class to provide a method to access the object's named ports.

At configuration time, pairs of ports on different MemObjects are connected as directed by the user's Python configuration script (described in Section 5.4). A MemObject communicates with its connected neighbors by passing a message structure to a send method on one of its ports; the message is passed to the corresponding receive method on the connected peer port, which invokes the appropriate method on the target MemObject to handle the message receipt. Through careful design, our ports implementation enables arbitrary MemObject connections while requiring only a single virtual function call per message. Port pairs always form a symmetric bidirectional connection between their associated MemObjects.

Some MemObjects have multiple distinct ports; for example, caches have a CPU-side port through which requests are accepted from CPUs or upstream caches, and a memory-side port through which requests are sent to downstream caches or main memory. These ports have distinct names, and messages received on these ports call distinct methods on the cache MemObject. Other MemObjects, such as buses, support multiple identical ports; in this situation, multiple connections to the same port name result in independent instances of the same port object, organized in a vector and distinguishable by index.

All messages communicated over ports are instances of the `Packet` class. A packet carries the command, size, address, and (optionally) data for a memory transaction. A packet also carries a pointer to a `Request` object, which contains information about the origin of the transaction, such as the program counter of the associated load or store instruction (if applicable). A complete memory transaction is always associated with a single immutable request, but may involve multiple different packets. For example, the CPU may send a packet to the L1 cache requesting a write of a single byte, which may in turn generate a different packet from the L1 cache to the L2 cache requesting a read of an exclusive copy of a full cache block.

M5's ports were inspired by ASIM's ports [3] and are similar to the sockets defined by the SystemC TLM-2.0 specification [6]. M5 ports differ from both of these in that they are used only for communicating requests and responses between arbitrary memory-system components, not arbitrary information between modules. M5's `Packet` class resembles TLM-2.0's generic payload, so an M5 port could be considered most similar to a TLM-2.0 socket that is constrained to communicate only generic payload objects. Unlike TLM-2.0 generic payloads, we have extended the `Packet` class to incorporate new features as we encounter new demands rather than defining a powerful but complex extension mechanism. Another distinction among these systems is that ASIM ports also model bandwidth and latency; M5 ports, like SystemC TLM-2.0 sockets, do not directly incorporate timing, but rely on interconnect objects such as buses for modeling timing where appropriate.

5.3.2 *Access Modes*

The M5 memory system supports three modes of access: timing, atomic, and functional. The same packet structure is used for all three; the sender chooses the access mode by calling distinct send methods on its port object. Timing and atomic accesses both simulate memory requests that occur in the modeled system, but provide different trade-offs between simulation accuracy and performance. Functional accesses provide "back door" access to the memory system for simulator-specific operations such as debugging and initialization.

Timing accesses are split transactions: a timing-mode send returns immediately, and the corresponding response (if any) will arrive via an independent receive method invocation at some future simulated time. This decoupling allows interconnect and cache modules to model realistic timing, using multiple events as necessary to model propagation delays and inter-packet contention. Of course, this realism has a cost in simulation performance. Atomic requests provide increased performance in exchange for decreased accuracy, such as in the fast-forwarding phase of a sampled simulation, by propagating a request from the originator to the final target in a sequence of nested calls, with the response (if any) transmitted back as part of the function return sequence. The atomic path supports a latency estimate achieved by summing component latencies along the path, but cannot account for inter-packet contention. However, the transaction is processed significantly more quickly because no additional events are scheduled. M5's timing and atomic modes correspond roughly to SystemC TLM-2.0's non-blocking and blocking transport interfaces, respectively [6].

To avoid complexity, component models such as caches are not required to handle interactions between timing-mode and atomic-mode accesses. As a result, all components in a simulated system must coordinate to use the same mode for simulated accesses at a given point in time. The `System` class contains a flag that indicates the current system mode, which other components can query to determine the appropriate access mode. This mode can vary across time (e.g., between fast-forwarding and detailed simulation phases) and among different systems in a multisystem simulation (e.g., between a load-generating client and a server under test, as described in Section 5.2.3).

Functional-mode accesses are required because memory system components (including caches) maintain functional state. Thus the current value of a memory location may exist only in the data storage of a simulated cache. For debugging and other accesses not directly related to simulated program execution, it is necessary to read and write memory locations regardless of their state in the cache hierarchy. Functional accesses propagate throughout the system to obtain the latest value (on a read) or update all cached copies (on a write). Functional accesses can be issued at any point in the simulation, so they must interact properly with outstanding timing-mode accesses.

5.3.3 I/O Devices

A unique feature of M5 is that I/O devices are treated as first-class memory objects. The I/O devices inherit from the same base classes as other objects in the memory system (e.g., caches) and communicate through the same data structures. Since the interface through which I/O devices communicate is the same as any other object (including the CPU), I/O devices are free to model arbitrarily complex timing and can be attached on any level of the memory hierarchy (e.g., between the first- and second-level caches).

Every platform has a combination of platform-specific memory mapped devices (e.g., interrupt controllers, timers) and platform-independent devices (e.g., Ethernet controllers, disk controllers). The platform-independent devices are normally much more complex than a timer or interrupt controller and also tend to be connected far away from the CPU on a peripheral bus such as PCI Express. Even though the devices vary in complexity and where they are connected, devices inherit from the same base classes. Devices that implement PCI inherit from a PCI base class that handles setting up PCI configuration space based on the requirements of the specific platform that is being simulated. By treating these devices as first-class memory objects, they can be attached where appropriate. Thus, a local interrupt controller or closely coupled NIC can be attached between the L1/L2 cache, while a serial port may be attached on a slow bus several bridges away from the CPU.

Devices needed solely for the platform to boot (e.g., cache configuration registers) typically do not require much timing fidelity. In these cases the devices can simply respond to memory requests without any timing of their own. On the other hand, an Ethernet controller is sufficiently complex that it needs internal timing to operate correctly. Here a state machine can be built that manages the scheduling of events for the device and all the interactions that are required by the system and the kernel.

5.3.4 Cache Coherence

Because the memory system combines function and timing, a multicore system with caches requires that the caches support a fully functional coherence protocol. M5 cache objects implement a broadcast-based MOESI snooping protocol. This protocol is complicated by the desire to support nearly arbitrary configurations containing multiple levels of both shared and private caches. To solve the complex race conditions created by multilevel bus hierarchies, M5 implements special "express" snooping requests that allow a cache to query and invalidate upstream caches atomically even when in timing mode. These requests provide timing similar to a cache that maintains duplicate copies of upstream cache tags (as in, for example, Sun's Niagara [4]), without requiring the cache object to know the configuration of caches above it.

Although M5's coherence protocol is adequate for many purposes, it is neither detailed nor flexible enough to support research into coherence protocols themselves. A previous attempt to decouple the coherence protocol from the cache model introduced complexity without providing significant advantages and was abandoned. Our current strategy for supporting coherence research (along with more general memory-system interconnect topologies) is based on integrating the Ruby memory-system simulator from the University of Wisconsin as described in Section 5.6.1.

5.4 Python Integration

M5 derives significant power from tight integration of Python into the simulator. While 85% of the simulator is written in C++, Python pervades all aspects of its operation. As mentioned in Section 5.2.1, all SimObjects are reflected in both Python and C++. The Python aspect provides initialization, configuration, and simulation control. The simulator begins executing Python code almost immediately on start-up; the standard `main()` function is written in Python, and all command-line processing and startup code is written in Python.

Initially, M5 was written almost entirely in C++. The configuration to simulate was specified via command-line arguments. This approach became unworkable once we started making the simulator more object oriented and wanted to specify per-object parameters flexibly. We then switched to reading the configuration from a user-provided `.ini`-format file. In this file, each section represented a single SimObject. Object configurations were repetitive, so we added a special parameter for inheriting values from another section; thus, for example, two CPUs could share the same parameter values. For additional flexibility we started running the `.ini` files through the C pre-processor before parsing them. This enabled the use of `#define` and `#ifdef` to generate families of configurations from a single `.ini` file.

When the limitations of the C preprocessor became constraining, we started writing Python scripts to generate the `.ini` files directly. We developed a framework in which every C++ SimObject had a corresponding Python class that knew how to generate the proper `.ini` section for that SimObject. The C preprocessor became unnecessary, and Python class inheritance replaced the special `.ini` parameter. This approach worked reasonably well, except that—since there was no explicit linkage between the Python and C++ code—adding parameters to an object required editing both Python and C++ in a coordinated fashion. Also, the C++ code did not support parameter inheritance, so parameter changes could require tedious and error-prone modification of multiple C++ classes.

To eliminate this redundancy, we extended the Python code to automatically generate the C++ code for parameter handling in the C++ SimObjects from the Python parameter descriptions. This step eliminated the manual maintenance of redundant C++ code and added parameter inheritance to C++ in one fell swoop. From there, it seemed obvious that generating and then parsing the intermediate `.ini` file was not particularly useful, so we embedded the Python interpreter in M5, allowing Python code to directly call C++ to create C++ SimObjects.

Once we embedded the Python interpreter, we found that we could leverage the expressive power of Python to simplify other areas of M5, such as initiating and restoring from checkpoints and controlling the main simulation loop. These additional features are supported by exposing C++ functions to Python using the open-source Simplified Wrapper and Interface Generator (SWIG). Our current belief is that only code that needs to be fast (e.g., run-time simulation models) should be in C++, and nearly everything else should be in Python. There is still a considerable amount of C++ code (85%) compared with Python (15%), but the ratio is constantly shifting toward Python.

5.4.1 Scripting Simulations

Nearly all aspects of an M5 simulation are controlled via a user-supplied Python script specified on the M5 command line. This script can use standard Python libraries to extract settings from additional command-line arguments or environment variables.

The bulk of a typical script is used to build the simulation configuration by constructing a tree of Python SimObject instances. In addition to directly instantiating the classes that correspond to C++ SimObjects, users can also subclass those classes to provide specialized default parameter settings. For example, the user's script can create an `L1Cache` class that is derived from the built-in `Cache` class but has a default size and associativity suitable for a first-level cache. Scripts can also "clone" instances to generate sets of identical objects without defining a new subclass.

Once the configuration tree has been defined in Python, the script invokes the `instantiate()` method on the root of the tree, which recursively creates the C++ counterparts for each of the Python SimObjects. The Python object parameter values are passed to the C++ object constructors using structures auto-generated from the parameters specified in the Python SimObject definition.

Once the C++ SimObjects have been created, control returns to the Python script. The script can invoke the C++ event loop for a specified number of simulation ticks via the `simulate()` method. It can also save state to, or restore state from, checkpoints stored on disk. A typical simulation script might have a loop that repeatedly invokes `simulate()` and then saves state, providing regular periodic checkpoints of simulation progress.

For a more detailed discussion and an example of a configuration script, see our earlier paper [2].

5.4.2 Build System

Before we started seriously using Python in the simulator itself, we abandoned GNU Make in favor of SCons, a build tool written entirely in Python. Our initial motivation for making the switch to SCons was simply that our `Makefile` was getting far too complicated. Our choice of SCons turned out to be a boon to the Python integration effort. Because SCons build files are just Python scripts, we can include the Python SimObject descriptions directly into our build files, simplifying the auto-generation of C++ parameter-handling code.

Having Python at our disposal at both compile time and run time allows us to extend M5's modularity to static configuration. For example, adding a new SimObject to M5 requires only a simple build file that identifies the Python and C++ files that contain the object's parameter declarations and model implementation, then including the directory containing that build file on the SCons command line.

5.5 ISA Independence

The majority of M5 is ISA independent, meaning it avoids making assumptions about the semantics or implementation of the ISA it will be used to simulate. Support for a particular ISA is provided by other components which are layered on top of this generic framework. This separation is a reflection of M5's modular design philosophy and has many of the same benefits. Separating out ISA support can be difficult, though, partially because it may not be clear how to distill out the truly ISA-dependent aspects of a system both conceptually and in an actual implementation. Also, once a particular ISA has been selected, it has to be able to work efficiently with the rest of the simulator across a generic interface. M5 achieves this using the architecture shown in Fig. 5.1.

At a high level, instruction behavior, instruction decoding, memory address translation, interrupt and fault handling, and some details of the simulated environment are considered ISA dependent. The following sections discuss each of these areas and the interfaces they use to interact with the rest of the simulator.

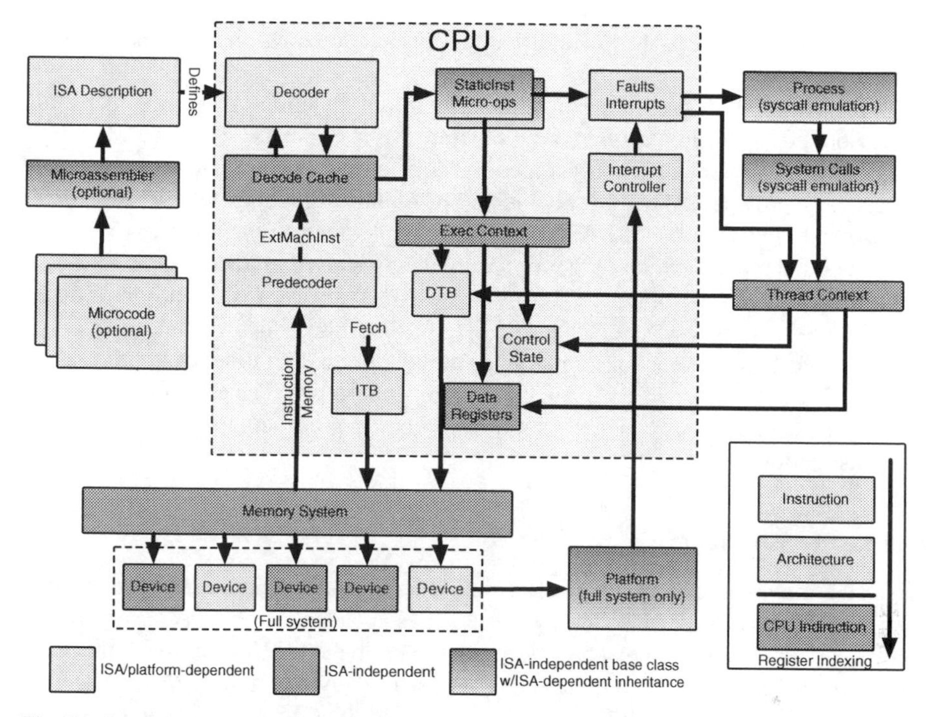

Fig. 5.1 Separation of the simulator into ISA/platform-dependent and -independent components

5.5.1 Instruction Behavior

The central task of a CPU is to execute instructions, and the ISA is largely defined by the set of instructions the CPU executes. Despite this close relationship, the CPU models themselves need to be fundamentally ISA independent to avoid having to implement the cross product of all ISAs and CPU types. To make that possible, two generic interfaces are used: one that allows a CPU to manage and manipulate instructions, and one that allows an instruction to interact with the CPU.

5.5.1.1 Static Instruction Objects

All instructions in M5 are represented by instances of `StaticInst` subclasses. Each instance provides member variables and functions that either describe the instruction in a standard format or manifest its behavior. These object's contents must be the same for equivalent instructions and cannot change after they are built. That allows them to be cached and reused, saving construction overhead. This is an especially effective technique since most execution time is spent in loops.

Each instruction object has an array of the register indices it accesses to allow the CPU to track data dependencies. It also provides several flags to describe itself and announce any special requirements. These specify, among other things, whether the instruction is integer or floating point, if it is serializing, or if it is a memory reference.

Every instruction provides either one or several functions that implement its behavior. For most instructions, this is simply the `execute` function. Memory instructions may work differently, though, if they need to wait on memory. In those cases, they provide `initiateAcc` and `completeAcc` functions which are called before and after their access. All instructions implement a function that returns a disassembled representation of themselves that can be used to trace execution.

For RISC ISAs, a single `StaticInst` can almost always realistically represent the behavior of an instruction. For CISC ISAs like x86 and for some especially complicated RISC instructions, microcode may be needed. In those cases, the main `StaticInst` sets a flag declaring that it contains micro-ops. The CPU fetches these micro-ops with an index called the micro-pc and processes them instead of the original instruction.

5.5.1.2 Register Indexing

Register indices are used to select small, fast storage locations an instruction accesses as it executes. Most simply, these act as an array, numbered consecutively and accessed directly. In a real processor implementing a real ISA, however, the process is significantly more complicated. There will likely be multiple independently indexed register files, and there may be complicated rules that change which storage location an index maps to over time. Even the existence of a unique storage location might be an illusion provided by, for instance, register renaming.

To provide a consistent, ISA-independent framework to describe and manage access to registers, these mechanisms can be restated in terms of "index spaces" and transformations between them. An index space is like an address space, but refers to register indices instead of memory addresses.

At a high level, an index space is implied by the instruction an index comes from and how it is used. In integer addition, for example, an index likely refers to an index space corresponding to the integer register file. There may be more than one index space for integer registers, though, if mechanisms like register windows are used. Each of these maps into a larger space, resolving aliases and unambiguously identifying the storage location requested by the instruction. SPARC's register windowing semantics [8] are relatively complex but can still be described this way.

At this point, the CPU is responsible for providing the illusion that this canonical index corresponds to a unique register. It is free to do that directly, or using register renaming or any other technique. Using this system, the roles of the ISA and the CPU model are clearly separated, allowing each to be implemented without being restricted by the specifics of the other.

5.5.1.3 Execution Contexts

To emulate an instruction's execution, a `StaticInst` object needs to access system state, typically registers or memory locations. Exactly where that state is stored and how it is accessed depends on the implementation of the CPU model. For example, a CPU model that uses register renaming will need to look up the physical register that corresponds with a logical index. M5 abstracts away these details by passing an "execution context object", or `ExecContext`, to the `StaticInst`'s `execute` method. The `ExecContext` defines a common interface for these operations, while allowing different CPU models to customize their implementation.

The simplest CPU model in M5, appropriately called the `SimpleCPU`, executes one instruction at a time and is not multithreaded. It maintains a single copy of architectural state which it presents directly to an instruction, so the `SimpleCPU` object itself is used as the `ExecContext`.

In more complicated models like our out-of-order CPU, called `O3`, each of the many instructions in flight sees a different view of the state of one of the CPU's several threads. In those cases, an object more specific than the entire CPU is needed. `O3` uses instances of a class called `DynInst` to contain the execution state for each dynamic instruction, including register remapping and any load/store ordering information. These objects are used as the `ExecContexts` for those CPUs.

In addition to `ExecContexts`, other classes exist that also help generalize the CPU models. `ThreadContexts` are very similar to `ExecContexts` but provide access to callers from outside the CPU like system calls. `ThreadState` classes centrally locate common state and allow sharing related declarations and code across implementations.

5.5.2 Instruction Decoding

5.5.2.1 ISA Description

Each ISA defines a function that accepts a contextualized blob of instruction memory and produces a fully decoded `StaticInst`. Defining that function and the potentially hundreds of classes that implement each different instruction is unfortunately very tedious and error prone. To make the process more manageable, M5 provides a custom high-level ISA description language. The ISA author describes what instructions exist, how they are encoded, and what they do in this concise representation, then a parser uses that description to generate C++ code which is compiled into M5.

The core element of the ISA description is the decoder specification, which looks like the following:

```
1  decode OPCODE {
2      0: Integer::add({{ Ra = Rb + Rc; }});
3      1: Integer::div({{
4          if (Rc == 0)
5              return new DivideByZeroFault();
6          Ra = Rb / Rc;
7      }});
8  }
```

In this example, the `OPCODE` bitfield (defined elsewhere) is used to switch among several instructions. Semantics for each instruction are defined inside double braces using a subset of C with minor extensions. The `div` instruction recognizes and reports a fault condition using `Fault` objects, which are described in Section 5.5.5. Decode blocks can contain other decode blocks to build up complex decode structures.

To process an instruction definition, the parser passes the arguments from the decode block (the C snippet and other optional flags) into a format function, in this case one called `Integer`. This function, written in Python, handles related batches of instructions. Much of the work is performed by a library of helper classes and functions that parse the C snippet to generate source and destination operand lists, set up default constructor code for the `StaticInst` subclass, etc. These instruction-specific code segments are then substituted into a C++ code template (typically specific to the format function), forming the final generated code.

The ISA description language also features `let` blocks, which are blobs of arbitrary Python code that are evaluated as they are parsed. These blobs can be used to define potentially very sophisticated helper functions and specialized mechanisms that can be used by other parts of the description. This feature is deceptively important and is the source of much of the power of these descriptions, allowing them to handle very complex ISAs like x86.

5.5.2.2 Predecoder

M5's decoding infrastructure expects to be fed information about one instruction at a time and to produce one instruction object at a time. For ISAs that force all instructions to be the same size and to be aligned, all that is needed is to fetch the appropriate amount of memory from the current PC and feed it into the decoding mechanism. For ISAs like x86 in which instructions can vary from a one to a dozen or more bytes long, that approach will not work. To address this problem, instruction memory must pass through a predecoder, defined by the ISA, before it goes into the normal decoder described above. The predecoder's job is to accept a stream of bytes of instruction memory and to produce a uniform, contextualized representation of an instruction called an `ExtMachInst`. The instruction will not be decoded, but all of its components will have been gathered and separated from the incoming stream and put in a uniform container.

5.5.3 Microcode Assembler

In ISAs like SPARC or ARM in which macro-ops are very uncommon and relatively simple, normal ISA description mechanisms are sufficient to build them. For ISAs like x86 in which some instructions behave like small subroutines, a microcode assembler has been developed that vastly simplifies defining them.

Like the ISA description parser itself, the microcode assembler is written in Python and interprets a custom language that leverages Python syntax. Macro-ops are defined as top-level constructs and are composed of micro-ops. The micro-ops are selected using a mnemonic and then a string of arguments in Python syntax. The assembler looks up a Python class in a dictionary using the mnemonic and passes the arguments to its constructor to create a representation of the micro-op. The macro-op is an instance of a user-defined Python class, and the micro-ops are added to it one at a time as they are encountered. The assembler can recognize directives mixed in with the micro-ops, written as a period followed by a name and optional parameters. The name is used to look up a function in another dictionary which is called with the parameters. Macro-ops also support labels and control flow. Together, it looks like the following:

```
1  def macroop exampleInst {
2      add r1, r2, r3
3      .useRegSize 2
4  end:
5      store r1, [r2, r3]
6  }
```

Here the second argument to the `store` takes advantage of the convenient similarity between Python's list syntax and traditional assembler memory address computation syntax.

The assembler also has a dictionary that acts as the variable scope when microops are constructed, directives are handled, etc. This allows defining arbitrary keywords for use in the microcode without having to teach the assembler how to recognize them. In the preceding example, the parameters `r1`, `r2`, etc. would be defined this way. They might be strings naming the index of a register like "`INTREG_R1`".

Once a macro-op object has been constructed, it is put in a dictionary keyed on its name. In the main part of the ISA description, the object can be recalled and used to fill in the portion of the decoder that goes with that instruction. The macro-op and micro-op Python objects know how to instantiate themselves in C++, so they are simply expanded where needed.

In addition to defining individual macro-ops, it is also possible to define a microcode ROM. This is done by using `def rom` instead of `def macroop` and can appear multiple times to continue appending to the same ROM object. This is to allow portions of a macro-op to be put into the ROM near where the rest is defined.

5.5.4 Memory Address Translation

While the concept of virtual memory is almost universal among ISAs, each implements it slightly differently. To support these differences, each ISA is responsible for implementing a translation lookaside buffer (TLB) object that performs instruction level address translation.

TLBs focus on paging, however, which is sometimes a limitation. CPUs assume the page offset of an address will be the same before and after translation. Because the size of a page is typically larger than any other size boundary an address might need to be aligned against, translation is assumed not to change alignment. In ISAs with different, less regular addressing mechanisms, like x86's segmentation, those assumptions may not hold. By carefully delegating each step of the process, however, even segmentation is supported in this way.

ISAs also handle TLB misses differently, usually in software, but sometimes in hardware in line with the original access. To support those semantics, translation can either complete immediately or finish later using a callback.

The idea of TLBs as translation caches with a collection of entries, a lookup mechanism, replacement policies, etc., is substantially the same across ISAs. Unfortunately, these ISA-independent elements are not yet separated from the translation process, and so the entire TLB is considered ISA dependent. Ideally, future versions of this system will pull those elements out into a common base TLB object.

5.5.5 Interrupts and Fault Handling

Faults are represented in M5 as instances of the ISA-independent `Fault` class. Because the cause, classification, and effects of faults are very ISA dependent, specific faults (e.g., page faults) are represented as subclasses of `Fault` by the ISA.

The specific architectural effects of a fault are implemented by the `invoke` virtual function, and the `Fault` subclasses are built into a class hierarchy so faults that behave the same share the same `invoke` function through inheritance. Because `Faults` are objects, they can also hold contextualizing information that affects their behavior (e.g., the address that caused a page fault or TLB miss).

`Fault` object pointers are returned by functions performing tasks that might fault, namely instruction execution and address translation. When appropriate, the CPU invokes any resulting fault object on the current architectural state, transforming it as defined by the ISA. Using this mechanism, the CPU is able to recognize faults without having to understand their semantics.

Because interrupts and faults are very similar and because the `Fault` object mechanism is very flexible, interrupts are just another type of `Fault` object in M5. Unlike faults, however, there are usually complicated rules that describe when an interrupt is recognized, how it gets from its source to a consuming CPU, and how it can be sent as an inter-processor interrupt (IPI). All of these behaviors are implemented by an `Interrupts` object defined by the ISA and instantiated as part of the CPU. Interrupts are communicated to that object instead of the CPU itself, and the CPU periodically polls it for an interrupt to service. Complicated systems like x86's xAPIC architecture can thus be emulated in detail without having to specialize the CPU model itself.

5.5.6 Platforms and Processes

While the choice of ISA primarily affects the CPUs that execute its instructions, real workloads expect certain properties from their environment that go beyond the ISA itself. For instance, in system-call emulation (SE) mode, the system calls available, their numbering, and how arguments are passed to them are a function of the ISA and OS being emulated. In full-system (FS) mode, the simulated OS code expects to be able to interact with a hardware environment and interface typical for that ISA.

In SE mode, the program M5 is simulating is represented internally by an instance of a `Process` class (a subclass of `SimObject`). Each supported ISA/OS combination provides a custom subclass of `Process` to translate requests from the simulated program into actions performed by M5. For example, a system call instruction requests a specific system call by providing a numeric index into the OS's system call table. M5 uses the appropriate `Process` object to translate that index into an internal function which will emulate that system call.

This abstraction also works in the opposite direction, allowing M5 to use a generic interface to interact with the simulated program. For example, when a real OS starts a process, it builds up an initial stack frame whose contents, layout, and location vary among ISAs and OSs. Other in-memory structures like Linux's vsyscall page or SPARC's register window spill and fill handlers might also need to be installed. The `Process` class provides virtual functions for initialization that are specialized by each subclass to handle these details. The `Process` class also provides a standard interface to retrieve system call arguments, allowing generic implementations of common system calls like `close` and `write` despite differing register- or stack-based argument-passing conventions across platforms.

Finally, `Process` objects provide a location for information about the current state of execution. The base `Process` class holds generic information such as the base and bounds of the process's heap area and the mapping from the process's file descriptors to M5 file descriptors. Subclasses can extend this information in ISA- or OS-dependent ways; for example, x86's `Process` subclass stores the location and size of the x86-specific global descriptor table (GDT) which is used by x86 Linux to implement thread-local storage. Subclasses can also add ISA- or OS-specific member functions; for example, SPARC's `Process` subclass implements a function to flush register window contents to memory.

In FS mode, M5 runs real OS code, which automatically takes care of all these process-level issues. The OS itself expects certain interfaces and hardware, though, so platform dependencies have not been eliminated, just pushed to a lower level. M5's device model and memory system are very modular and flexible, so providing the hardware expected for a particular platform ideally involves just implementing any missing devices and hooking everything up at the right address. The commonality of the PCI device model allows most of M5's devices and device-configuration infrastructure to be reused across platforms. The exact mechanisms used to access devices can differ, though. In some cases, like Alpha and MIPS, devices are mapped directly into memory. In others, like x86, there may be a separate I/O address space accessed with dedicated I/O instructions. M5 defines a partition of the physical address space for these I/O operations, supported by special features in the simulated x86 TLB. I/O instructions like `IN` and `OUT` are implemented as loads and stores within this address range.

5.6 Future Directions

M5 is by no means a finished product. The simulator is constantly being enhanced to support additional features and component models. An increasing number of enhancements are being contributed by users outside the group of initial developers; for example, during the course of writing this chapter, someone unexpectedly contributed an implementation of the POWER ISA. Two current projects that will greatly enhance M5 are integration with the Ruby memory system simulator and parallelization.

5.6.1 Ruby Integration

M5's bus-based cache coherence protocol (Section 5.3.4) is adequate for a large class of experiments. However, as users look to study on-chip networks, they will need support for more sophisticated interconnects and coherence protocols. Instead of trying to evolve our memory system to add these capabilities, we chose to leverage existing efforts by forming a joint project with the GEMS [5] simulator team. GEMS was designed specifically for memory-system research and has a detailed memory system model called Ruby. Ruby models caches, memories, and networks and uses a domain-specific language called SLICC to specify coherence protocols. As of this writing, the integration of Ruby and M5 is in its initial stages, with a rough but working initial version available for limited use.

5.6.2 Parallelization

Although architecture research has increasingly focused on multicore systems, most simulators are still single threaded. As the number of simulated cores increases, so too does the time to complete a simulation of that system. We have begun the process of parallelizing M5 to make use of multicore processors. Our first steps include the ability to assign SimObjects to separate event queues, enabling a transition from a single global event queue to multiple per-thread queues. Initially, we plan to simulate independent systems on different threads, a capability we have already demonstrated in prototype form. Ultimately, we hope to allow for the simulation of multiple cores within the same system on a multicore processor.

Acknowledgments We thank all the users of and contributors to M5 who have made the simulator successful. We also thank the institutions that have supported its development, including the National Science Foundation (under Grant Nos. CCR-0105503 and CCR-0219640), AMD, Hewlett-Packard, IBM, Intel, MIPS, and Sun. Thanks to Brad Beckmann for reviewing a draft of this chapter. Any opinions, findings, and conclusions or recommendations expressed in this chapter are those of the author(s) and do not necessarily reflect the views of the National Science Foundation (NSF) or any other sponsors.

References

1. Advanced Micro Devices, Inc.: AMD SimNow™ simulator. URL http://developer.amd.com/cpu/simnow
2. Binkert, N.L., Dreslinski, R.G., Hsu, L.R., Lim, K.T., Saidi, A.G., Reinhardt, S.K.: The M5 simulator: Modeling networked systems. IEEE Micro **26**(4), 52–60 (2006).
3. Emer, J., Ahuja, P., Borch, E., Klauser, A., Luk, C.K., Manne, S., Mukherjee, S.S., Patil, H., Wallace, S., Binkert, N., Espasa, R., Juan, T.: Asim: A performance model framework. IEEE Comp **35**(2), 68–76 (2002).
4. Kongetira, P., Aingaran, K., Olukotun, K.: Niagara: A 32-way multithreaded SPARC processor. IEEE Micro **25**(2), 21–29 (2005).
5. Martin, M.M.K., Sorin, D.J., Beckmann, B.M., Marty, M.R., Xu, M., Alameldeen, A.R., Moore, K.E., Hill, M.D., Wood, D.A.: Multifacet's general execution-driven multiprocessor simulator (GEMS) toolset. SIGARCH Comput Archit News **33**(4), 92–99 (2005). DOI http://doi.acm.org/10.1145/1105734.1105747
6. Open SystemC Initiative: OSCI TLM-2.0 Language Reference Manual (2009).
7. Saidi, A.G., Binkert, N.L., Hsu, L.R., Reinhardt, S.K.: Performance validation of network-intensive workloads on a full-system simulator. In: *Proceedings of the 2005 Workshop on Interaction between Operating System and Computer Architecture (IOSCA)*, pp. 33–38 Austin, TX, Oct (2005).
8. Weaver, D.L., Germond, T. (eds.): The SPARC Architecture Manual (Version 9). PTR Prentice Hall (1994).

Chapter 6
Structural Simulation for Architecture Exploration

David August, Veerle Desmet, Sylvain Girbal, Daniel Gracia Pérez, and Olivier Temam

Abstract While processor architecture design is currently more an art than a systematic process, growing complexity and more stringent time-to-market constraints are strong incentives for streamlining architecture design into a more systematic process. Methods have emerged for quickly scanning the large ranges of hardware block parameters. But, at the moment, systematic exploration rarely goes beyond such *parametric* design space exploration. We want to show that it is possible to move beyond parametric exploration to *structural* exploration, where different architecture blocks are automatically composed together, largely broadening the scope of design space exploration. For that purpose, we introduce a simulation environment, called UNISIM, which is geared toward interoperability. UNISIM achieves this goal with a combination of modular software development, distributed communication protocols, a set of simulator service APIs, architecture communications interfaces (ACIs), and an open library/repository for providing a consistent set of simulator components. We illustrate the approach with the design exploration of the on-chip memory subsystem of an embedded processor target. Besides design space exploration, we also show that structural simulation can significantly ease the process of fairly and quantitatively comparing research ideas and illustrate that point with the comparison of state-of-the-art cache techniques. Finally, we disseminate the whole approach for both facilitating design space exploration and the fair comparison of research ideas through ArchExplorer, an atypical web-based infrastructure, where researchers and engineers can contribute and evaluate architecture ideas.

6.1 Structural Simulation

Simulator development is a huge burden for architecture design groups because of long coding time and the increasing complexity of architectures; homogeneous and heterogeneous multi-cores only make matters worse. Structural simulation aims at

O. Temam (✉)
INRIA Saclay, Batiment N, Parc Club Universite, rue Jean Rostand, 91893 Orsay Cedex, France
e-mail: oliver.temam@inria.fr

R. Leupers, O. Temam (eds.), *Processor and System-on-Chip Simulation*,
DOI 10.1007/978-1-4419-6175-4_6,

improving simulator development productivity by breaking down architectures into components roughly corresponding to hardware blocks and then allowing to reuse components within or across research groups. We use the term "structural simulation" rather than "modular simulation" (and "component" rather than "module") in order to highlight that the approach implies not only software modularity but also a software breakdown which reflects the hardware blocks. We start by discussing how structural simulation can facilitate simulator design; we survey several prominent structural simulation environments and then we show, through the UNISIM environment, how to further improve simulator interoperability.

6.1.1 Structural Simulation Environments

Low-level design languages like Verilog or VHDL are structural, but they are too slow for rapid architecture prototyping or cycle-level simulation. Conversely, simulators for rapid prototyping and cycle-level modeling are abstract models of architectures and typically suffer from two weaknesses as a consequence: the lack of design constraints may breed insufficiently precise if not unrealistic simulator models, and these models may lack the inherent modularity of hardware blocks, which is key for reuse, interoperability, and development productivity. Structural high-level (cycle-level or higher) simulators aim at bridging the gap between Verilog/VHDL and ad hoc high-level simulators.

One of the first instances of such a high-level structural simulation framework is HASE [5], which dates back to 1995 and provides many of the functionalities of modern structural simulation environments: structural design, hierarchical description of the architectures, and a discrete event engine, at the expense of simulation speed. ASIM [10], developed at Intel, is a structural simulation framework which implements port-based communication interfaces between components, a clean separation between instruction feeders (traces, emulators) and the performance model, together with a set of tools. While ASIM was not publicly disseminated outside Intel, it is considered as one of the first successful application of high-level structural simulation. SimFlex [13] is a public simulation environment based on ASIM concepts and implementation, which extends ASIM with advanced functionalities such as sampling and checkpointing. The LSE (Liberty Simulation Environment) [31] is a structural simulation framework which aims at implementing a rigorous communication interface between components in the form of a three-signal handshaking protocol. This approach improves reuse by distributing control between components at the level of component ports and helps implement an efficient component wake-up strategy. MicroLib [24] proposes a similar handshaking protocol, though it is based on the widely adopted SystemC simulation framework. SystemC [23] is an open source C++ library for the structural development of computer architectures (especially embedded architectures) proposed by the OSCI group and now an IEEE standard. While SystemC provides only elementary support for cycle-level communications, insufficient to ensure compatibility and reuse between components, it is geared toward multi-abstraction-level simulation, from cycle-level to transaction-level modeling (TLM). SystemC has become an industry standard and numerous commercial tools like the ARM RealView fast simulators [1], VaST [32], or Virtutech Simics (see Chapter 3) propose SystemC interfaces.

Like SystemC and Virtutech, an increasing number of simulation environments are expanding to system-level simulation in order to encompass the full hardware system and to accommodate the operating system impact of multi-cores. Other simulation infrastructures reflect that trend, such as the M5 full-system modular simulation infrastructure, see Chapter 5, the GEMS [21] environment for exploring cache coherence protocols, or even the HP COTSON infrastructure, see Chapter 4, for exploring data centers.

6.1.2 UNISIM

UNISIM is designed to rationalize simulator development by making it possible and efficient to distribute the overall effort over multiple research groups, even without direct cooperation. The following four observations drive the development of UNISIM:

1. Many existing simulators are monolithic or designed for a single architecture. As a result, it is difficult to extract a micro-architecture component from the simulator for reuse, sharing, or comparison.
2. The lack of interoperability among existing simulators hampers reuse, sharing, and comparison of ideas and hinders the take-up of academic results by companies which cannot afford the effort to integrate multiple academic simulators.
3. The most popular and open structural simulation environment, SystemC, does not impose strict enough design guidelines to ensure the interoperability of components.
4. Due to the growing impact of technology on architecture design, simulators require many more features than just performance evaluation: power and temperature modeling and other technology constraints detailed in Chapter 14, sampling for simulation speed, debugging support for parallel computing, etc. These features are currently implemented in an ad hoc manner which makes them hard to integrate in a vast array of simulators.

6.1.2.1 Control

While mapping hardware blocks to simulator components is both attractive and intuitive, it can conflict with the objective of reusing simulator components. The key difficulty is the reuse of *control logic*. Control is often implemented, or simply represented, as centralized or only partially decentralized; a symptomatic example is the DLX block diagram from the famous textbook [14], where control is centralized into one block, see Fig. 6.4. While this implementation is structural, it is difficult to reuse: any modification in any of the hardware blocks would require a modification of the control block. And while control logic may correspond to a small share of transistors, it often corresponds to the largest share of simulator code. For instance, a cache bank can account for many more transistors than the cache controller, but it is just an array declaration in the simulator, while the cache controller can correspond to several hundred simulator lines.

In order to address these conflicting decomposition/reuse objectives, UNISIM uses a solution pioneered by LSE [31] and the MicroLib environment [24]. This solution provides a customizable representation of control which lends itself well to reuse and structural design. This control *abstraction* takes the form of a handshaking mechanism between components: a component makes no assumption about the other components beyond its incoming and outgoing signals. All control logic corresponding to interactions with other components are embedded in these signals. This approach has two key benefits: (1) by construction, it provides a *distributed* simulator implementation of control, see Fig. 6.4 and (2) it defines a rigorous and standardized interface between components, which, in turn, improves components' interoperability.

6.1.2.2 Simulator Services

Beyond executing the code and measuring the number of cycles, simulators must provide an increasing number of services. Technology issues are imposed to precisely evaluate delays, power, and temperature [28], among other metrics. Simulator speed constraints may require to plug in sampling techniques [12, 27, 34], see Chapters 10 and 11, and checkpointing techniques [33] for replaying or speeding up executions. System and CMP simulators are complex to debug and thus require non-trivial debugging support. While there exist tools for each functionality, they are often implemented in an ad hoc manner, sometimes even embedded in a simulator, and cannot be easily plugged to other simulators.

In order to both increase the functionality of simulators and provide a more efficient way to leverage the work of different research groups, UNISIM implements a set of *Service APIs*. These APIs are essentially standardized function calls; any simulator component implementing these function calls automatically benefits from the corresponding services. For instance, a cache component providing statistics on its activity can get an evaluation of power consumption, provided a power model is plugged into the engine. For instance, in Fig. 6.1 we show the components of a full-system Mac simulator, each with ports for accessing a debugging API. Using the debugging API, we could plug the standard ddd debugger into UNISIM. This is the second benefit of the *Service APIs* approach: the services are plugged at the engine level. Therefore, not only can any simulator benefit from a service as long as it implements the corresponding API, but it is also easy to replace a service tool with another, provided again that it is API compliant. For instance, two power models can be easily compared on any simulator with that approach. For the aforementioned ddd example, a small adapter (44 lines) must be developed to make a UNISIM component gdb compliant.

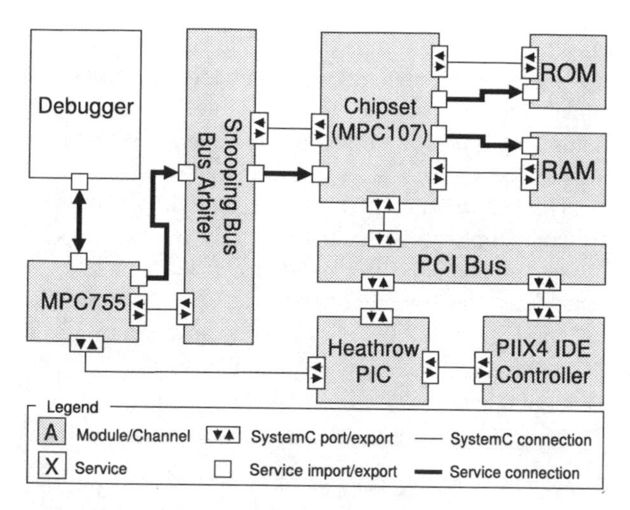

Fig. 6.1 Full-system PowerMac G3 simulator with debugging services

6.1.2.3 Simulator Interoperability

Not only is simulator functionality difficult to reuse and inter-operate, but the simulator implementations of different hardware blocks themselves can be difficult to extract from a simulator and reuse in another one. For instance, most of the SystemC simulators are not interoperable because of insufficiently clear communication and development guidelines. Besides enforcing a stricter communication protocol, as mentioned in Section 6.1.2.1, UNISIM takes three additional steps for achieving interoperability.

Wrap and reuse. The first step consists in acknowledging that simulator development is a huge effort, and any group which has invested several man-years on a tool will not drop that effort easily. As a result, UNISIM is designed to create heterogeneous simulators by *wrapping* existing simulators within a UNISIM component, allowing existing simulators to interact with UNISIM components, or even with other simulators. Besides the syntactic wrapping, an adapter must sometimes be developed to translate the simulator *Model of Computation* [8] into the UNISIM/SystemC one, i.e., the method through which the different components or parts of the simulator are called or woken up. In order to illustrate that this approach is pragmatic, we have wrapped the full SimpleScalar simulator into a UNISIM component, stripped it off its simple memory model, and connected it to another UNISIM component which contains a detailed SDRAM component developed into a third simulation environment. Only 50 source lines of SimpleScalar had to be modified in order to break the pipeline loop, and the resulting simulator is only 2.5 times slower despite the more complex memory model and the wrapping.

Open repository. The second interoperability action is to build an open library or repository providing a set of consistent and interoperable models and components. The repository maintains information about inter-component compatibility and component history and allows users to easily locate components meeting their needs, thus improving the reuse of components. Finally, this library is open, allowing anyone to upload components, while retaining intellectual property rights to the components and applying a license of the author's choice. We show an application of this open repository for design space exploration in Section 6.3.

Architecture Communications Interfaces. The third step is to ensure that architecture blocks are compatible at the *hardware* level; for that purpose, they should agree on a set of input and output control and data signals. This set of signals forms a communication interface which we term *Architecture Communications Interfaces*, as an analogy to software-level APIs (Application Programming Interfaces). For instance, a processor/memory interface, depicted in Fig. 6.2, enables to connect processors with a large range of cache mechanisms and also to compose arbitrarily deep cache hierarchies with interface-abiding cache, see Fig. 6.3.

`address`	bi-directional, 32 bits Memory request address.
`data`	bi-directional, path width Data for read and write requests.
`size`	bi-directional, $log_2(max(\#bytes))$ bits Request size in bytes.
`command`	processor → cache, 3 bits
proc./L1	Request type (`read, write, evict, prefetch`).
L1/L2	Request type (`read, write, evict, prefetch, readx, flush`).
proc./mem.	Request type (`read, write, evict, prefetch, readx, flush`).
`cachable`	processor → cache, 1 bit Whether or not the requested address is cachable.

Fig. 6.2 Processor/memory interface

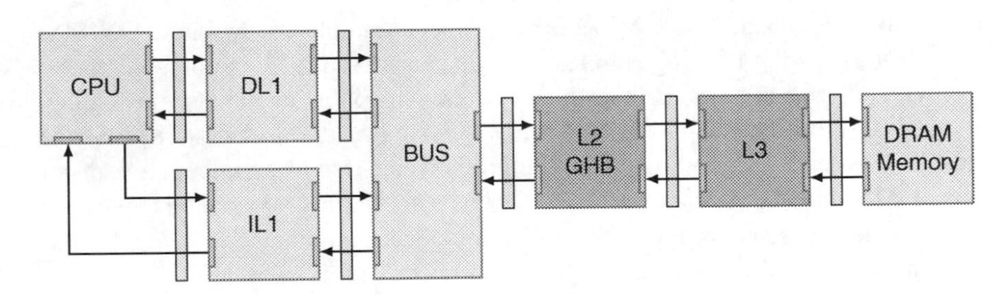

Fig. 6.3 Composing architectures

ACIs raise two main questions: (1) Do we need to design a new ACI for each new hardware block variation? (2) Which hardware blocks, typically studied in architecture research, are eligible for an ACI definition?

Extending ACIs to accommodate new mechanisms. For most data cache mechanisms, the innovations proposed are *internal* to the block and these innovations have little or no impact on the interface with the processor and the memory in many,

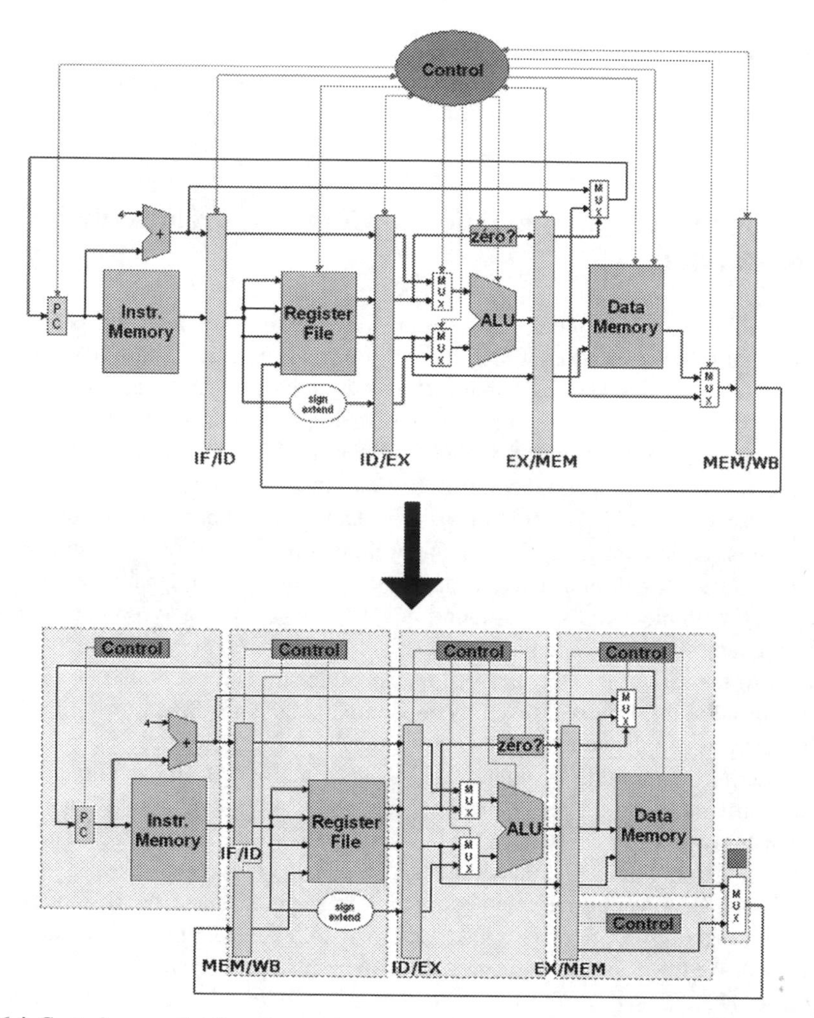

Fig. 6.4 Central versus distributed control

though not all, cases. Moreover, an ACI can be extended with the necessary signals, without affecting backward compatibility.

Hardware blocks eligible for ACIs. It can also be argued that processor/memory is a special form of interface, more clearly defined than for other hardware blocks in the system.

Some hardware blocks, such as the commit stage of a superscalar processor, effectively have a large set of connections with the rest of the architecture, which change across architectures, and are thus difficult to consider in isolation. However, there are quite a few hardware blocks considered as "domains" of computer architecture, which have good modularity properties, and for which it would be possible to define an interface with the rest of the system. A non-exhaustive list of such hardware blocks includes instruction caches, TLBs, prefetchers, branch predictors, interconnects (buses and network on chips) and network topology, main memory (DRAMs, SRAMs, RDRAM, etc.), any co-processor in a heterogeneous multi-core, cores themselves in multi-cores, and functional units.

6.2 Using Structural Simulation to Compare Architecture Research Ideas

While novel architecture ideas can be appreciated purely for the new insight they provide, empirical evaluation of architecture ideas through simulation has become, rightfully or not, a fixture of any architecture research or development work. Novel ideas are expected to bring some quantitative improvement over state-of-the-art mechanism(s). Many domains of sciences, such as biology or physics, often request that research articles provide the ability to reproduce experiments in order to confirm the researchers' claims; sometimes, reproducing experiments is even part of the reviewing process. However, reproducing architecture ideas of other researchers or engineers, often based on articles, white papers, or on sparsely available, unmaintained, or incomplete tools, is a daunting task, and it is largely comprehensible that reproducibility and systematic comparison has not become mainstream in our domain. At the same time, the lack of interoperable simulators and the fact simulators are usually not disclosed (even by academics) make it difficult, if not impossible, to fairly assess the benefit of research ideas.

In order to illustrate these points, and the potential benefit of designing simulators in a structural and interoperable way, we made a comparison of 10 data cache mechanisms over a 15-year period. We have collected research articles on data cache architectures from the 2000–2004 editions of the main conferences (ISCA, MICRO, ASPLOS, HPCA). We have implemented most of the mechanisms corresponding to pure hardware optimizations (we have not tried to reverse-engineer software optimizations). We have also implemented the older but widely referenced mechanisms (*Victim Cache*, *Tagged Prefetching*, and *Stride Prefetching*). The different mechanisms, a short description, and the corresponding reference are listed in Table 6.1.

We have implemented all cache techniques as independent components, plugged into the popular superscalar simulator SimpleScalar 3.0d [2, 4] in order to illustrate across-simulator interoperability at the same time; we refer the reader to [24] for the exact simulator configuration in these experiments. We have compared the mechanisms using the SPEC CPU2000 benchmark suite [30] and extracted traces using SimPoint [12, 27], see Chapter 10.

6.2.1 Validating Re-implementations

The lack of publicly disclosed and interoperable simulators makes the task of comparing against prior art exceedingly difficult and time consuming. For several mechanisms, there was no easy way to do an IPC validation. The metric used in *FVC*

Table 6.1 Target data cache optimizations

Acronym	Mechanism	Description
VC	Victim Cache [19] (L1)	A small fully associative cache for storing evicted lines; limits the impact of conflict misses without (or in addition to) using associativity
FVC	Frequent Value Cache [37] (L1)	A small additional cache that behaves like a victim cache, except that it is just used for storing frequently used values in a compressed form (as indexes to a frequent values table). The technique has also been applied, in other studies [35, 36], to prefetching and energy reduction
TK	Timekeeping [15] (L1)	Determines when a cache line will no longer be used, records replacement sequences, and uses both information for a timely prefetch of the replacement line
TKVC	Timekeeping Victim Cache [15] (L1)	Determines if a (victim) cache line will again be used, and if so, decides to store it in the victim cache
Markov	Markov Prefetcher [18] (L1)	Records the most probable sequence of addresses and uses that information for target address prediction
TP	Tagged Prefetching [29] (L2)	One of the very first prefetching techniques: prefetches next cache line on a miss or on a hit on a prefetched line
SP	Stride Prefetching [3] (L2)	An extension of tagged prefetching that detects the access stride of load instructions and prefetches accordingly
CDP	Content-Directed Data Prefetching [6] (L2)	A prefetch mechanism for pointer-based data structures that attempts to determine if a fetched line contains addresses and, if so, prefetches them immediately
CDPSP	CDP + SP (L2)	A combination of CDP and SP as proposed in [6]
TCP	Tag Correlating Prefetching [16] (L2)	Records miss patterns and prefetches according to the most likely miss pattern
DBCP	Dead-Block Correlating Prefetcher [20] (L1)	Same as TCP, but records also hits for a better identification of the appropriate pattern
GHB	Global History Buffer [22] (L2)	Same as stride prefetching, but tolerates varying strides within a stream

and *Markov* is miss ratio, so only a miss ratio-based validation was possible. *VC*, *Tag*, and *SP* have been proposed several years ago, so the benchmarks and the processor model differed significantly. *CDP* and *CDPSP* used an internal Intel simulator and their own benchmarks. For all the above mechanisms, the validation consisted in ensuring that absolute performance values were in the same range, and that trends were often similar (relative performance difference of architecture parameters, among benchmarks, etc). For *TK*, *TKVC*, *TCP*, and *DBCP*, we used the performance graphs provided in the articles for the validation.

Because one of the key points is to argue that research articles may not provide sufficient information on experiments and methodology, we decided, on purpose, not to contact the authors in a first step, in order to assess how much we could dig from the research articles only. Later on, we have either contacted the authors or have been contacted by authors and tried to fix or further validate the implementation of their mechanisms.

For several of the mechanisms, some of the implementation details were missing in the article or the interaction between the mechanisms and other components wash not sufficiently described, so we had to second-guess them. We illustrate the potential impact of such omissions with *TCP*; the article properly describes the mechanism, how addresses are predicted, but it gives few details on how and when prefetch requests are sent to memory. Among the many different possibilities, prefetch requests can be buffered in a queue until the bus is idle and a request can be sent. Assuming this buffer effectively exists, a new parameter is the buffer size; it can be either 1 or a large number, and the buffer size is a trade-off, since a too short buffer size will result in the loss of many prefetch requests (they have to be discarded) and a too large one may excessively delay some prefetch requests. Figure 6.5 shows the performance difference for a 128-entry and a 1-entry buffer. All possible cases are found: for some benchmarks like *mgrid* and *applu*, the performance difference is tiny, while it is dramatic for *art, lucas*, and *galgel*. For instance, the performance of *lucas* decreases (with a 128-buffer).

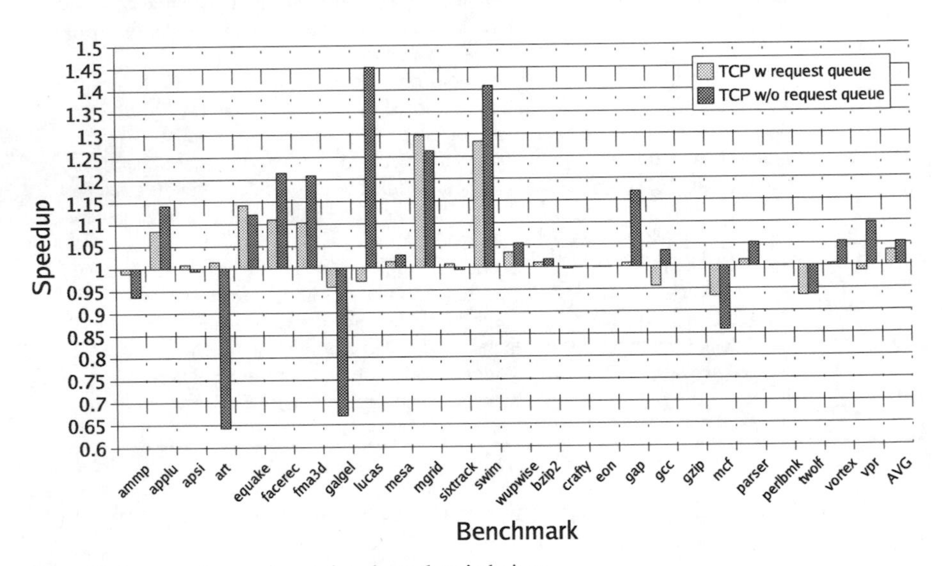

Fig. 6.5 Impact of second-guessing the authors' choices

6.2.2 A Quantitative Comparison of Data Cache Techniques

How has data cache performance improved over years? Figure 6.6 shows the average IPC speedup over the 26 benchmarks for the different mechanisms with respect

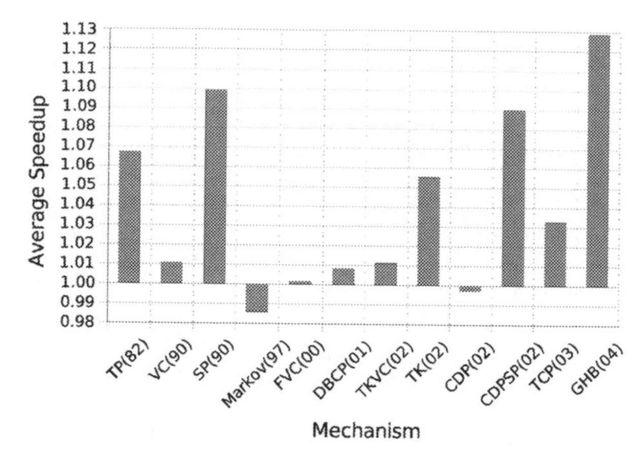

Fig. 6.6 Speedup of the different cache mechanisms, ranked by chronological order

to a standard cache of the same size. We find that the best mechanism is *GHB* (published in 2004), an evolution of *SP*, an idea originally published in 1990, and which is the second best performing mechanism, then followed by *TK*, proposed in 2002. While which mechanism is the best very much depends on industrial applications (e.g., cost and power in embedded processors versus performance and power in general-purpose processors), it is still fair to say that the progress of data cache research over the past 15 years has been all but regular.

A common simulation platform but also a common methodology is important.

As an example of the impact of methodology, we want to show the impact of varying the benchmarks. Even picking different subsets of the same benchmark suite can lead to radically different design decisions.

We have ranked the different mechanisms for every possible benchmark combination of the SPEC CPU2000 benchmark suite from 1 to 26 benchmarks (there are 26 SPEC benchmarks). First, we have observed that for any number of benchmarks less than or equal to 23, i.e., the average IPC is computed over 23 benchmarks or less, there is always *more than one winner*, i.e., it is always possible to find two benchmark selections with different winners. In Fig. 6.7, we have indicated how often a mechanism can be a winner for any number N of benchmarks from $N = 1$ to $N = 26$ (i.e., is there an N-benchmark selection where the mechanism is the winner?). The table shows that, for any selection of 23 benchmarks or less, the selection can lead to research articles with opposite conclusions as to which cache mechanism performs best. Note that not using the full benchmark suite can also be detrimental to the proposed mechanism. For instance, *GHB* performs better when considering all 26 benchmarks rather than the benchmark selection considered in the article. So not only architecture ideas should be compared on a common simulation platform, but the comparison methodology should be precisely defined, lest wrong conclusions are drawn. We refer the reader to [25] for a more detailed analysis and comparison.

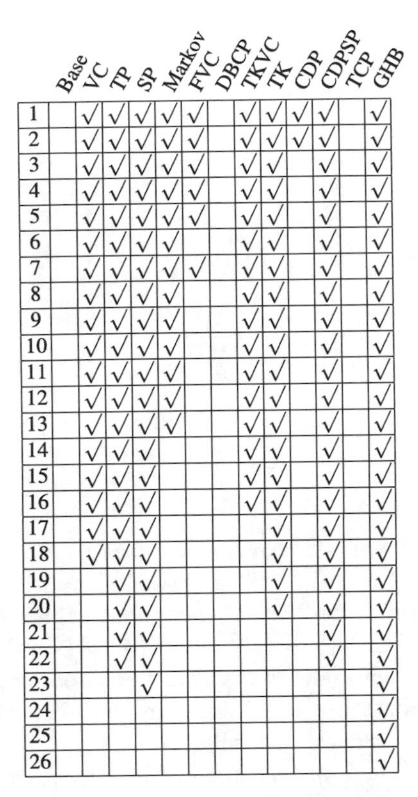

	Base	VC	TP	SP	Markov	FVC	DBCP	TKVC	TK	CDP	CDPSP	TCP	GHB
1		√	√	√	√	√		√	√	√	√		√
2		√	√	√	√	√		√	√	√	√		√
3		√	√	√	√	√		√	√		√		√
4		√	√	√	√	√		√	√		√		√
5		√	√	√	√	√		√	√		√		√
6		√	√	√	√			√	√		√		√
7		√	√	√	√	√		√	√		√		√
8		√	√	√	√			√	√		√		√
9		√	√	√	√			√	√		√		√
10		√	√	√	√			√	√		√		√
11		√	√	√	√			√	√		√		√
12		√	√	√	√			√	√		√		√
13		√	√	√	√			√	√		√		√
14		√	√	√				√	√		√		√
15		√	√	√				√	√		√		√
16		√	√	√				√	√		√		√
17		√	√	√					√		√		√
18		√	√	√					√		√		√
19			√	√					√		√		√
20			√	√					√		√		√
21			√	√							√		√
22			√	√							√		√
23				√									√
24													√
25													√
26													√

Fig. 6.7 Which mechanism can be the winner with N benchmarks?

6.3 Structural, Open, and Continuous Design Space Exploration

The architect usually relies on a trial-and-error process where intuition and experience often drive the creation and selection of appropriate designs. However, as architecture complexity increases, and thus the number of possible architecture options similarly increases, it is no longer obvious that experience and intuition are always the sole and best drivers for architecture design decisions. The cache mechanisms comparison of Section 6.2.2 illustrates this concern by suggesting that the progress of research may not always be regular over time, in large part because our current methodology does not emphasize comparison of research results. There is probably an architecture complexity tipping point where human insight would be more productive if combined with systematic architecture design space exploration. In order to address this issue we have combined structural simulation with an atypical web-based permanent and open design space exploration framework into ArchExplorer.

ArchExplorer mitigates the *practical* hurdles which prevent a researcher from performing a broad exploration and fair comparison of architecture ideas, especially the time and complexity involved in reimplementing other researchers' works.

ArchExplorer can be summarized as a framework for an *open and permanent exploration of the architecture design space*. Instead of requesting a researcher to find, download, and run the simulators of competing mechanisms, we provide a remote environment where the researcher can upload his/her own simulator and compare the mechanism against all other existing and previously uploaded mechanisms. The continuous exploration serves two purposes: allow to explore a huge design space over a long period of time and progressively build over time a large database of results that will further speed up any future comparison.

After uploading the mechanism, the whole exploration process is *automatic*: from plugging the mechanism into an architecture target to retuning the compiler for that target, statistically exploring the design space and publicly disseminating exploration results.

We have implemented this whole process at `archexplorer.org` using several of the data cache mechanisms mentioned in Section 6.2.2 plus a few others, for the memory system of an embedded processor. Because the comparison is done not only for a fixed processor baseline but by varying all possible architecture parameters and components, it casts an even more accurate, and fairly different, light on data cache research.

The overall methodology and approach is summarized in Fig. 6.8. In short, a researcher adapts (wraps) his/her custom simulator to make it compatible with the UNISIM environment described in Section 6.1.2.3, uploads the simulator together with a specification of valid and potentially interesting parameter ranges, and the mechanism is immediately added to the continuously explored design space. The architecture design points are statistically selected/explored, and for each architecture design point, the compiler is automatically retuned for a truly meaningful comparison with other architecture design points, and the benchmarks recompiled accordingly. After the set of benchmarks has been simulated for this design point, performance results are accumulated in the database, and the ranking of the mechanisms is automatically updated and publicly disseminated on the web site.

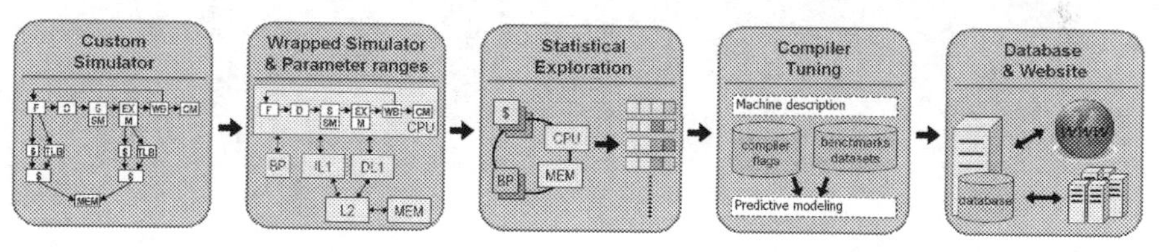

Fig. 6.8 Overview of ArchExplorer

Our embedded processor target is IBM PowerPC405 [17], which is a simple 32-bit embedded RISC processor core including a 5-stage pipeline and 32 registers. The core has no floating-point unit. We consider a 70 nm version running at the maximum frequency of 800 MHz (to date, the PowerPC405 has been taped out at 90 nm with a maximum clock frequency of 667 MHz); the observed average memory latency is 85 cycles over all benchmarks. We use seven of the data cache mechanisms listed in Table 6.1 (*VC*, *TKVC*, *TP*, *SP*, *CDP*, *CDPSP*, *GHB*) plus a skewed associative cache [26] with acronym *SKEW*. The architecture design points are compared using 11 MiBench [11] embedded benchmarks (`bitcount`, `qsort`, `susan_e`, `patricia`, `stringsearch`, `blowfish_d`, `blowfish_e`, `rijndael_d`, `rijndael_e`, `adpcm_c`, `adpcm_d`) and compiled using GCC.

6.3.1 Automatically Tuning the Compiler for the Architecture Design Point

The potential impact of the compiler on architecture design decisions is often overlooked. However, the performance benefits of compiler optimizations are similar, sometimes even higher, than the performance benefits of architecture optimizations. Consider, for instance, the impact of tuning the GCC compiler for PowerPC405 in Fig. 6.9 over the `-O3` optimization.

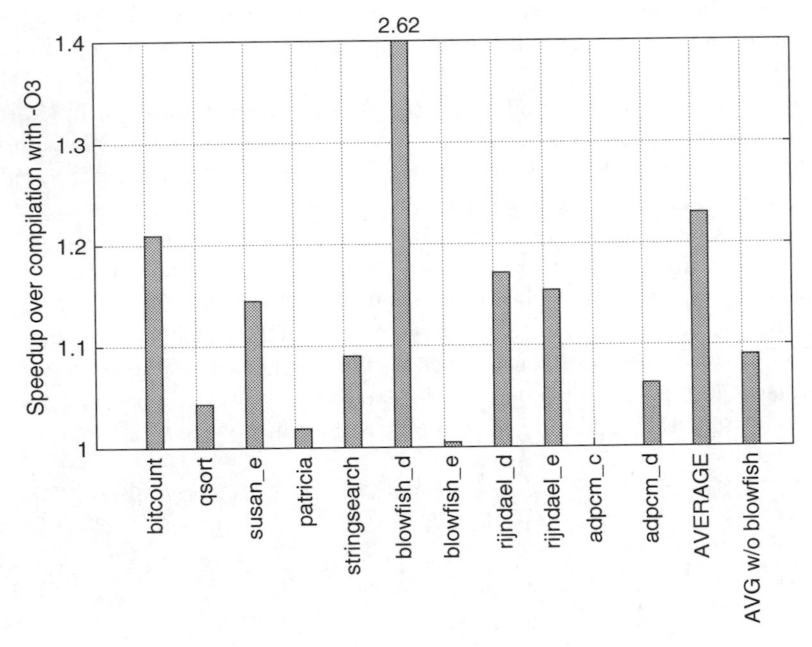

Fig. 6.9 Speedup after tuning the compiler on the reference architecture

Usually, two architecture design points P_1 and P_2 are compared using benchmarks compiled with a compiler tuned for the baseline architecture. In reality, programs will be run with a compiler tuned for the final architecture. So while P_1 may perform better than P_2 using the baseline-tuned compiler, that relative performance could be *reversed* if the compiler is first tuned to each design point and the benchmarks recompiled accordingly. In Fig. 6.10, we report the fraction of design exploration decisions that are reversed when the compiler is tuned versus when the compiler is not tuned.

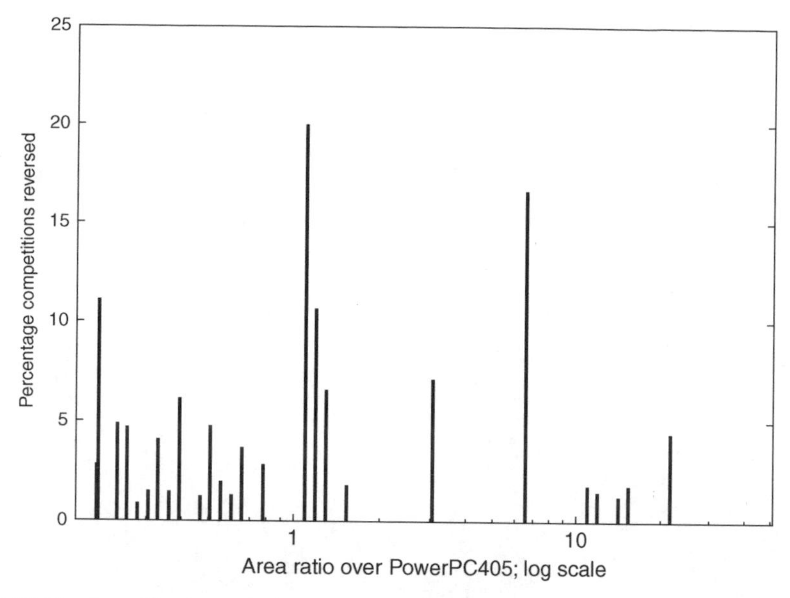

Fig. 6.10 Fraction of hardware design decisions reversed by software exploration

As a result, for each architecture design point, we first tune the compiler by statistically exploring combinations of compiler optimizations. While manual compiler tuning is a tedious process, recent research in iterative compilation [7] has shown that it is possible to automatically tune a compiler for a given architecture platform.

6.3.2 Statistical Exploration of the Design Space

One of the pitfalls of systematic exploration is the huge size of the design space. Expanding design space exploration from parametric to structural exploration only makes matters worse, e.g., 10^{24} design points in our example. As a result, it has become standard practice to stochastically explore the design space [7]. Because we do not just explore parameters, but compose architectures using various architecture components, we resort to stochastic exploration akin to genetic programming. The principle is that each design point corresponds to a large set of parameter values, and each parameter can be considered as a gene. We in fact distinguish two gene levels: genes describing components (nature and number, e.g., depth of a cache hierarchy), and for each component/gene, the sub-genes describing components' parameter values. The genetic mutations first occur at the component level (swapping a component for another compatible one) and then at the parameter level. The database stores all gene combinations tested so far and the corresponding results.

In spite of the sheer size of the design space, we can show that exploration converges relatively quickly due to the many design points with similar performance.

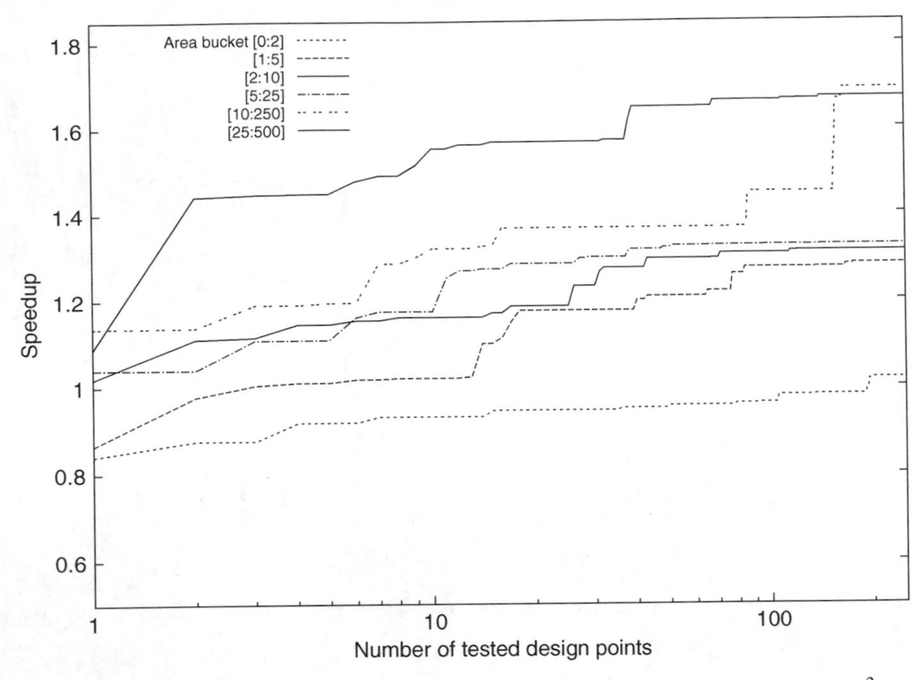

Fig. 6.11 Convergence speed of statistical exploration for different area size intervals (in mm^2)

We illustrate this fast convergence in Fig. 6.11 for different chip area sizes. We refer the reader to [9] for further details.

6.3.3 Combining Quantitative Comparison and Exploration

We now put to task the whole process of the exploration of a processor memory subsystem. We want to compare the behavior of all aforementioned data cache architectures for a large set of area budgets.

Data cache mechanisms versus tuned reference architecture. In Fig. 6.12, we compare the performance achieved using standard data cache architectures against the performance achieved using the data cache techniques of Table 6.1. In the former case, we only vary the typical processor parameters, hence the term *parametric* exploration, and the curve represents the performance achieved for each design area size. In the latter case, we vary the data cache structure and name this *structural* exploration, and Fig. 6.12 distinguishes between the different cache mechanisms. The performance bump of the parametric envelope, around an area ratio of 40, denotes that, for the parameter ranges used below that area size, there always exists a standard cache parametrization which outperforms a data cache hierarchy.

We find that all data cache mechanisms only moderately outperform the standard data cache architecture, when it is duly explored, in terms of both performance and

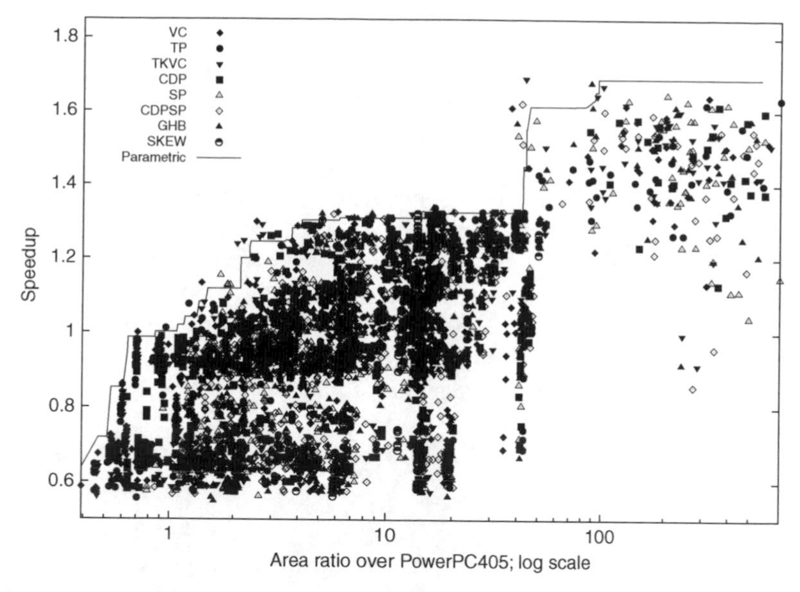

Fig. 6.12 Parametric versus structural exploration

energy, see Fig. 6.13. While these conclusions are naturally dependent on the target benchmarks and architectures, they paint a rather unexpected picture on the actual benefits of sophisticated cache architectures.

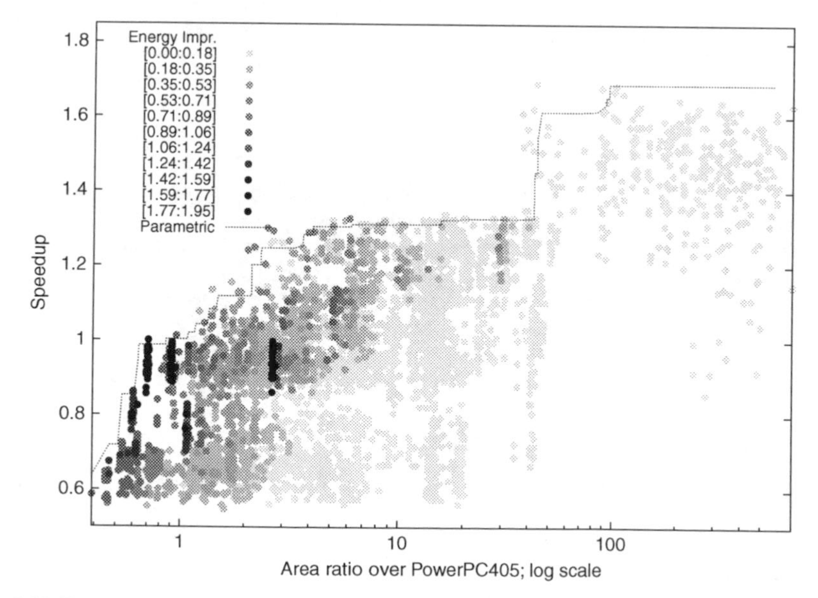

Fig. 6.13 Energy improvements

Best data cache mechanisms as a function of area budget. In the comparison of Section 6.2.2, *GHB* was found to be the best cache mechanism. In Fig. 6.12, we have varied the parameters of the reference architecture and the parameters specific to each mechanism in order to assess the relative merits of these mechanisms over a broad design space. While *GHB* still appears to outperform competing mechanisms for certain area sizes, almost every other mechanism also emerges as the winner for at least one area size. And in fact, there is no clearly dominant mechanism; the best mechanism varies widely with the target area size. For instance, *SKEW* and *CDPSP* perform better for large area budgets, *TKVC* works well for small budgets, and *VC* performs well across the spectrum. Overall, the conclusions are quite different from the conclusions of Section 6.2.2, which show that the design space must be broadly explored to truly assess the relative quantitative merits of architecture mechanisms.

6.4 Conclusions

In this chapter, we have shown that the benefit of structural simulation is not only improved simulation development productivity, or more rigorous simulator development: it can also form the foundation for a design space exploration infrastructure that can significantly improve the efficiency of architecture design in two ways. By facilitating the quantitative assessment and comparison of architecture innovations and by providing the means for an automated and systematic (albeit stochastic) exploration of the design space. We have also suggested that an atypical "cloud-based" approach can overcome the gruesome but key practical hurdles to the adoption of structural simulation, comparison of architecture ideas, and broad design space exploration.

References

1. ARM. Soc designer with maxsim technology, information available at www.arm/com/products/devtools/maxsim.html. Technical report, ARM.
2. Austin, T., Larson, E., Ernst, D.: Simplescalar: An infrastructure for computer system modeling. *Computer*, **35**(2), 59–67, February (2002).
3. Baer, J.-L., Chen, T.-F.: "An effective on-chip preloading scheme to reduce data access penalty." In: *Proceedings of the 1991 ACM International Conference on Supercomputing*, pp. 176–186, (1991) Cologne, West Germany, June 17–21, (1991).
4. Burger, D., Austin, T.: The simplescalar tool set, version 2.0. Technical Report CS-TR-97-1342, Department of Computer Sciences, University of Wisconsin, June (1997).
5. Coe, P., Howell, F.W., Ibbett, R., Williams, L.: A hierarchical computer architecture design and simulation environment. *ACM Trans Model Comput Simul*, 8, 431–446, (1998).
6. Cooksey, R., Jourdan, S., Grunwald, D.: "A stateless, content-directed data prefetching mechanism." In: *Proceedings of the 10th International Conference on Architectural Support for Programming Languages and Operating Systems (ASPLOS-X)*, pp. 279–290, San Jose, CA, October (2002).
7. Cooper, K., Subramanian, D., Torczon, L.: Adaptive optimizing compilers for the 21st century. *J Supercomput*, **23**(1), 7–22, (2002).

8. Davis, J., Goel, M., Hylands, C., Kienhuis, B., Lee, E., Liu, J., Liu, X., Muliadi, L., Neuerdorffer, S., Reekie, J., Smyth, N., Tsay, J., Xiong, Y.: Overview of the Ptolemy project. Technical Report UCB/ERL No. M99/37, EECS, University of California at Berkeley, (1999).
9. Desmet, V., Girbal, S., Temam, O.: "A methodology for facilitating a fair comparison of architecture research ideas." In *2010 IEEE International Symposium on Performance Analysis of Systems and Software (ISPASS)*. IEEE Computer Society, 9: White Plains, NY, March 28–30, (2010).
10. Emer, J., Ahuja, P., Borch, E., Klauser, A., Luk, C.-K., Manne, S., Mukkerjee, S.S., Patil, H., Wallace, S., Binkert, N., Juan, T.: ASIM: A performance model framework. *IEEE Comput, 35(2)* 68–76, February (2002).
11. Guthaus, M.R., Ringenberg, J.S., Ernst, D., Austin, T.M., Mudge, T., Brown, R.B.: "Mibench: A free, commercially representative embedded benchmark suite." In: *IEEE 4th Annual Workshop on Workload Characterization*, pp. 3–14, Austin, TX December 2 (2001).
12. Hamerly, G., Perelman, E., Lau, J., Calder, B.: Simpoint 3.0: faster and more flexible program analysis. *MOBS '05: Workshop on Modeling, Benchmarking and Simulation*, Madison, WI, USA, June 4 (2005).
13. Hardavellas, N., Somogyi, S., Wenisch, T.F., Wunderlich, R.E., Chen, S., Kim, J., Falsafi, B., Hoe, J.C., Nowatzyk, A.G.: Simflex: a fast, accurate, flexible full-system simulation framework, for performance evaluation of server architecture. *SIGMETRICS Perform Eval Rev*, **31**(4), 31–34, March (2004).
14. Hennessy, J.L., Patterson, D.A.: *Computer Architecture: A Quantitative Approach*. Morgan Kaufmann, San Francisco, CA (1996).
15. Hu, Z., Kaxiras, S., Martonosi, M.: Timekeeping in the memory system: predicting and optimizing memory behavior. In: *Proceedings of the 29th Annual International Symposium on Computer Architecture (ISCA)*, pp. 209–220, Anchorage, Alaska, May (2002).
16. Hu, Z., Martonosi, M., Kaxiras, S.: "TCP: Tag correlating prefetchers." In: *Proceedings of the 9th International Symposium on High Performance Computer Architecture (HPCA)*, Anaheim, CA, February (2003).
17. IBM. PowerPC 405 CPU Core. September (2006).
18. Joseph, D., Grunwald, D.: "Prefetching using Markov predictors." In: *Proceedings of the 24th Annual International Symposium on Computer Architecture (ISCA)*, pp. 252–263, Denver, CO, June (1997).
19. Jouppi, N.P.: Improving direct-mapped cache performance by the addition of a small fully-associative cache and prefetch buffers. *Technical Report, Digital, Western Research Laboratory, Palo Alto*, CA, March (1990).
20. Lai, A.-C., Fide, C., Falsafi, B.: "Dead-block prediction & dead-block correlating prefetchers." In: *Proceedings of the 28th Annual International Symposium on Computer Architecture*, pp. 144–154, Göteborg, Sweden, June 30–July 04, (2001).
21. Martin, M.M.K., Sorin, D.J., Beckmann, B.M., Marty, M.R., Xu, M., Alameldeen, A.R., Moore, K.E., Hill, M.D., Wood, D.A.: Multifacet's general execution-driven multiprocessor simulator (gems) toolset. *SIGARCH Comput Architect News*, **33**(4), 92–99 (2005).
22. Nesbit, K.J., Smith, J.E.: "Data cache prefetching using a global history buffer." In: *Proceedings of the 10th International Symposium on High Performance Computer Architecture (HPCA)*, p. 96, Madrid, Spain, February (2004).
23. OSCI. SystemC, OSC Initiative. Technical Report, OSCI, (2003).
24. Pérez, D.G., Mouchard, G., Temam, O.: "Microlib: A case for the quantitative comparison of micro-architecture mechanisms." In: *International Symposium on Microarchitecture*. ACM, Portland, OR, December 4–8, (2004).
25. Pérez, D.G., Mouchard, G., Temam, O.: "Microlib: A case for the quantitative comparison of micro-architecture mechanisms." In: *37th Annual International Symposium on Microarchitecture (MICRO-37 2004), 4–8 December 2004, Portland, OR*, pp. 43–54. IEEE Computer Society (2004).
26. Seznec, A.: "A case for two-way skewed-associative caches." In: *Proceedings of the 20th Annual International Symposium on Computer Architecture*, pp. 169–178, San Diego, CA, USA, May 16–19, (1993).

27. Sherwood, T., Perelman, E., Hamerly, G., Calder, B.: "Automatically characterizing large scale program behavior." In: *Tenth International Conference on Architectural Support for Programming Languages and Operating Systems (ASPLOS-X)*, pp. 45–57. ACM Press, (2002).
28. Shivakumar, P., Jouppi, N.P.: CACTI 3.0: An integrated cache timing, power and area model. *Technical Report, HP Laboratories Palo Alto*, CA, August (2001).
29. Smith, A.J. Cache memories. *Comput Surveys*, **14**(3), 1473–530, September (1982).
30. SPEC. SPEC2000. http://www.spec.org (2000).
31. Vachharajani, M., Vachharajani, N., Penry, D.A., Blome, J.A., August, D.I.: Microarchitectural Exploration with Liberty. In: *Proceedings of the 34th Annual International Symp. on Microarchitecture*, pp. 150–159 Austin, TX, December (2001).
32. VaST. System engineering tools for the simulation and modeling of embedded systems. http://www.vastsystems.com. Technical Report, VaST systems (1999).
33. Wenisch, T.F., Wunderlich, R.E., Falsafi, B., Hoe, J.C.: TurboSMARTS: Accurate Microarchitecture Simulation Sampling in Minutes. *SIGMETRICS '05*, Banff, Alberta, Canada, June 6–10, (2005).
34. Wunderlich, R.E., Wenisch, T.F., Falsafi, B., Hoe, J.C.: "Smarts: accelerating microarchitecture simulation via rigorous statistical sampling." In: *Proceedings of the 30th Annual International Symposium on Computer Architecture*, pp. 84–97. ACM Press, San Diego, CA, June 9–11, (2003).
35. Yang, J., Gupta, R.: "Energy efficient frequent value data cache design." In: *Proceedings of the 35th International Symposium on Microarchitecture (MICRO)*, pp. 197–207, Istanbul, Turkey, November (2002).
36. Zhang Y., Gupta, R.: Enabling partial cache line prefetching through data compression. In: *International Conference on Parallel Processing (ICPP)*, Kaohsiung, Taiwan, October (2003).
37. Zhang, Y., Yang, J., Gupta, R.: "Frequent value locality and value-centric data cache design." In: *Proceedings of the 9th International Conference on Architectural Support for Programming Languages and Operating Systems (ASPLOS-IX)*, pp. 150–159, Cambridge, MA, November (2000).

Part II
Fast Simulation

Chapter 7
Accelerating Simulation with FPGAs

Michael Pellauer, Michael Adler, Angshuman Parashar, and Joel Emer

Abstract This chapter presents an approach to accelerating processor simulation using FPGAs. This is distinguished from traditional uses of FPGAs, and the increased development effort from using FPGAs is discussed. Techniques are presented to efficiently control a highly distributed simulation on an FPGA. Time-multiplexing the simulator is presented in order to simulate numerous virtual cores while improving utilization of functional units.

7.1 Introduction

In order to influence the design of a real microprocessor, architects must gather data from simulators that are accurate enough to convince their skeptical colleagues. This precision comes at the cost of simulator performance. Industrial-grade simulators often run in the range of 10–100 kHz (simulated clock cycles per second) [1]. Such low performance can limit the variety and length of benchmark runs, thus further reducing confidence in architectural conclusions, especially for more radical proposals.

Although parallelizing the simulator can help improve performance, the arrival of multicore processors has actually exacerbated the problem. This is because of three main factors. First, simulating four cores is at least four times the work of simulating one core, but running the simulator on a four-core host machine does not in practice result in a 4$\times$ speedup, due to communication overheads. Second, next-generation multicores typically increase the number of cores, so that architects often find themselves simulating six- or eight-core target machines on a four-core host. Third, the on-chip interconnect network grows in complexity as the number of cores increases, requiring the simulation of more complex topologies and communication protocols.

Recently there has been a trend to use reconfigurable logic architectures as accelerators for computation. To facilitate this several companies have begun producing

J. Emer (✉)
Intel/Massachusetts Institute of Technology, 77 Reed Road, Hudson, MA 01749, USA
e-mail: joel.emer@intel.com

R. Leupers, O. Temam (eds.), *Processor and System-on-Chip Simulation*,
DOI 10.1007/978-1-4419-6175-4_7, © Springer Science+Business Media, LLC 2010

products that allow a field programmable gate array (FPGA) to be added to a general-purpose computer via a fast link such as PCIe [2], Hypertransport [3], or Intel Front-Side Bus [4]. As FPGAs can take advantage of the fine-grained parallelism available in a processor simulation, it is natural to explore whether such products can help address the simulator performance problem. A simple back-of-the-envelope calculation shows that an FPGA running at 100 MHz could take 10 FPGA cycles to simulate one target cycle and still achieve a simulation speed of 10 MIPS, a large improvement over software simulators. Contemporary efforts to explore FPGA-accelerated processor simulation include Liberty [5], UT-FAST [1, 6], ProtoFlex [7], RAMP Gold [8], and HAsim [9, 10]. Collaboration between these groups is facilitated by the RAMP project [11].

Although FPGAs can improve simulator execution speed, the process of designing a simulator on an FPGA is more complex than designing a simulator in software. FPGAs are configured with hardware description languages and are not integrated into most modern debugging environments. There is a danger that increased simulator development time will offset any benefit to execution time. No discussion of FPGA-accelerated simulators is complete without presenting techniques that address this problem. Additionally, FPGAs impose a *space constraint*: the simulator must fit within the FPGA's capacity. Thus to be successful a simulator that uses FPGAs must

1. keep development time short enough so that architects remain ahead of the processor design cycle;
2. model cycle-by-cycle behavior with the same accuracy as a software simulator, while taking advantage of the fine-grained parallelism of an FPGA in order to improve simulation speed:
3. fit on the host FPGA, while maintaining an acceptable level of detail and scaling to support interesting experiments.

This chapter presents an approach to solving these problems. Section 7.2 discusses how using FPGAs for architectural simulation differs from traditional uses such as circuit prototyping. Section 7.3 presents several techniques to reduce development effort. Section 7.4 explores implementing the simulator in such a way as to take advantage of the fine-grained parallelism offered by FPGA. Finally, Section 7.5 discusses time-multiplexing a simulator in order to achieve more efficient utilization of an FPGA for multicore simulation.

7.2 FPGAs as Architectural Simulators

Traditionally FPGAs have occupied three positions in the circuit design flow. The first is to distribute pre-configured FPGAs in place of a custom-fabricated chip. In this case the register-transfer level (RTL) description of the circuit is designed with FPGAs in mind; thus it can take advantage of FPGA-specific structures such as Xilinx synchronous block RAM resources.

The second use is circuit prototyping, where FPGAs are used to aid verification before the costly step of fabrication. In this use the RTL was designed with ASICs in mind and thus may contain circuit structures—such as multi-ported register files or content-addressable memories—that are appropriate for ASICs, but result in inefficient FPGA configurations.

The third use is functional emulation, where a design is created which implements the functionality of the final system, but contains no information on the expected timings of the various components. Usually the goal of such an emulator is to produce a version which is functionally correct with minimal design effort. These designs may use FPGA-specific structures and avoid FPGA-inefficient ones, as they are under no burden to create a circuit related to the final ASIC.

Using an FPGA to aid in architectural simulation occupies something of a middle ground. A simulator helps the architect to make key architectural decisions via exploration, thus it must combine the functionality of the system with some notion of expected timings of its final components, but these timings may vary widely from those of the FPGA substrate. Similarly, if the target is expected to be implemented as an ASIC, then FPGA-inefficient structures should not be ruled out.

The key insight is that an FPGA-accelerated simulator must be able to correctly *model* the timing of all structures, but does not have to accomplish this by directly configuring the FPGA into those structures. The FPGA is used only as a general-purpose, programmable substrate that implements the model. This allows the architectural simulator to simulate FPGA-inefficient structures using FPGA-specific components, while *pretending* that their timings match their ASIC counterparts.

To illustrate this, consider the example in Fig. 7.1. The architect wishes to simulate a target processor which contains a register file with two read ports and two write ports (7.1A). Read values appear on the same target cycle as an address is asserted. External logic guarantees that two writes to the same address are never asserted on the same model clock cycle.

Directly configuring the FPGA into this structure would be space inefficient because it cannot use built-in block RAM resources. Block RAM typically only has two total ports and has different timing characteristics than the proposed register file—read values appear on the next FPGA cycle after an address is asserted. Thus a direct emulation would use individual registers and multiplexers (7.1B), which can be quite expensive.

An FPGA-based simulator separates the FPGA clock from the simulated *model clock*. Such a simulator could use a block RAM paired with a small finite state machine (FSM) to model the target behavior (7.1C). In this scheme the current cycle of the simulated target clock is tracked by a counter. The FSM ensures that the cycle counter is not incremented until two reads and two writes have been performed. Thus we are able to design a simulator with a high frequency and low area, at the expense of now taking 3 FPGA cycles to simulate one model cycle.[1]

---

[1] This is 3 instead of 4 because the simulator can perform the first write on the same FPGA cycle as the second read to the synchronous block RAM.

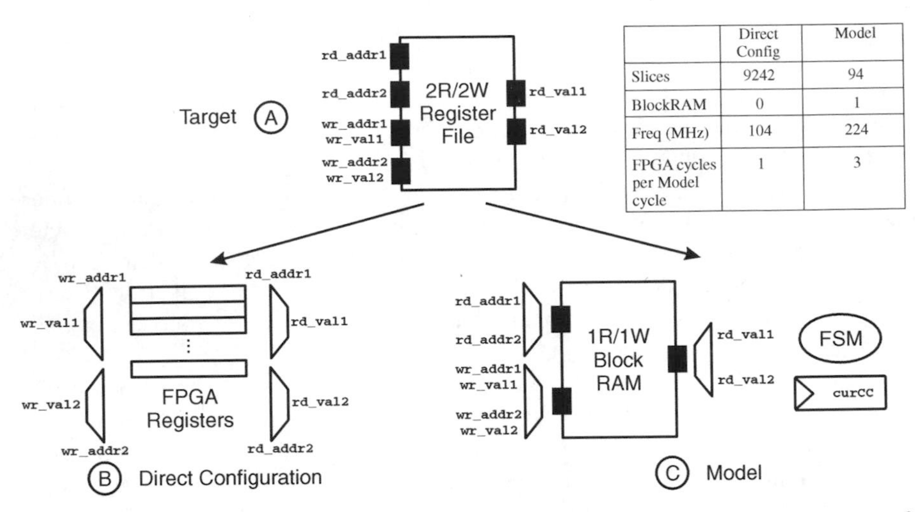

	Direct Config	Model
Slices	9242	94
BlockRAM	0	1
Freq (MHz)	104	224
FPGA cycles per Model cycle	1	3

Fig. 7.1 Separating the model cycle from the FPGA cycle helps the simulator to take advantage of synchronous block RAM

In order to reason about the performance of such simulators we introduce a new term: FPGA-cycles-to-model-cycles ratio (FMR). FMR is similar to the microprocessor performance metric cycles per instruction (CPI) in that one can observe the FMR of a run, a region, or a particular class of instructions. The FMR of a simulator combined with its FPGA clock rate gives simulation rate:

$$\text{frequency}_{\text{simulator}} = \frac{\text{frequency}_{\text{fpga}}}{\text{FMR}_{\text{overall}}}$$

This can be plugged into the traditional calculation for simulated instructions per second (IPS):

$$\text{IPS}_{\text{simulator}} = \frac{\text{frequency}_{\text{fpga}}}{\text{CPI}_{\text{model}} \times \text{FMR}_{\text{overall}}}$$

Finally, it is important to note one way in which FPGA simulators share the same restriction as software simulators: they give little insight into the physical properties of the target design. Because the RTL used to configure the FPGA into the simulator makes many accommodations for FPGAs, its device utilization and critical path are unlikely to give insight into those characteristics of the target design.[2]

[2] There are scenarios in which the FPGA simulator characteristics may give some degree of insight into the corresponding characteristics of the final design. However, this should not be one of the assumptions when using an FPGA for architectural simulation.

7.3 Addressing Development Effort

Unfortunately, developing an FPGA simulator can require significantly more effort than a traditional software simulator. Accelerator FPGAs are configured using hardware description languages and are currently not integrated into debugging environments. Additionally, there is no equivalent of software's standard library infrastructure, which complicates printout-oriented debugging. Furthermore, the long running times of FPGA synthesis and place-and-route tools can lengthen the compile–run–debug loop.

Despite these obstacles, FPGA accelerator development does remain significantly simpler than ASIC hardware description, for several reasons. First, FPGA developers do not have to worry about physical circuit characteristics, such as parasitic capacitance, that complicate ASIC development. Second, the reconfigurable nature of the FPGA means that it is easier to take an iterative approach to development, rather than working toward a final irrevocable tapeout. Finally, because FPGA accelerators are added to existing general-purpose computers, the capabilities of the host computer can be used to aid in debugging and to perform functions too complicated to implement on an FPGA.

Altogether, we find that the development effort can be made tractable with the application of good software engineering techniques—modular interfaces, libraries of reusable code, layering of functionality—along with a few specific considerations which are described in this section.

7.3.1 High-Level Hardware Description Languages

Structural hardware description languages such as VHDL and Verilog give designers precise control over their microarchitectures. However, this control often comes at the cost of complex, low-level code that is unportable and difficult to maintain.

When using an FPGA as a simulator the architect is not describing a final product, and thus does not need such exacting control. Thus high-level hardware description languages such as Bluespec [12], HandelC [13], or SystemC [14] can be a good fit.[3]

The HAsim [9], UT-FAST [1], and ProtoFlex [7] FPGA simulators are all written in Bluespec SystemVerilog. The benefits of Bluespec are similar to those for using high-level languages in software development: raising the level of abstraction improves code development time and reuse potential, while simultaneously eliminating many low-level bugs caused by incorrect block interfacing. Bluespec also features a powerful *static elaborator* that allows the designer to write polymorphic hardware modules which are instantiated at compile time with distinct types. This

[3] It should be noted that high-level hardware description languages do not necessarily result in worse FPGA utilization. There are cases where high-level knowledge exposes optimization opportunities [20].

brings many of the benefits of software languages' high-level datatype systems into hardware development.

7.3.2 Building a Virtual Platform

Development efforts can be further eased by adopting a standardized set of interfaces for the FPGA to talk to the outside world. This *virtual platform* provides a set of virtualized device abstractions to FPGA developers, enabling them to focus on implementing core functionality without spending time and energy debugging low-level device drivers. Furthermore, most FPGA-based simulators are likely to be hosted on hybrid compute platforms comprising one or more FPGAs and one or more CPUs. Extending well-understood communication protocols such as remote procedure call (RPC) and shared memory to the hybrid CPU/FPGA environment makes the platform more approachable for sharing responsibilities between the FPGA and the CPU.

Figure 7.2 illustrates the structure of the virtual platform implemented as part of the HAsim simulator. The primary interfaces between the simulator and the platform are a set of virtual devices and an RPC-like communication protocol called remote request–response (RRR) that enables multiple distributed services on the CPU and the FPGA to converse with each other [15]. The primary benefit of this approach is portability—the virtual platform can be ported to a new physical FPGA platform without altering the application. Only low-level device drivers must be rewritten. Additionally, the virtual platform is general enough to apply to other FPGA-accelerated applications beyond simulating processors.

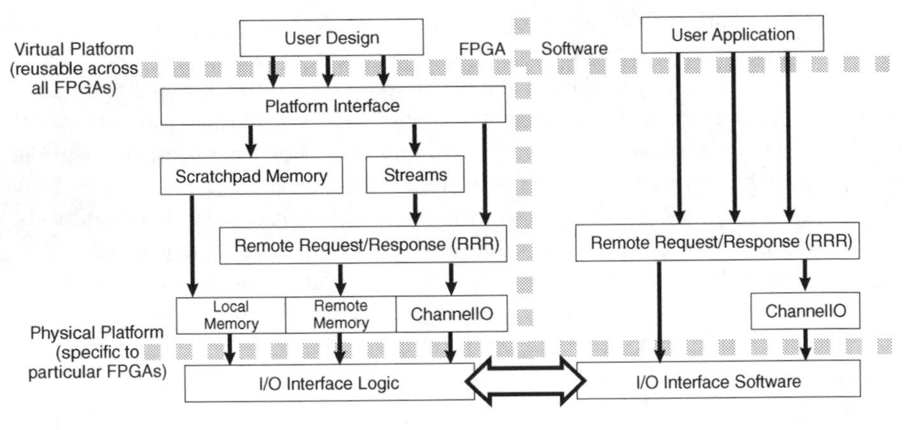

Fig. 7.2 HAsim's virtual platform

In addition to generalized interaction, it is useful to build a layer of specific services on top of the virtual platform. Example services include communicating command-line parameters from software, enabling print-outs and debugging state dumps from hardware, and communicating dynamic assertion failures in hardware.

Once this capability is in place, the problem becomes coordinating many distributed modules on the FPGA interacting with the off-chip communication controller. A methodology for abstracting on-FPGA communication details in a modular fashion is presented in [10].

7.3.3 Interacting with a Software Simulator

With the virtual platform in place hardware/software interaction becomes a tool which can significantly ease FPGA simulator development. Although any functionality can be partitioned on the FPGA or in software, it makes sense to take simulator events which are rare but difficult to handle—such as system calls or page faults—and handle them in software. Thus the common case should be simulating on the FPGA, with occasional invocations of software.

HAsim uses the M5 full-system simulator (Chapter 5) as a software backer to handle system calls and page faults. The modular nature of M5 allows HAsim to tie directly into M5's memory subsystem while ignoring its CPU timing models. When the FPGA detects a system call it transfers the architectural state of the simulated processor to HAsim's software, which invokes M5. After the state is updated, it is transmitted back to the FPGA, at which point simulation on the FPGA can resume. The ProtoFlex project applied these principles to emulation [7]. They demonstrated that if these events are rare enough the impact on performance can be minimized, while still resulting in significant gains in development effort.

7.3.4 Timing-Directed Simulation

Timing-directed simulation is another technique for reducing simulator development effort. In such a scheme the simulator is divided into separate *functional* and *timing* partitions which interact in order to form a complete simulation. The functional partition is responsible for correct ISA-level execution of the instruction stream. The timing partition, or *timing model*, is responsible for driving the functional partition to simulate the cycle-by-cycle behavior of a particular microarchitecture. The functional partition handles responsibilities common to all architectures, such as decoding instructions, updating simulator memory, or guaranteeing that floating-point operations conform to standards. The timing partition is responsible for the tasks of simulating a specific microarchitecture, such as deciding what instruction to issue next, tracking branch mispredictions, and recording that floating-point multiply instructions take 5 clock cycles to execute.

The functional partition might be complex to implement, optimize, and verify, but once it is complete it can be reused across many different timing models. Additionally, the timing models themselves are significantly simpler to implement than simulators written from scratch: they do not need to worry about ISA functional correctness, but only track microarchitecture-specific timing details. Often structures can be excluded from the timing model partially or completely, as their behavior is handled by the functional partition. A common example of this is a timing model of a cache, which need only track tags and status bits—it need not store the actual instructions or data as these are not relevant to deciding whether a particular load hits or misses.

Traditionally, simulators implement both partitions in software. The HAsim FPGA simulator demonstrated that the functional partition could be implemented on the FPGA. The HAsim microarchitecture defines eight operations (Fig. 7.3): load a new in-flight instruction (`getInstruction`), test how that instruction relates to other in-flight instructions (`getDependencies`), execute the operation (`getResults`), perform memory side effects (`doLoads`, `doStores`), and finally remove the instruction from being in-flight, possibly making its results visible (`commitResults`, `commitStores`, `rewind`). Details of how timing models can use these operations to simulate features like out-of-order issue and speculative execution are presented in [9].

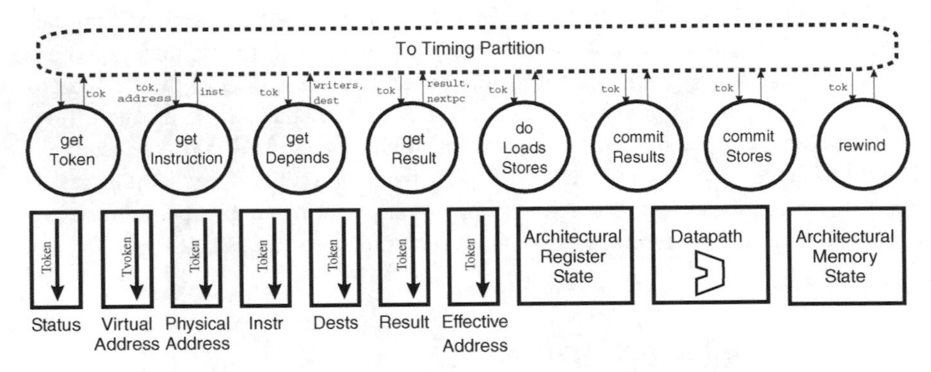

Fig. 7.3 Overview of the microarchitectural implementation of the HAsim simulator's functional partition

The UT-FAST simulator combines the hardware–software partitioning approach with the partitioned simulator approach [1]. In the UT-FAST approach the functional partition is a software emulator such as QEMU. It generates a trace of instruction execution that is fed into a timing model on the FPGA. A key benefit of this scheme is the ability to use an existing functional emulator to generate the trace. (Modifications are required to support rollback of the emulator when the timing model goes down a different path from that of the generated trace.) For a detailed discussion of this approach see [1, 6].

7.4 Fine-Grained Distributed Simulation

Section 7.2 demonstrated that separating the FPGA clock from the simulated model clock can result in significant benefits. The problem becomes taking many such modules—representing the various functions of a target processor—and composing them together in a manner that results in a consistent notion of model time. This section examines how to construct a simulator while enabling the FPGA to take advantage of the fine-grained parallelism available in the target design.

7.4.1 Unit-Delay Simulation

One straightforward way to coordinate distributed modules is to assign each module n FPGA cycles to simulate one model cycle. This is *unit-delay* simulation, historically used in projects such as the Yorktown Simulation Engine [16]. Recall that the register file in Fig. 7.1 had an FMR of 3. If this was the slowest module, then every other module would have 3 cycles to finish simulation, and the overall FMR of the simulator would be 3.

The advantage of the unit-delay scheme is that there is very low overhead. It is straightforward to code and can result in good performance when n is low. The disadvantage is that in practice there are processor events such as exceptions that are rare, but take large numbers of FPGA cycles to simulate. Additionally, it can be difficult to bind n if the simulator uses off-chip communication or to alter all of the modules if n changes.

7.4.2 Dynamic Barrier Synchronization

An alternative is to have the FMR determined dynamically. This is *dynamic barrier* synchronization, where the modules communicate to a central controller when they are done simulating a model cycle. When all modules have finished, the controller increments the model cycle counter and alerts all modules to proceed. The number of FPGA cycles required for each model cycle is now the dynamic worst case, which can result in a significant improvement in overall FMR compared to unit-delay simulation.

The main problem with barrier synchronization is the scalability of signals to and from the central controller, which can impose a large burden on the FPGA place-and-route tools. When exploring this effect, it was observed that quadrupling the number of modules in a system resulted in a 39% loss of clock speed due to the controller alone [17]. Thus a dynamic barrier scheme is best suited to situations where the distributed number of modules is small and the dynamic average FMR is low.

7.4.3 Port-Based Modeling

Port-based modeling is a technique used in structural simulators such as Intel's Asim [18]. This section discusses port-based models and shows that their implementation on FPGAs can overcome the disadvantages of the above simulation techniques. For a more thorough treatment of this subject see [17].

In a port-based model the simulator is decomposed into several modules as the architect finds convenient. The modules themselves have no notion of model time—conceptually we consider their operations to be infinitely fast. All communication between modules must go through *ports*, which are essentially FIFOs of message type t. The ports are statically annotated with a latency l. A message placed into the port on model cycle n will emerge on cycle $n + l$. (Note that ports do not represent queues in the target processor, which would be represented within the modules themselves.) Ports of latency zero are allowed but may not be arranged into "combinational loops"—a familiar restriction to hardware designers.[4]

A module now simulates a model cycle as follows:

1. Check all input ports for arriving messages.
2. Perform all local calculations and state updates.
3. Write any output messages to outgoing ports.

Figure 7.4 shows an example of modeling a target processor with ports. This is a four-way superscalar processor, with four execution units of varying capabilities, and thus can issue up to four instructions per cycle under ideal circumstances. To support this the register file has seven read ports and four write ports (the jump unit only requires one read port). When the target is recast as a port-based model the system is partitioned into modules using the pipeline stages as a general guideline. Pipeline registers are replaced by ports of latency 1, such as those connecting Fetch and Decode. The instruction and data memories are represented as simple static latencies, which is unrealistic but illustrative for this example.

The target ALU being considered uses a two-stage pipeline for the simple operations and a four-stage pipeline for the multiplier. Note that the simulator is not necessarily implemented with pipeline stages at the same locations, but can instead use any FPGA-optimized circuit, as the port ensures that the operation appears to consume 4 model cycles. Divide operations are handled differently—in the target they take a varying amount of time. Thus they cannot be represented by a static port latency—the latency is fixed to 1, but operation completions must be modeled using a higher-level communication protocol.

7.4.4 A-Ports: Implementing Port-Based Models on FPGAs

A sequential software simulation of a port-based model traditionally statically schedules the modules so that each module is simulated once every model cycle, based on a topological sort of the ports. (Such a sort is guaranteed to exist thanks

[4] Sometimes port-based models contain "false" combinational loops between modules. In this case, a module's simulation loop must be loosened to correctly handle the false dependencies, as presented in [21].

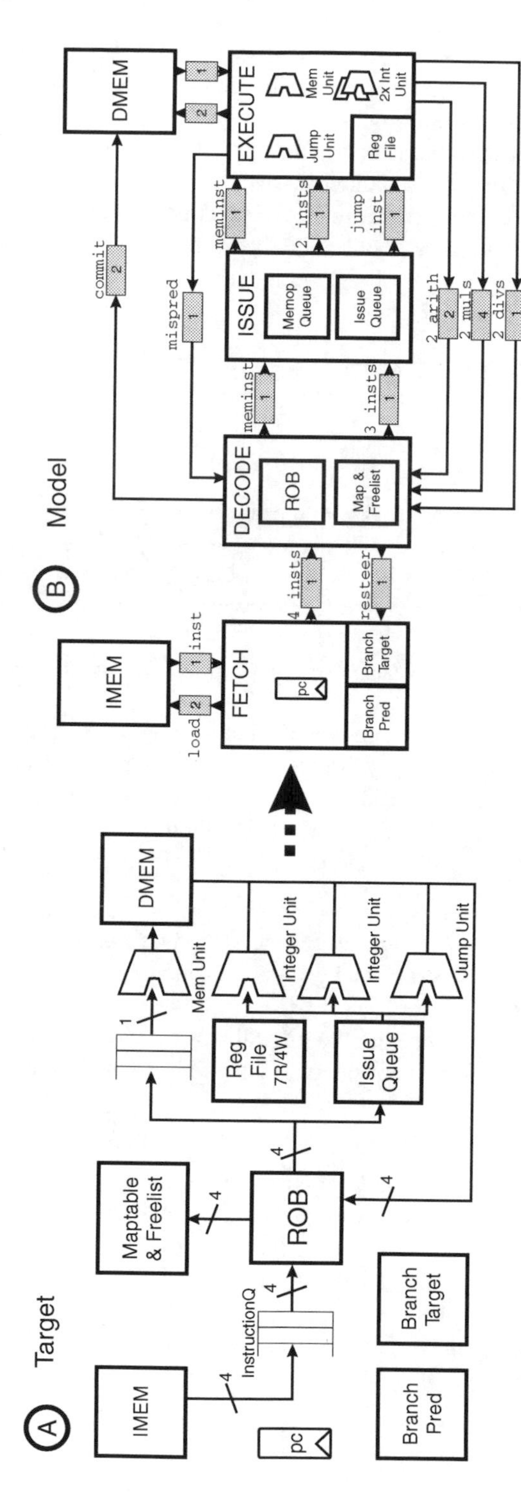

Fig. 7.4 Turning a target superscalar processor into a port-based model

to the "no combinational loops" restriction.) Barr et al. [19] showed that such a simulator can be parallelized to run on a multicore host by cutting the graph at pre-determined ports and distributing the resulting groups of modules to the various host cores. This is because the large number of modules which are ready to simulate in parallel can overwhelm today's multicore hosts.

An FPGA is better able to take advantage of this fine-grained parallelism, so the barrier can be removed and each module can make a local decision when to proceed to the next model cycle. One way to accomplish this is to use customized FIFOs called A-Ports (Fig. 7.5) [10, 17].[5]

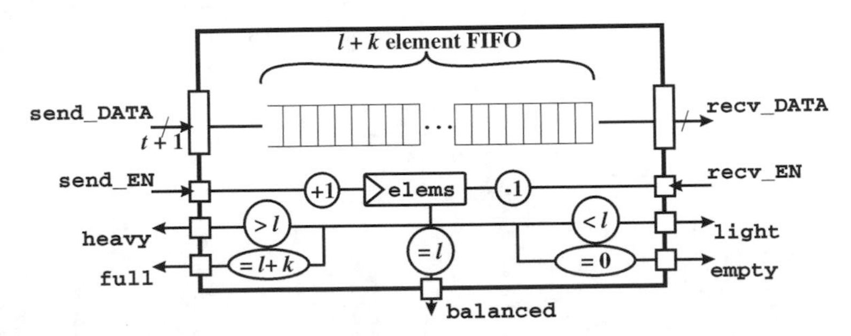

Fig. 7.5 An A-Port is a FIFO that maintains a notion of utilization relative to its latency

In order to distribute the computation of model time, we restrict each module so that it must write a value to every output port on every model cycle—ports may not be conditionally written. Instead, the domain of every port's message type is extended to include a special `NoMessage` value, which represents the absence of a message in the target design. Thus the complete distributed simulation loop is as follows:

- When all incoming A-Ports are not empty, a module may begin computation. Note that some of its inputs may be `NoMessage`, and that this is explicitly different from an empty port.
- When computation is complete, the module must write all of its outgoing A-Ports. It may write `NoMessage` or some other value, but must write all of them exactly once.
- One message is consumed from each incoming A-Port and the loop repeats.

The net effect of this simulation loop is to allow every module in the system to produce and consume data at any wall-clock rate, while still maintaining a local notion of a model clock step. The current model cycle is simply how many times a module has executed the simulation loop. As a consequence, the simulator can enter a state where adjacent modules are concurrently simulating different model cycles. This is called simulator *slip*. A producer may run into the future, pre-computing

[5] The name A-Ports reflects that they are a generalization of ports in the Asim simulator.

values as fast as long as output buffering is available. Similarly, a fast consumer can drain its input buffers.

We say an A-Port of latency l is *balanced* when it contains exactly l messages. When an A-Port contains more than l elements it is *heavy*, and similarly it is *light* when it contains fewer than l elements. Observe the following:

- When an A-Port is balanced, the modules it connects are simulating the same model cycle.
- When an A-Port is heavy, the producer module is simulating into the future relative to the receiving module.
- When an A-Port is light, the situation is reversed.

This slipping does not alter results of simulation, but it can improve simulation rate over barrier synchronization, as demonstrated in Fig. 7.6. In this example, instructions a and c take more FPGA time to simulate compared to b and d. Observe that on FPGA cycle 6 module A is simulating model cycle 3, whereas module B is simulating model cycle 2.

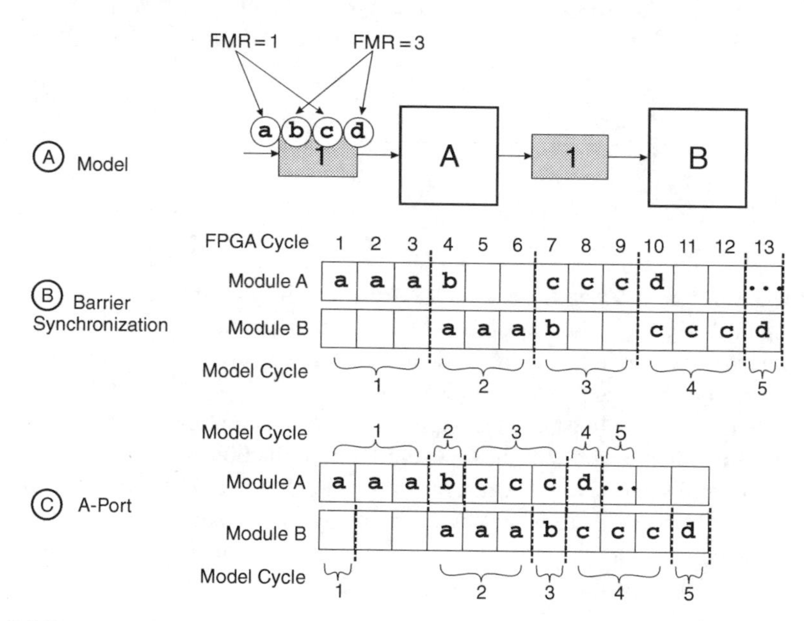

Fig. 7.6 Demonstrating how an A-Port implementation can result in a performance improvement over barrier synchronization

Obtaining a snapshot of relevant architectural state in the A-Ports scheme is complicated by the fact that the modules may have slipped in time. A simple distributed technique to re-synchronize slipped modules as well as an assessment of the performance improvement granted by slip is presented in [17].

7.5 Fine-Grained Time-Division Multiplexing

A common architectural experiment involves scaling the number of cores in a multicore configuration and observing the effect on performance. When creating an FPGA-accelerated simulator, FPGA utilization can artificially constrain how far these experiments can be scaled.

One technique that can aid models with scaling to more cores is to time-division multiplex the simulator components. In this scheme the logic which implements the target cores is not duplicated directly. Instead, a single *physical implementation* is shared to simulate multiple *virtual instances* of the cores in the system, as shown in Fig. 7.7.

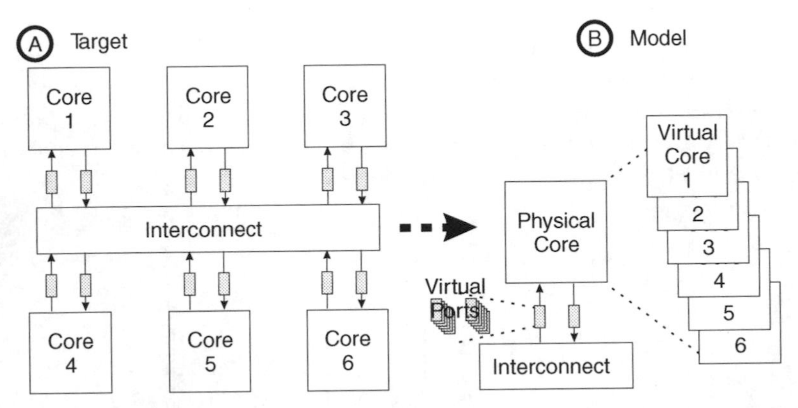

Fig. 7.7 Multiplexing a physical core between many virtual cores in order to simulate a multicore processor

One straightforward way to perform multiplexing would be to have the physical core simulate a given virtual instance until a slow event occurs, such as off-chip communication, at which point the physical core could switch to simulating a different virtual instance.[6] Such a technique would save area with respect to duplication, as the state of the virtual instances is duplicated, but the combinational logic is not. However, this scheme would result in a significant reduction in simulation rate, as only one virtual instance would be active in the simulator at a time. Furthermore, on any given FPGA cycle there would likely be many idle modules in the system, as faster modules wait for slower modules to finish. Lastly, this kind of multiplexing may reduce the fidelity of simulation as the timings of interactions between the cores become less precise.

By making the multiplexing decision on a finer granularity we can achieve better performance by putting idle modules to work while simultaneously maintaining precise fidelity.

[6] This kind of multiplexing bears a resemblance to multi-threading in real microprocessors, but it is important to distinguish that this is a simulator technique, not a technique in the target architecture. The cores being multiplexed do not have to support multi-threading.

7.5.1 Module-Based Multiplexing

One practical way to time-division multiplex a module is to have the physical implementation simulate each virtual instance in a round-robin fashion. In this scheme local module state is duplicated, but the input and output ports between modules are not. Instead, the ports themselves are initialized to contain the messages for different virtual instances in a round-robin order (Fig. 7.8). The simulation loop for each module becomes the following:

- When all incoming ports are not empty, they all contain messages of the same virtual instance. The module may simulate the next model cycle for that instance.
- Any output messages it produces will be the output for that instance.

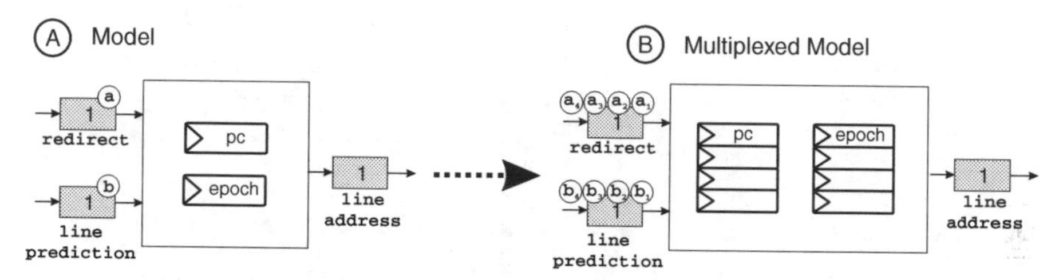

Fig. 7.8 Round-robin multiplexing cycles between the virtual instances in a fixed order

The original port-based simulation scheme (Section 7.4.4) allowed adjacent modules to be simulating different module clock cycles. In the time-division multiplexed scheme, adjacent modules may now also be simulating different virtual instances. This helps keep modules busy, as they are more likely to have work in the queue if an upstream module encounters a rare-but-slow event.

Performance in this scheme can be limited by the overly restrictive round-robin ordering. One solution would be to allow a ready virtual instance to bypass an idle one at the head of the queue, but so far this scheme has proven too expensive to implement on an FPGA. We have found it to be more expensive than direct module duplication.

7.5.2 Pipelining the Modules

A more practical approach to minimize idle virtual instances is to pipeline the module implementation.[7] Under ideal cases pipelining the simulator modules can entirely eliminate the multiplexing performance penalty, achieving the performance of the original duplicated modules.

---

[7] Again, note that this refers to altering the implementation of the modules on the FPGA, not altering the timing characteristics of the target circuit, which are preserved by the ports.

To understand why, consider the situation in Fig. 7.9A. Module *A* is faster than module *B*, which has an FMR of 4 (7.9B). Multiplexing the system 4 ways without pipelining would decrease the overall FMR to $4 \times 4 = 16$ (7.9C). However, if module *B* can be pipelined into four stages, then overall FMR can be reduced back to the original 4 (7.9D). Note that the fourth stage of module *B* finishes each model cycle for the first core on the same FPGA cycle as the original non-multiplexed design.

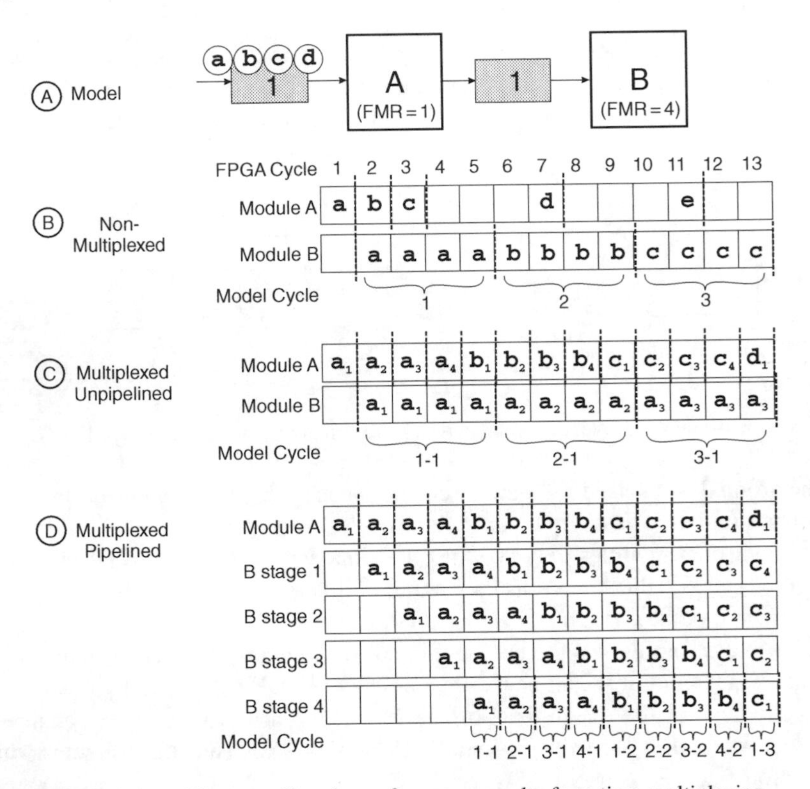

Fig. 7.9 Pipelining modules can offset the performance penalty from time-multiplexing

Of course, such a scheme will rarely be operating under ideal conditions, and pipeline bubbles will be introduced. Such bubbles can be minimized if there are always ready virtual instances waiting for simulation. Thus a key to achieving simulator performance is keeping module pipeline depths less than or equal to the number of virtual instances.

7.5.3 Modeling Interconnect Networks

The above multiplexing scheme only works because there is no communication between the various virtual instances of the target multicore. The simulation of model cycle k for Core x is independent from cycle k for Core y. But of course this is an oversimplification: the target cores can influence each other's behavior through the *on-chip interconnect*. Therefore the multiplexing of the interconnect must be handled slightly differently than that of the cores themselves.

Figure 7.10 shows three standard interconnect topologies: a crossbar, a ring, and a torus. Implementing time-multiplexed models of these systems is complicated by the fact that the networks are not connected to n separate cores, but to one physical implementation which is time-division multiplexed n ways, as shown in Fig. 7.11. The networks are then simulated as follows:

1. Crossbar: The crossbar module simulates a model cycle by reading the port from the core n times in sequence and then reading the incoming port from the memory controller once. It then chooses winners, writes the core response port n times in sequence, and writes the memory controller response port (Fig. 7.11A).
2. Ring: Rather than duplicate the ring stops, we can multiplex them along with the cores. Now there is only one physical ring stop implementation. We model communication between the ring stops by connecting the physical output port back in as the input (Fig. 7.11B). Note that when virtual instance 1 simulates a model cycle the output it produces should be the input for the memory controller on the *next* model cycle. Similarly the output of instance n should be the input for instance $n-1$. We can accomplish this by inserting logic to do a small reordering of the messages which pass through the port (Fig. 7.11C).
3. Torus: A torus with width w and height h acts as a two-dimensional version of the ring. The West output port is connected to the East input port, and every wth item is reordered w spots (Fig. 7.11D). The North output port is connected to the South input port, and the last h items are reordered to the front (Fig. 7.11E).

This technique extends to bidirectional topologies by adding backward facing ports which have different reorderings. Furthermore, topologies whose edges are subsets of these graphs may be modeled by always sending a `NoMessage` value along edges that do not exist. For example, the torus could simulate a mesh by always populating the wraparound edges `NoMessage`.

Using this implementation technique the simulation of the interconnect network can be overlapped with the simulation of the cores themselves, which can aid in scaling. Interestingly, we have found that doubling the number of cores does not halve simulator performance, as it would in a traditional sequential simulator. In fact, the overall FMR per core goes down by roughly 40%. This is partially because adding a more realistic on-chip network to the model means that cores spend more model cycles starved, waiting for responses from the memory controller. As a consequence the FPGA can simulate them faster, as pipeline bubbles are easy to simulate. This inverse relationship between target IPC and simulation rate has often been noted in software simulators.

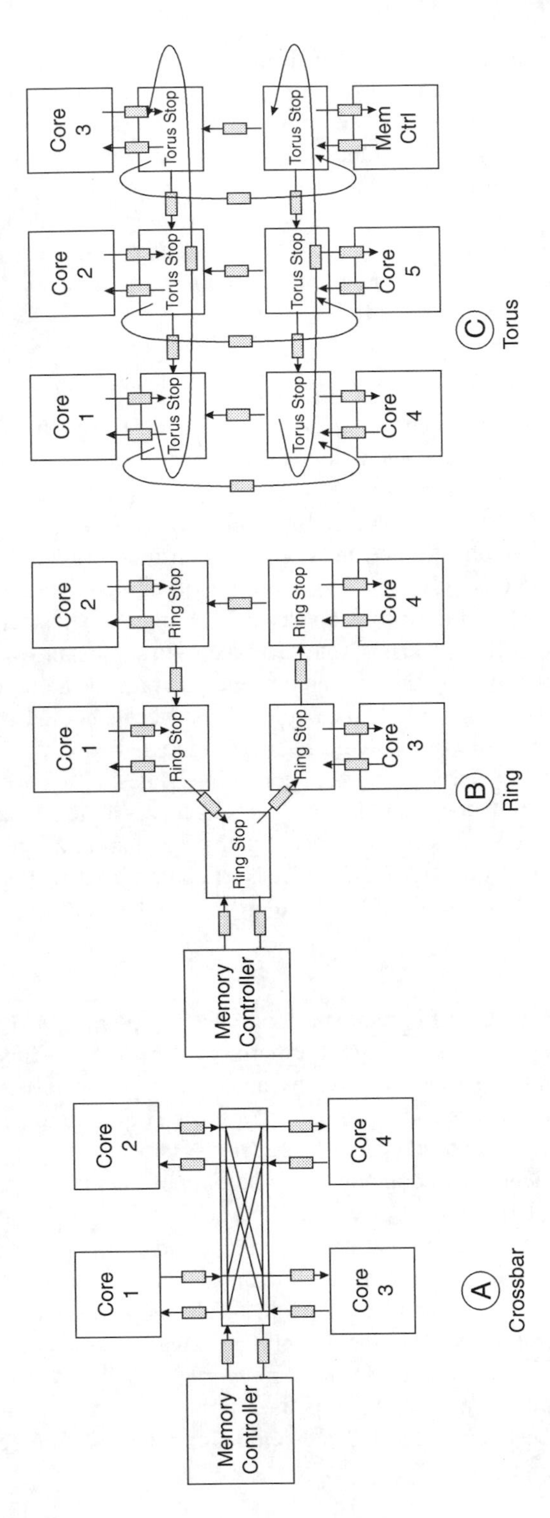

Fig. 7.10 Three example target multicore interconnects

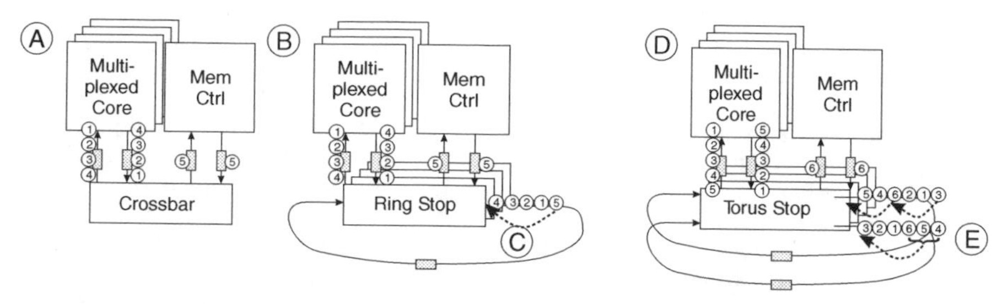

Fig. 7.11 Modeling the interconnects involves interacting with a time-division multiplexed core, so the FPGA models do not resemble real interconnect topologies, and their usage characteristics are nearly identical

7.6 Conclusion

Using FPGAs to accelerate architectural simulation represents a novel approach—both to how FPGAs are used and to how simulators are constructed. Development effort remains the major barrier to wider adoption, and ongoing efforts to reduce implementation time are critical.

If reconfigurable logic platforms are able to show success in the processor-simulation domain, then there is hope that they may prove to be of interest as general-purpose computation accelerators. FPGAs have already proven themselves as ASIC-replacement computation engines. The community hopes that the development of a standardized I/O framework and development infrastructure would allow a wide audience to explore using them as integrated computation accelerators.

References

1. Chiou, D., Sunwoo, D., Kim, J., Patil, N.A., Reinhart, W.H., Johnson, D.E., Keefe, J., Angepat H.: "FPGA-accelerated simulation technologies FAST: Fast, full-system, cycle-accurate simulators." In: *Proceedings of the 40th Annual IEEE/ACM International Symposium on Microarchitecture (MICRO)*, Chicago, IL, pp. 249–261 December (2007).
2. HiTech Global Design and Distribution, LLC. `http://www.hitechglobal.com` (2009).
3. DRC Computer Corp. `http://www.drccomputer.com` (2009).
4. Nallatech, Inc. `http://www.nallatech.com` (2009).
5. Penry, D.A., Fay, D., Hodgdon, D., Wells, R., Schelle, G., August, D.I., Connors, D.: "Exploiting parallelism and structure to accelerate the simulation of chip multi-processors." In: *The 12th International Symposium on High-Performance Computer Architecture (HPCA)*, Salt Lake City, UT, pp. 29–40 February (2008).
6. Chiou, D., Sunwoo, D., Kim, J., Patil, N.A., Reinhart, W.H., Johnson, D.E., Xu, Z.: "The fast methodology for high-speed soc/computer simulation." In: *International Conference on Computer-Aided Design (ICCAD)*, San Jose, CA, pp. 295–302, November (2007).
7. Chung, E., Nurvitadhi, E., Hoe, J., Mai, K., Falsafi, B.: "Accelerating architectural-level, full-system multiprocessor simulations using FPGAs." In: *FPGA '08 Proceedings of the Eleventh International Symposium on Field Programmable Gate Arrays* pp. 77–86, Monterey, CA, February (2008).

8. Tan, Z., Waterman, A., Cook, H., Asanovic, K., Patterson, D.: Ramp gold: An FPGA-based architecture simulator for multiprocessors. In: *Proceedings of the 47th Design Automation Conference (DAC)*, Anaheim, CA, June (2010).
9. Pellauer, M., Vijayaraghavan, M., Arvind, M.A., Emer, J.: "Quick performance models quickly: Closely-coupled timing-directed simulation on fpgas." In *IEEE International Symposium on Performance Analysis of Systems and Software (ISPASS)*, pp. 1–10, Austin, TX, April (2008).
10. Pellauer, M., Vijayaraghavan, M., Arvind, M.A., Emer, J.: "A-ports: An efficient abstraction for cycle-accurate performance models on FPGAS." In *IEEE International Symposium on Performance Analysis of Systems and Software (ISPASS)*, pp. 87–96, Monterey, CA, February (2008).
11. Wawrzynek, J., Patterson, D., Oskin, M., Lu, S.L., Kozyrakis, C., Hoe, J. C., Chiou D., Asanovic, K. Ramp: A research accelerator for multiple processors. *IEEE Micro* March/April 27(2):46–57 (2007).
12. Bluespec Inc. `http://www.bluespec.com` (2008).
13. Page. I: Constructing hardware-software systems from a single description. *J VLSI Process*, 12, 87–107 (1996).
14. OSCI. SystemC language reference manual version 2.1.
15. Parashar, A., Adler, M., Pellauer, M., Emer, J.: Hybrid CPU/FPGA performance models. In: *Workshop on Architectural Research Prototyping (WARP)*, pp. 1–2, Beijing, China, June (2008).
16. Pfister, G.: "The Yorktown simulation engine." In: *19th Conference on Design Automation (DAC)*, pp. 51–60, Las Vegas, NV, June (1982).
17. Pellauer, M., Vijayaraghavan, M., Arvind, M.A., Emer, J.: A-port networks: Preserving the timed behavior of synchronous systems for modeling on fpgas. *ACM Trans Reconfigurable Technol Syst*, September 2(3):Article 16 (2009).
18. Emer, J., Ahuja, P., Borch, E., Klauser, A., Luk, C.K., Manne, S., Mukherjee, S.S., Patil, H., Wallace, S., Binkert, N., Espasa, R., Juan, T.: Asim: A performance model framework. *Computer*, February 35(2):68–76 (2002).
19. Barr, K.C., Matas-Navarro, R., Weaver, C., Juan, T., Emer, J.: Simulating a chip multiprocessor with a symmetric multiprocessor. In: *Boston Area Architecture Workshop (BARC)*, Boston, MA, January (2005).
20. Interra Systems. Bluespec Testing Results: Comparing RTL Tool Output to Hand-designed RTL. `http://www.bluespec.com/images/pdfs/InterraReport042604.pdf`, April (2004).
21. Vijayaraghavan, M., Arvind, M.A.: "Bounded Dataflow Networks and Latency-Insensitive Circuits." In: *Proceedings of Formal Methods and Models for Codesign (MEMOCODE)*, pp. 171–180, Cambridge, MA, July (2009).

Chapter 8
Scalable Simulation for MPSoC Software and Architectures

Rainer Leupers, Stefan Kraemer, Lei Gao, and Christoph Schumacher

Abstract *Multi-processor systems-on-chip* (MPSoCs) are gaining a lot of attraction due to their good performance to power ratio. In order to cope with the complexity of such systems, early availability of full system simulation is of high importance. The simulation time is increasing with the growing number of processors. Therefore, scalable simulation techniques are required to mitigate this problem. Two new concepts to increase the simulation speed are becoming popular. First, raising the abstraction level increases simulation speed at the expense of a lower simulation accuracy. Second, exploiting all available processor cores in today's host systems increases the simulation speed without sacrificing the accuracy. Depending on the individual use case, one technique alone or a mixture of both techniques can be applied to create a fast simulation environment which is suitable for design space exploration, software development, performance estimation, and debugging.

8.1 Introduction

The increasing popularity of multi-processor systems-on-chip (MPSoC) in the embedded domain poses big obstacles for system designers, software developers, and simulation developers. This chapter focuses on the challenges imposed by this trend on system simulation and the corresponding simulation techniques.

It is expected that the performance of a single core of a simulation host will show only a limited growth within the next years, instead the total number of cores will increase. However, even commercial state-of-the-art simulation engines are not yet capable of exploiting multi-processor hardware efficiently. In order to cope with the growing complexity of the simulated systems, new and scalable simulation approaches are required.

Traditionally, simulations have been mainly modeled to reflect the needs of hardware developers and system programmers. Use cases for simulators have been early

R. Leupers (✉)
Software for Systems on Silicon (SSS), RWTH Aachen University, SSS-611910,
Templergraben 55, 52056 Aachen, Germany
e-mail: leupers@iss.rwth-aachen.de

R. Leupers, O. Temam (eds.), *Processor and System-on-Chip Simulation*,
DOI 10.1007/978-1-4419-6175-4_8,

design space exploration (DSE), performance evaluation, profiling, and testing. However, the growing complexity of simulated systems also poses a big challenge for the application developers. The early availability of a simulation environment can help to manage the software development risk imposed by the complex and sometimes even heterogeneous multi-processor structure of modern systems. Therefore, system simulators that take the specific requirements of application software developers into account are of great importance for a smooth software development process. In contrast to hardware development, for software development it is possible to trade accuracy for simulation speed. Otherwise it would not be possible to create fast software prototypes that are usable interactively.

Processor simulation constitutes the central building block of system simulation. Therefore, a lot of research effort has been carried out in this field. Section 8.2 focuses on the automatic generation of a processor simulator based on a high-level processor description language. Furthermore, statically and dynamically compiled simulation techniques are presented in the context of generated processor simulations. Section 8.3 describes how the simulation speed of a processor simulator can be further improved by using a hybrid simulation approach. Especially complete system simulation suffers from the simulation speed issue. Since the simulation speed usually decreases approximately linearly with the number of simulated processor cores, abstract simulation is applied to maintain reasonable simulation performance.

In addition to the approaches mentioned above, the available parallelism in the host system can be exploited to increase the simulation speed. However, parallelizing system simulations is a complex task due to the limited coarse grained parallelism directly exploitable. Section 8.4 gives a survey of the most important approaches in parallelizing simulations.

A case study of a scalable multi-processor system and its impact on the simulation environment is presented in Section 8.5.

8.2 Retargetable Instruction Set Simulation

*Instruction set simulator*s (ISSs) are of utmost importance in both architecture and software design. Nonetheless, it is a complex task to develop a high-quality simulator that is correct, accurate, and fast; therefore, providing a retargetable implementation approach is of profound interest. This section first discusses the retargetable simulator generation approach, then introduces a *just-in-time cache compiled simulation* (JIT-CCS) technique that allows self-modifying code to be supported.

8.2.1 Retargetable Simulator Generation

The emergence of *Architecture description language*s (ADLs) drastically improves the efficiency of describing a processor architecture and developing the toolkit thereof. LISA [21] is one of the most powerful ADLs available for description of in-order pipelined architectures. A LISA model consists of descriptions of processor resources and instructions. An excerpt of a LISA model that features a classic DLX [9] style data path (line 4 of the figure) is shown in Fig. 8.1. The resources encompass memories (line 2), registers (3), and pipeline registers (5). Coding (12) and syntax (13) definitions are parts of the instruction description and so is the behavior definition, which is divided into operations and distributed among the corresponding pipeline stages (14, 24). These operations are triggered in order (17, 25), and the pipeline registers are used to pass values between them (15, 24).

```
1  RESOURCE {
2    RAM uint32 data_mem { SIZE(0xFFFF); BLOCKSIZE(32); FLAGS(R|W); };
3    REGISTER uint32 R[0..15];
4    PIPELINE pipe = { FE ; DC ; EX ; MEM; WB };
5    PIPELINE_REGISTER IN pipe { uint32 src1; uint32 src2; uint32 dst; };
6  }
7  INSTRUCTION ADD IN pipe.DC {
8    DECLARE {
9      LABEL s1, s2, d;
10     INSTANCE ADD_ex;
11   }
12   CODING { /*Opcode*/ 0b0001 /*Operands*/ d=0bx[4] s1=0bx[4] s2=obx[4] };
13   SYNTAX { "add" "r" ~d "," "r" ~s1 "," "r" ~s2 }
14   BEHAVIOR {
15     OUT.src1 = R[s1]; OUT.src2 = R[r2];
16   }
17   ACTIVATION { ADD_ex }
18 }
19 OPERATION ADD_ex IN pipe.EX {
20   DECLARE {
21     REFERENCE s1, s2, d; //Use the labels defined in the DC stage.
22     INSTANCE Writeback;  //An operation in the WB stage.
23   }
24   BEHAVIOR { OUT.dst = IN.src1 + IN.src2; }
25   ACTIVATION { Writeback }
26 }
```

Fig. 8.1 Excerpt of a LISA model

Both interpretive and compiled simulators [2] can be generated from LISA models. As an example, the code of a compiled simulator generated from this LISA model is presented in Fig. 8.2. Assuming the program is accessible a priori and is not subject to change, each instruction can be pre-decoded to define the operations that need to be executed. The results of decoding are stored in a two-dimensional array (`operations` of Fig. 8.2), which consists of function pointers to operations. Since the architecture features a pipelined data path, a separate *program counter* (PC) is defined for each pipeline stage to denote the operation currently executing at this pipeline stage. At run-time, these PCs are advanced in a pipelined way, except for the first one – the `FE_PC`, which is assigned to the address of either the next instruction or the branch destination if a branch is taken. Since only the PC of the last pipeline stage (write back, aka `WB`) is visible to the programmers, the `WB_PC` is an alias of the program counter of the programmers' view.

PC	FE	DC	EX	MEM	WB
0x8000	Fetch	ADD r3, r2, r1	ADD_ex	Nop	Writeback
0x8004	Fetch	LD r7, 0x40	Nop	LD_mem	Writeback
0x8008	Fetch	BR r15	Nop	Nop	Nop

Simulation Compiler

```
typedef void (*operation_pointer)();
operation_pointer operations[PROGRAM_SIZE][DATAPATH_LENGTH] = {
  ...
  /*0x8000*/ Fetch, ADD, ADD_ex, Nop,    Writeback,
  /*0x8004*/ Fetch, LD,  Nop,    LD_mem, Writeback,
  /*0x8008*/ Fetch, BR,  Nop,    Nop,    Nop,
  ...
};
void run() {
  int FE_PC = 0; int DC_PC = 0; int EX_PC = 0; int MEM_PC = 0;
  do {
    // Advance pipeline
    PC/*Aka WB_PC*/ = MEM_PC; MEM_PC = EX_PC; EX_PC = DC_PC; DC_PC = FE_PC;
    FE_PC = isBranch? DEST_PC: (FE_PC + 4);
    // Execute operations
    operations[FE_PC][0](); operations[DC_PC][1]();
    operations[EX_PC][2](); operations[MEM_PC][3]();
    operations[PC][4]();
  } while (PC != EXIT);
}
```

Fig. 8.2 Example of compiled simulation

In comparison with interpretive simulation, this approach moves the effort of decoding to compile time as much as possible, so that the simulation speed can be significantly improved. It assumes the program is not changeable at run-time; therefore, it is known as *statically compiled simulation*. Experimental results show that a speedup of more than three orders of magnitude compared to a commercial interpretive simulator can be achieved [20].

8.2.2 Just-In-Time Cache Compiled Simulation

Statically compiled simulation assumes that the program memory is not subject to change at run-time, which is normally reasonable for application simulation. However, operating systems often modify the program memory dynamically, hence they cannot be supported by this technique.

A *dynamically compiled simulation* technique called *just-in-time cache compiled simulation* is introduced to address this problem. As shown in Fig. 8.3, instead of directly decoding the program memory, an instruction cache is created that is called a *JIT cache*. The decoding is only applied to the instructions of this artificial cache, as opposed to the statically compiled counterpart elaborated in Fig. 8.2, where the entire program memory is decoded statically at compile time. In every cycle, the simulator fetches a new instruction by accessing the program memory location indicated by the `FE_PC`. The instruction is stored to the JIT cache, which is organized in a direct mapped way, in order to provide unique resolution to cache locations in

```
extern int ProgramMemory[PROGRAM_SIZE];
int JitCache[CACHE_SIZE];
typedef void (*operation_pointer)();
operation_pointer operations[CACHE_SIZE][DATAPATH_LENGTH];
void run() {
  int FE_PC = 0; int DC_PC = 0; int EX_PC = 0; int MEM_PC = 0;
  do {
    // Advance pipeline
    PC/*Aka WB_PC*/ = MEM_PC; MEM_PC = EX_PC; EX_PC = DC_PC; DC_PC = FE_PC;
    FE_PC = isBranch? DEST_PC: (FE_PC + 4);
    // Check JIT-Cache, compile if cache miss is encountered
    if (JitCache[FE_PC%CACHE_SIZE] != ProgramMemory[FE_PC])
      JitSimulationCompiler(FE_PC);
    // Execute operations
    operations[FE_PC%CACHE_SIZE][0](); operations[DC_PC%CACHE_SIZE][1]();
    operations[EX_PC%CACHE_SIZE][2](); operations[MEM_PC%CACHE_SIZE][3]();
    operations[PC%CACHE_SIZE][4]();
  } while (PC != EXIT);
}
```

Fig. 8.3 Example of just-in-time cache compiled simulation

favor of quick access. The content of the cache is compared with the fetched instruction. If they are equal, the prior result of decoding can be used directly, otherwise, the JIT simulation compiler has to be employed to perform the decoding.

It is also important to determine the size of the JIT cache. If it is too small, the simulation compiler has to be invoked very frequently, which creates a significant overhead. On the other hand, an excessively large cache will unnecessarily increase the memory requirements of the simulator itself. For simulation of very large applications, an extra advantage of JIT-CCS is that the application does not have to be decoded entirely; therefore, the decoding results put less pressure on the memory. As presented in [2], a JIT-CCS system reaches a very close performance to a statically compiled simulator with a JIT cache of modest size (of 8192 entries).

Note that although the JIT-CCS systems are still slower than binary translators [Chapter 6], JIT-CCS technique has two important advantages. First, with JIT-CCS, retargetable simulator generation can be supported with ease. Second, it can be used to implement cycle-accurate simulators, which are important in hardware design and verification.

8.3 Hybrid Simulation

Although compiled simulation techniques have improved the execution speed of ISSs significantly, the speed is still limited compared to native execution of the application on a PC. Many researches (e.g., [13, 28]) attempt to provide alternative solutions by annotating timing information back to the application's source code to do performance estimation. As explained in Fig. 8.4, the application's source code is usually optimized and transformed to a low-level representation. Provided that the cycle cost of each basic statement (see lowered code of Fig. 8.4) is given, the timing of each basic block of the low-level representation can be estimated. After the timing information has been annotated, the low-level representation can be compiled

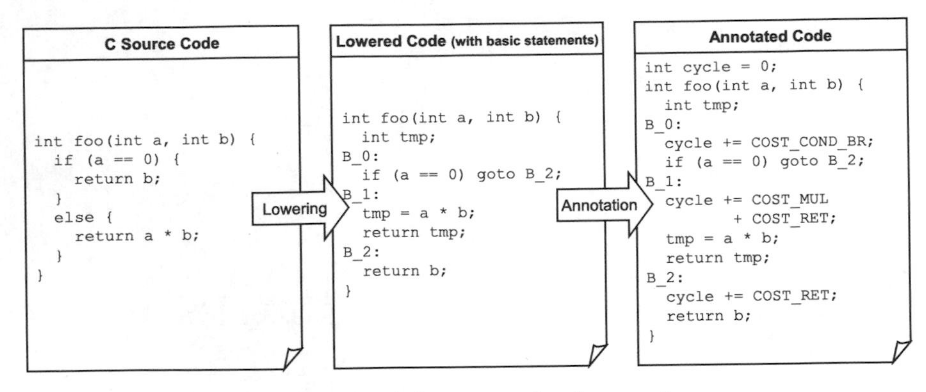

Fig. 8.4 Performance estimation using timing-annotated native execution

natively. During the course of execution, the performance of the application can be assessed. This approach is also known as *source-level performance estimation.*

Even though the execution speed of this approach is extremely high, two limitations are inevitable. First, because the entire course of a compiler backend that encompasses code selection, register allocation, etc. is not taken into account, the accuracy of performance estimation is limited, especially when dynamic delays (such as that introduced by memory access through an interconnect) have a big impact on the performance. Second, it cannot support applications that contain assembly or include library functions with no source code available. To address these issues, hybrid simulation is proposed, which allows employment of *instruction set simulator*s when simulating critical functions or functions without C source code are available.

This section will first introduce a way to integrate timing-annotated native execution with instruction set simulation, then discuss an on-demand profiling paradigm to use hybrid simulation.

8.3.1 Dynamic Execution Engine Switch of HySim

HySim [7] is a hybrid simulation framework featuring two simulation engines – a fast but less accurate *virtual simulator* (VS)[1] and an accurate but relatively slow instruction set simulator. In the embedded domain, an application usually consists of many C functions, some of which are executed many times. HySim allows each of these function calls to be either executed using the VS or the ISS and the decisions can be made separately.

The technical challenge HySim faces is that a *shared execution context* must be maintained so that it can be accessed by the functions executed on both the VS and the ISS. To address it, a novel transformation named *C virtualization* [7] is

[1] Also known as *virtual coprocessor* in [7]

introduced. Figure 8.5 shows an example of C virtualization and hybrid simulation. The example application consists of three C functions. It is compiled and loaded into the ISS, which contains resources visible to the programmer such as program memory, data memory, and registers. The source code of the functions `foo` and `bar` is transformed using the C virtualization approach. After transformation, all the accesses via pointers and accesses of global variables are virtualized to accesses of the ISS resources utilizing support functions. Additionally, stub functions are generated to support the execution engine switch. Since the output of C virtualization is still C code, it can be directly compiled with a native compiler such as GCC.

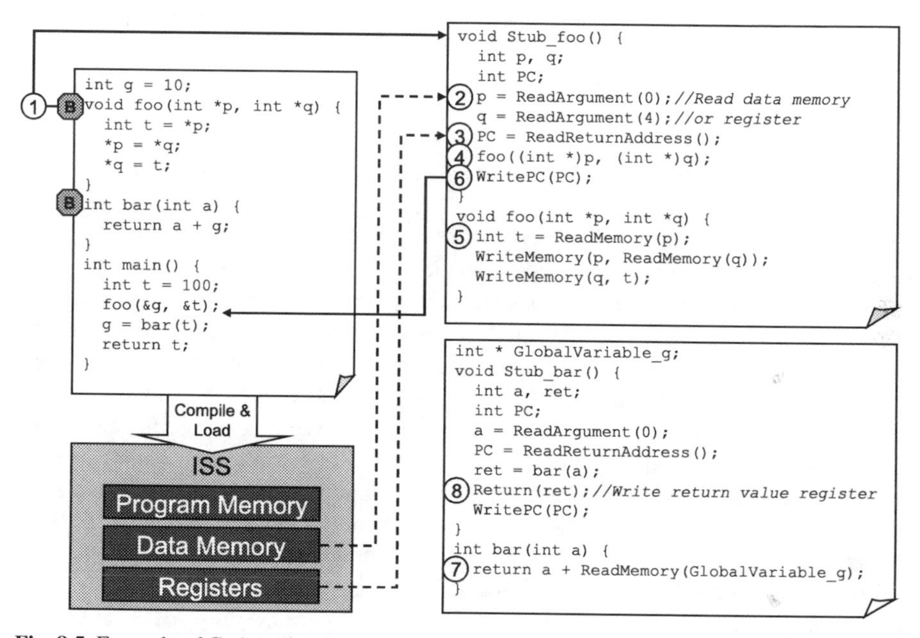

Fig. 8.5 Example of C virtualization and hybrid simulation

When the example application is loaded, a breakpoint is set on each virtualized function. The simulation begins with the `main` function, which passes the addresses of a global variable `g` and a local variable `t` to the subroutine `foo` it calls. Once `foo` is called, the breakpoint set at its beginning is hit; therefore, HySim can take over the control to decide whether the execution should be launched at the ISS or the VS. For the former case, the simulation simply continues. Suppose the execution of `foo` is launched at the VS, the stub function `Stub_foo` is invoked (①). According to the calling conventions, this function prepares the incoming arguments for the virtualized `foo` (②). It also retrieves the return address of the ISS and saves it to a local variable (③), because it indicates where to continue the simulation once `foo` returns. Afterward, function `foo` is called (④), which can read and write both the global variable `g` and the local variable `t` located at the data memory of the ISS using support functions (⑤). Note, that the data memory and the registers of the

ISS, by this means, become a shared context of the ISS and the VS. After the execution of `foo`, the PC is overwritten with the saved return address and the simulation continues at the ISS (⑥).

Using the same approach, the call to `bar` can be handled. The function `bar` accesses the global variable `g` directly. A global pointer (`GlobalVariable_g`) is defined to support the virtualization, and the address of `g` at the ISS is assigned to this pointer when the application is loaded. Using this pointer as a handle, `bar` can access the memory at the ISS where `g` is located (⑦). Similar to incoming arguments, the return value of `bar` can be handled with a support function (⑧).

Switching is not limited to the direction from the ISS to the VS. In a similar way, switching of the opposite direction can also be supported.

8.3.2 On-Demand Profiling-Based Performance Estimation

Timing information can be analyzed from the source code and annotated to the virtualized code to provide a rough performance estimation. However, since HySim offers two execution modes, it is possible to efficiently utilize them to provide fast simulation without sacrificing too much accuracy. Suppose a function has already been simulated using the ISS, hence the timing information is known. If the function is executed again with the same control path being taken, it can be executed with the VS and the already known timing information can be used to advance the clock. In other words, if each unique control path of a function is defined as a *scenario*, this approach *caches* the timing information of the profiled scenarios.

Nevertheless, to realize on-demand profiling, one technical difficulty is that the scenario of a function is only known after it has been executed, whereas this information is needed before simulating it. To address this problem, HySim uses a lightweight checkpointing-replay technique called *CrossReplay* [6]. Figure 8.6 shows the flowchart of CrossReplay. When a function is about to be simulated, the VS is speculatively chosen as the execution engine. During the execution, a lightweight checkpoint and a signature are generated. The signature can be anything that indicates a scenario, e.g., a vector of executed basic blocks. After querying a database, if a matching signature is found, it indicates that previously profiled timing information can be utilized. Therefore, CrossReplay retrieves the timing information from the database and advances cycles according to it. Conversely, if the query does not provide a matching record, the function must be replayed on a dedicated replay ISS by loading the lightweight checkpoint. After replaying, the profiled timing information is stored into the database, the clock is advanced, and the simulation continues.

The speed of HySim heavily depends on the proportion of execution mapped to the VS to that of the ISS. C virtualization and CrossReplay also create some overhead to the native execution of the VS. As shown in a case study [6], with negligible inaccuracy introduced, the simulation of a MPSoC can be sped up by 2–5 times compared to instruction set simulation.

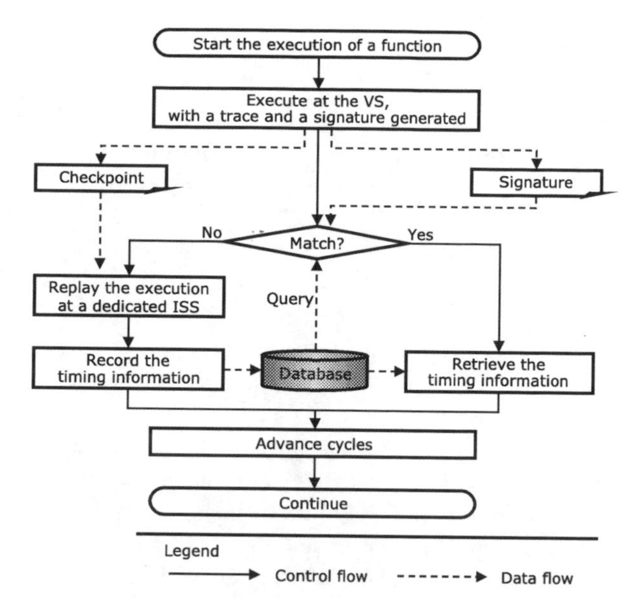

Fig. 8.6 Using CrossReplay for on-demand profiling

HySim is similar to sampling/statistical simulation [Chapters 11 and 13] in the sense that both techniques improve simulation speed by partially applying detailed simulation. However, the use cases of HySim and sampling/statistical simulation are different. The former can be used for application development, whereas the source code of applications is subject to change, since it does not need lengthy preprocessing. The latter is more suitable for architecture design.

8.4 Parallel Simulation

When simulating a multi-core system, the user usually witnesses a superlinear increase in simulation time with growing number of processing elements to simulate. The simulation host CPU must simulate more objects, and the overall efficiency decreases since the simulation OS process memory working set also grows. Therefore, there is more pressure on caches and memory interface of the simulation host. The effect is also observed in [22], an important example of parallel simulation approaches. Accordingly, opting for a parallelized simulation in order to prevent simulation speed from deteriorating seems to be self-evident.

Paradoxically, although MPSoCs offer plenty of inherent parallelism, it is not easily exploited in simulation. The main problem is to maintain the causality of events [5]. To illustrate the problem, consider a system consisting of two CPUs attached to a shared memory, as displayed in Fig. 8.7. To keep it simple, assume that all memory accesses happen instantaneously, that the CPUs share a common

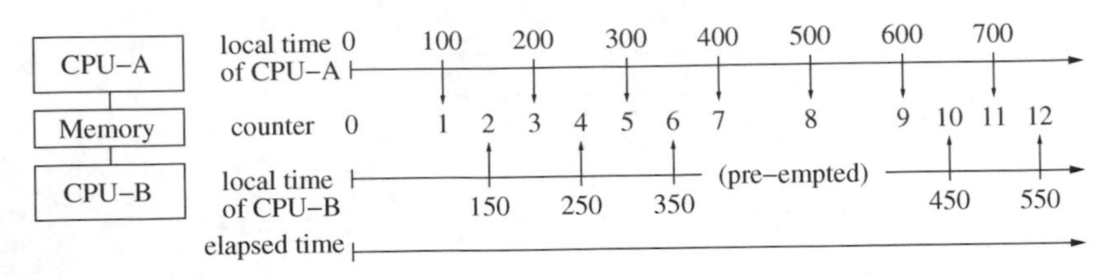

Fig. 8.7 Causality problem: example system

clock, and that accesses to the same memory location do not happen at the same time. The example uses processors to illustrate the causality problem, but it applies to parallel simulation in general.

Traditional sequential simulation would advance time usually by alternatingly stepping both CPUs. A straightforward strategy to simulate such a system in parallel is to assign the simulation of each CPU to its own simulation host core and to let the two host processors step their respective CPUs within a loop. Without special means, it is certain that the local time of the processors will drift apart. Reasons for this include preemption of either thread due to the host's OS, different simulation speeds or type of CPUs, or different instruction mixes on the simulated CPUs.

The result of such a parallel simulation is most likely to be different compared to the traditional sequential simulation approach described above, if not functionally wrong.

Imagine both cores of the example system increment a common counter in regular intervals of 100 cycles. In sequential simulation, both CPUs would then observe an increment of one made by the other CPU every 100 cycles. This is not necessarily the case with the simple parallel simulation scheme described above, where events from the future can influence the past.

Say that at cycle 375 of CPU-B's local time, its simulation thread gets preempted. While the thread is suspended, CPU-A increments the counter three times at its local times 400, 500, and 600. After CPU-B's thread wakes up, CPU-B increments the common counter again at its cycle 450 and observes that it has seemingly been incremented three times by CPU-A, although only 100 cycles have passed in CPU-B's local time frame. Causality is violated and the CPUs could observe any positive number of increments during a single interval.

8.4.1 Synchronization Strategies

To cope with the causality problem, a number of synchronization algorithms for parallel discrete event simulation (PDES) [5] have been suggested, which are often categorized similar to the following [1]:

Oblivious parallel simulation strategies do not take into account to which stimuli processes react. Therefore, all elements of a simulation have to be executed at every time step, even if there is no direct need for execution. This kind of algorithm has been suggested for digital logic simulation given that communication and synchronization overhead is the primary concern.

Synchronous strategies execute activity in the same order as sequential simulation except if multiple parts of the simulation model are to be executed at exactly the same time-stamp. In the latter case, execution is conducted in parallel. The result can be guaranteed to match one of the sequential executions, if the modeling style has the property that activity at a certain time cannot influence activity at exactly the same time. VHDL, Verilog, and SystemC signals implement such a property using *delta-cycles*.

Given the model in Fig. 8.7, this would mean that both cores would always be stepped exactly one cycle, followed by a synchronization of the executing threads. However, depending on the underlying hardware and the size of the model, lock-step barrier synchronization can be very costly.

Conservative asynchronous algorithms try to improve the performance of parallel simulation over synchronous algorithms by exploiting knowledge of the time it takes for individual parts to influence each other and thereby executing activity earlier than the synchronous paradigm dictates.

The downside is that the necessary information regarding timing is usually not readily available, which may cause a simulation engine to fall back to lock-step synchronization as explained above.

Optimistic asynchronous algorithms try to carry the idea of the conservative asynchronous algorithms even further. In case it is not known when parts of the simulation will communicate, the simulation engine will try to predict if or assume that there is no communication for a time. If it later turns out that the prediction or guess was incorrect, the simulation is rolled back to the point of time where the break in causality occurred.

Although this kind of algorithm provides the greatest possible potential for speedup, it relies on correct predictions of communication behavior and the availability of an inexpensive rollback mechanism. Both features are difficult to implement in today's production simulation environments for MPSoC.

Simulation without strict synchronization might also be an option, because perfect accuracy is not always required. For example in functional models that focus on behavior, it may be tolerable that causality is violated. The parts of the simulation that do require synchronization must still be carefully identified and implemented (e.g., bus locking).

8.4.2 Further Reading

Issues with PDES are well known and are studied for many decades. References [1, 4, 5, 12] explain the theoretical background. Accordingly, a number of simulators have been implemented. In the beginning, simulators were targeted to a comparatively narrow range of target systems as [23]. What followed were many attempts to widen the range of possible target systems and to deal with simulator construction in a more abstract manner [16]. These insights were then applied to construct parallel engines executing simulation languages like VHDL [15] or Liberty [22].

Other approaches take measures beyond classical PDES, like augmenting the simulation engine with knowledge about the OS software running on the simulated platform [30] or executing different simulation time segments in parallel [8].

8.5 Case Study—Scalable MPSoC Simulation Environment

MPSoCs are becoming a viable alternative to the traditional monolithic processor system implementations. Especially, heterogeneous MPSoCs are of great interest because of their good performance to energy ratio.

Two major challenges can be foreseen for future CMOS technologies: first, the minimization of the wire delay [3, 10] and second, the management of the constantly increasing design complexity.

The tiled architectures approach [27] addresses these problems by combining a high number of simple processing tiles into a computing system instead of designing a single, monolithic processing architecture. The simplicity of the individual tiles leads to the minimization of the wire delay problem and also helps to reduce the design complexity. The inter-tile communication is realized by a distributed routing fabric based on packet switching. These kinds of architectures are designed to be highly scalable and therefore they are suitable for massively parallel applications.

In the following the tiled architecture developed in the context of the European SHAPES project [17, 19, 24] is presented. The goal of this project is to design a Teraflops board that can be used as building block for Petaflop-class systems. Due to its flexibility and scalability the SHAPES architecture is intended to cover the entire range from commodity market applications (4–8 tiles), classical signal processing applications such as audio video processing, ultra-sound, and radar (2K tiles) up to numerically demanding, massively parallel scientific simulations (32K tiles). Two applications that fit this kind of architecture well have already been ported in the course of this project: wave field synthesis (WFS) [25] and lattice quantum chromo dynamics (LQCD) [29].

8.5.1 Overview of the SHAPES Platform

Figure 8.8 shows the overall structure of the SHAPES multi-tile platform. The main hardware building block of this platform is a dual-core tile consisting of a general purpose RISC processor and a VLIW DSP. Figure 8.9 depicts the schematic of such a tile and its key components. An ARM926EJ-S serves as general purpose processor for control dominated code, while a VLIW mAgic [18] processor is used for efficient execution of the signal processing parts of an application.

All processing elements and the on-tile memory are interconnected using a multi-layer AMBA bus for high throughput. Each RDT tile also provides a rich set of peripherals which are connected via an AMBA peripheral bus (APB). Tiles with different configurations, i.e., number of CPUs, number and type of peripherals and

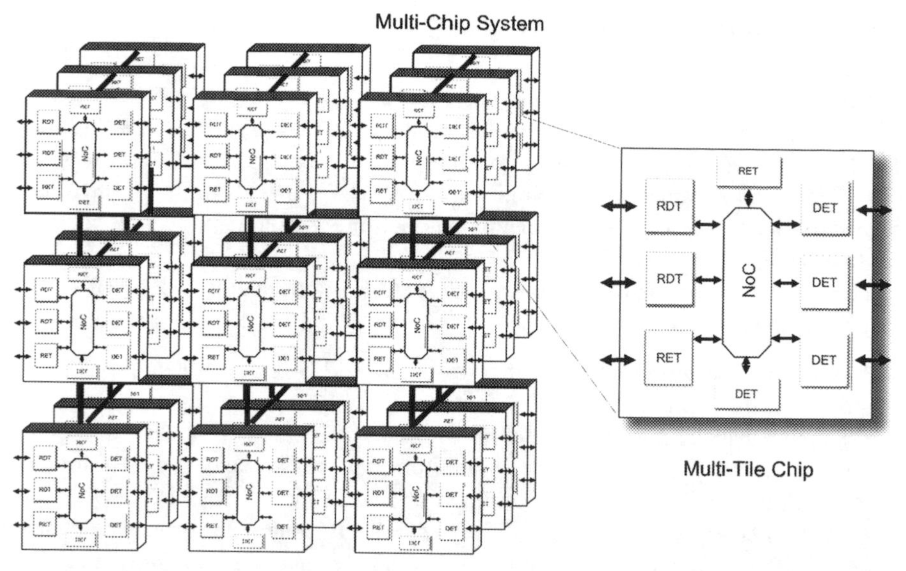

Fig. 8.8 Multi-Tile SHAPES platform

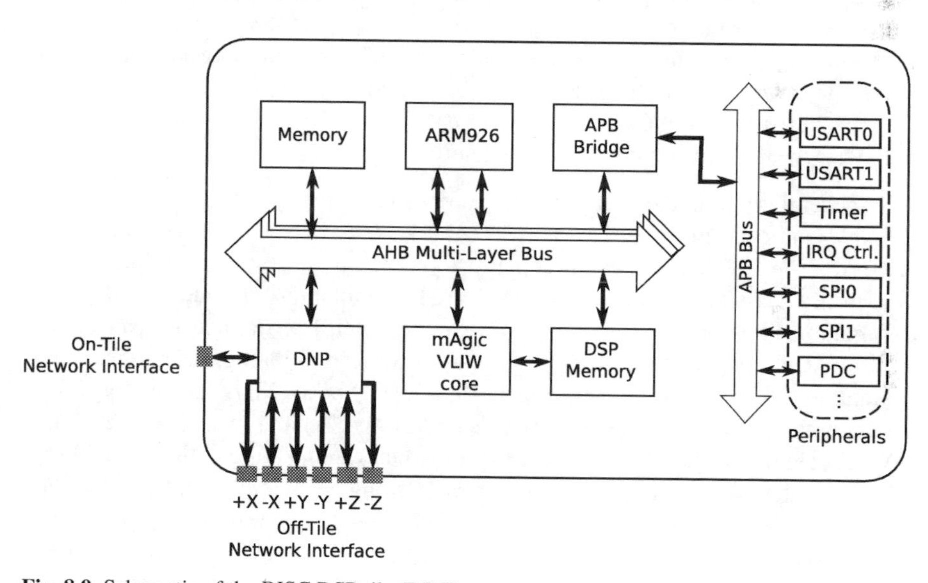

Fig. 8.9 Schematic of the RISC DSP tile (RDT)

memory size, can be obtained by stripping down the RDT tile. Up to eight tiles can be combined into a single chip. These multi-tile chips can then be arranged in a three-dimensional mesh topology (see Fig. 8.8). In case of a multi-tile chip a network-on-chip (NoC) is used as on-chip interconnect. The inter-tile communication is realized by the *distributed network processor* (DNP) [17], which serves as a generalized DMA controller.

8.5.2 SHAPES Simulation Environment

Due to the high development effort and costs of hardware prototypes, system simulations are becoming a viable alternative during the early design stages of the MPSoC development. This is especially true for the multi-tile SHAPES platform. In order to test different parameter configurations a *virtual SHAPES platform* (VSP) has been developed. The key aspects of this simulation environment are presented in the following.

For a scalable MPSoC platform, such as SHAPES, four main use cases of simulation can be identified:

- *Design space exploration*: With the virtual SHAPES platform it is possible to quickly evaluate different combinations of tiles. Hence, promising candidates for a detailed analysis can be identified.
- *Software development*: Having a simulation platform at hand, it is possible to start software development and optimization before the hardware is ready.
- *Performance estimation*: The possibility to collect detailed performance characteristics for a given application, e.g., context switching overhead, allows the user to optimize the task to processor mapping, to optimize the application itself, and to optimize the interprocess communication.
- *Debugging*: The main advantage compared to debugging a software application on a hardware board is the possibility to access virtually all hardware resources in a non-intrusive fashion. This increased visibility is of big help for understanding the hardware–software interaction in detail.

In general, the simulation speed plays a critical role for the usability of a simulation environment. This is particularly true for complex architectures as the SHAPES platform. The simulation speed of a platform is mainly determined by the abstraction level. Classical register transfer level (RTL) simulations provide very detailed information about the simulated system. However, their low simulation speeds prohibit efficient design space exploration and software development. Therefore, the simulation level needs to be much higher than RTL. For the SHAPES platform two levels of abstraction are considered: *cycle accurate* (CA) and *instruction accurate* (IA). CA simulation has the advantage that the micro-architecture of the processor is fully simulated, thus leading to a very accurate timing behavior. In contrast, IA simulation provides much higher simulation speed by neglecting the micro-architectural features. Thus, CA simulation provides a higher accuracy while IA simulation permits shorter design cycles. Depending on the use case the developer can choose between both simulation modes in order to obtain the best balance between simulation speed and accuracy.

The virtual SHAPES platform is modeled in SystemC using the Synopsys virtual platform environment [26]. The ARM processor and the mAgic processor are modeled as ISS at IA and CA levels. The ARM processor model includes a cache model and is taken from the Synopsys IP library, whereas the floating point VLIW DSP is modeled using the architecture description language LISA [11]. Building upon the single tile simulator it is possible to automatically generate a multi-tile simulator using a script-based generation approach. Based on input parameters like tile types, number of tiles, and type of interconnect, the multi-tile simulator is automatically assembled and built.

The created simulation environment is an important tool for hardware-dependent software developers due to its non-intrusive debugging capabilities. Moreover, the simulation helps to identify good mappings of application tasks to processors and allows collecting a variety of performance characteristics.

8.5.3 Advanced Simulation Techniques Applied in SHAPES

Using the presented VSP, the simulation time increases approximately linearly with the number of tiles, because most of the simulation time is spent for processor simulation. ISS-based processor simulation techniques have been pushed to their limits and not much speed up can be expected in the future. More advanced techniques, e.g., binary translation [Chapter 9], can be applied to ameliorate the situation. However, in case of the complex VLIW DSP used in SHAPES, binary translation does not pay off, because it relies on similarities between simulation host and target which are not present for VLIW DSPs. Therefore, a hybrid simulation approach (see Section 8.3) is selected. Based on the observation that the user requires detailed simulation only for a fraction of the simulation time, it is possible to temporally switch between a fast but inaccurate and a slow but detailed simulation. To further reduce the number of simulation runs, checkpointing [14] can be applied to skip the time consuming but repetitive booting of a simulated OS.

Figure 8.10 shows the speedups achieved with different hybrid simulation configurations. The baseline of this measurement is the simulation of the SHAPES platform utilizing only ISS-based processor simulation. For comparison reasons

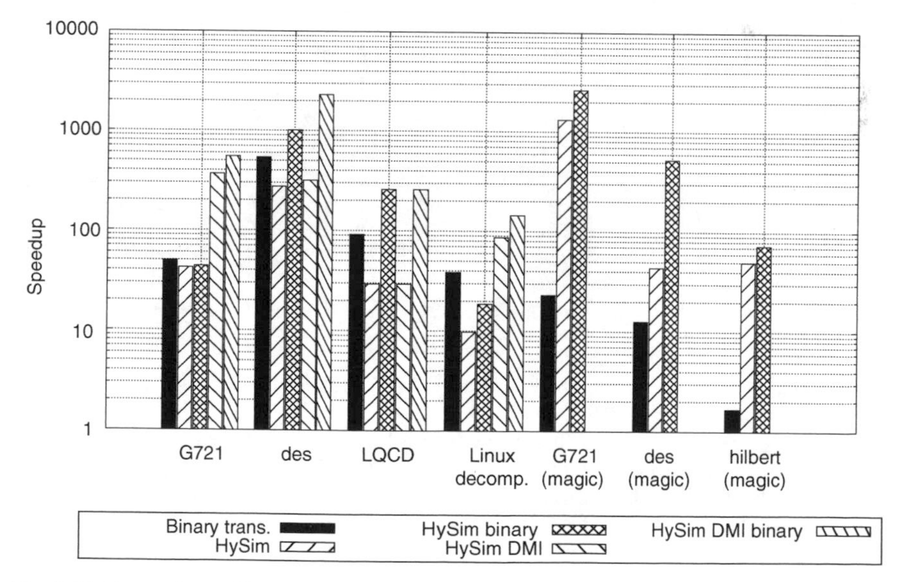

Fig. 8.10 Comparison of binary translation and hybrid simulation for the SHAPES platform

the speedup reachable with binary translation is also shown. Four different hybrid simulations are used in this experiment. The first configuration utilizes the HySim framework to switch between the VS and the ISS-based simulation. The second configuration combines binary translation with the hybrid simulation. Configuration three and four are identical to the first two configuration except that they make use of the DMI interface to access the global memory.

Applying these kinds of advanced simulation techniques will ensure sufficiently high simulation performance as required for software development on large-scale simulators.

8.6 Conclusion

The emergence of MPSoCs has triggered a number of changes in the embedded systems domain. Raising the number of processors in an MPSoC increases the simulation time and poses a big challenge for simulation developers. To mitigate this problem new processor simulation techniques are required since the commonly used ISS-based processor simulation has been pushed to its limits and not much improvement can be expected from this technique. The simulation environment developed for the SHAPES platform shows by using one of those techniques or a combination of them it is possible to develop a sufficiently fast simulator for such a complex platform. This will also scale with future SHAPES platforms with substantially more cores than the current SHAPES platform simulation. It can be foreseen for the near future that two techniques are gaining interest for simulation developers. It can be expected that raising the abstraction level will be the simulation technique of primary choice until a versatile parallel simulation environment will be available.

Acknowledgments We thank Pees, Braun, Nohl, Hoffmann, and all the people who have contributed to the development of these techniques [2, 20].

References

1. Bailey, M., Briner, J., Chamberlain, R. Parallel logic simulation of VLSI systems. *ACM Comput Surv* **26**(3), 255–294, (1994).
2. Braun, G., Nohl, A., Hoffmann, A., et al. A universal technique for fast and flexible instruction-set architecture simulation. *Comp-Aid Des Inte Circuits Syst, IEEE Trans* **23**(12), 1625–1639, (2004).
3. Carloni, L.P., Sangiovanni-Vincentelli, A.L.: Coping with latency in SoC design. *IEEE Micro* **22**(5), 24–35, Sep./Oct. (2002).
4. Chandy, K., and Misra, J.: Distributed simulation: A case study in design and verification of distributed programs. *Software Engineering, IEEE Trans* **SE-5**(5), 440–452, Sept. (1979).
5. Fujimoto, R. Parallel discrete event simulation. *Commun ACM* **33**(10), 30–53, (1990).
6. Gao, L., Karuri, K., Kraemer, S., et al.: Multiprocessor performance estimation using hybrid simulation. In: *DAC '08: Proceedings of the 45th annual conference on Design automation*, pp. 325–330, New York, NY, USA, June, AMC (2008), Anaheim, CA.

7. Gao, L., Kraemer, S., Leupers, R., et al.: A fast and generic hybrid simulation approach using C virtual machine. In: *CASES '07: Proceedings of the 2007 International Conference on Compilers, Architecture, and Synthesis for Embedded Systems*, pp. 3–12, Salzburg, Austria (2007) New York, NY, USA, Sept, ACM.
8. Girbal, S., Mouchard, G., Cohen, A., et al.: dist: a simple, reliable and scalable method to significantly reduce processor architecture simulation time. In: *SIGMETRICS '03: Proceedings of the 2003 ACM SIGMETRICS international conference on measurement and modeling of computer systems*, pp. 1–12, New York, NY, USA, 2003. ACM, (2003).
9. Hennessy, J.L., Patterson, D.A.: San Francisco, CA, USA, *Computer architecture: A quantitative approach.* Morgan Kaufmann Publishers Inc., (2003).
10. Ho, R., Mai, K.W., Horowitz, M.A.: The future of wires. *IEEE Trans Comp-Aid Des* **89**(4), 490–504, Apr (2001).
11. Hoffmann, A., Kogel, T., Nohl, A., Braun, G., Schliebusch, O., Wahlen, O., Wieferink, A., Meyr, H.: A novel methodology for the design of application-specific instruction-set processors (ASIPs) using a machine description language. *IEEE Trans Comp-Aid Des* **20**(11), 1338–1354, (2001).
12. Jefferson, D.: Virtual time. *ACM Trans Program Lang Syst* **7**(3), 404–425 (1985).
13. Karuri, K., Faruque, M.A.A., Kraemer, S., et al.: Fine-grained application source code profiling for ASIP design. In *DAC '05*, pp. 329–334, New York, NY, USA, June, ACM, (2005) Anaheim, CA.
14. Kraemer, S., Leupers, R., Petras, D., Philipp, T.: A checkpoint/restore framework for SystemC-based virtual platforms. In *System-on-Chip, 2009. SOC 2009. International Symposium on*, pp. 161–167, Tampere, Finland (2009) Piscataway, NJ, USA, Oct, IEEE Press.
15. Lungeanu D., Shi, C.: Parallel and distributed VHDL simulation. In: *DATE '00: Proceedings of the conference on design, automation and test in Europe*, pp. 658–662, New York, NY, USA, 2000. ACM (2000).
16. Nicol, D., Heidelberger, P.: Parallel execution for serial simulators. *ACM Trans Model Comput Simul* **6**(3), 210–242, (1996).
17. Paolucci, P.S., Jerraya, A.A., Leupers, R., Thiele, L., Vicini, P.: SHAPES: A tiled scalable software hardware architecture platform for embedded systems. In: *Proceedings of the International Conference on Hardware/Software Co-design and System Synthesis (CODES+ISSS)*, pp. 167–172, Seoul, Korea (2006) New York, NY, USA, Oct, ACM.
18. Paolucci, P.S., Kajfasz, P., Bonnot, P., Candaele, B., Maufroid, D., Pastorelli, E., Ricciardi, A., Fusella, Y., Guarino, E.: mAgic-FPU and MADE: A customizable VLIW core and the modular VLIW processor architecture description environment. *Comp Phys Communi* **139**(1), 132–143 (2001).
19. Paolucci, P.S.: The diopsis multiprocessor tile of SHAPES. In: *6th International Forum on Application-Specific Multi-Processor SoC MPSOC'06*, Colorado, USA, August (2006).
20. Pees, S., Hoffmann, A., Meyr, H.: Retargetable compiled simulation of embedded processors using a machine description language. *ACM Trans Des Autom Electron Syst* **5**(4), 815–834 (2000).
21. Pees, S., Hoffmann, A., Zivojnovic, V., et al.: LISA—machine description language for cycle-accurate models of programmable DSP architectures. In: *DAC '99: Proceedings of the 36th annual ACM/IEEE Design Automation Conference*, pp. 933–938, New York, NY, USA, 1999 ACM (1999).
22. Penry, D., Fay, D., Hodgdon, D., et al.: Exploiting parallelism and structure to accelerate the simulation of chip multi-processors. In *HPCA*, pp. 29–40 IEEE Computer Society, Austin, TX Feb (2006).
23. Reinhardt, S., Hill, M., Larus, J., et al.: The wisconsin wind tunnel: virtual prototyping of parallel computers. *SIGMETRICS Perform Eval Rev* **21**(1), 48–60 (1993).
24. SHAPES – Scalable Software Hardware computing Architecture Platform for Embedded Systems. http://www.shapes-p.org.

25. Sporer, T., Beckinger, M., Franck, A., Bacivarov, I., Haid, W., Huang, K., Thiele, L., Paolucci, P.S., Bazzana, P., Vicini, P., Ceng, J., Kraemer, S., Leupers, R.: SHAPES – a scalable parallel hw/sw architecture applied to wave field synthesis. In *Proceedings of the AES 32nd International Conference*, pp. 175–187, Hillerød, Copenhagen, Denmark, September (2007).
26. Synopsys Inc. "Synopsys Virtual Platforms", Mountain View, CA. 2010 http://www.synopsys.com/Tools/SLD/VirtualPrototyping/
27. Taylor, M.B., Kim, J., Miller, J., Wentzlaff, D., et al. The raw microprocessor: A computational fabric for software circuits and general-purpose programs. *IEEE Micro* **22**(2), 25–35 (2002).
28. Wang, Z,. Herkersdorf, A.: An efficient approach for system-level timing simulation of compiler-optimized embedded software. In: *DAC '09*, pp. 220–225, New York, NY, USA, July (2009) ACM, San Franscisco, CA.
29. Wilson,. K.G.: Confinement of quarks. *Phys Rev D* **10**(8), 2445–2459, Oct (1974).
30. Yi, Y., Kim, D., Ha, S. : Fast and accurate cosimulation of MPSoC using trace-driven virtual synchronization. *Comp-Aid Des Integ Circuits Syst, IEEE Trans*, **26**(12), 2186–2200 Dec. (2007).

Chapter 9
Adaptive High-Speed Processor Simulation

Nigel Topham, Björn Franke, Daniel Jones, and Daniel Powell

Abstract Instruction set simulators are essential tools in all forms of microprocessor design; simulators play a key role in activities ranging from ASIP design-space exploration to hardware–software co-verification and software development. Simulation speed is the primary concern when functional simulators are used as CPU emulators for software development. Conversely, the ability to measure performance is of critical importance during the exploratory phases of co-design, whereas the ability to use a simulator as a golden reference model is important for hardware–software co-verification. A key challenge is to provide the highest level of performance, for the different observability and performance measuring demands of each use-case. In this chapter, we describe an adaptive simulator designed to meet these diverse requirements. Adaptation takes two forms: first, the simulator has a high-speed JIT compilation capability allowing it to be extended dynamically according to simulated program behavior; and second, it is able to learn how to model the timing behavior of the target processor and thereby deliver approximate performance figures with very low overhead. The simulator maintains a precise model of the architectural state of the processor it simulates, enabling it to be used also as a back-end target for a debugger, to assist in software development, as well as providing a Golden Reference Model to a co-simulation environment. Through the use of these performance-enhancing dynamic adaptations, the simulator is capable of simulating an embedded system at speeds approaching, or even exceeding, real time.

9.1 Introduction

This chapter discusses how various forms of *adaptation* in simulators can provide fast, accurate, and observable modeling of a processor. Accuracy and observability are also possible to achieve, by simulating at a low level, but such simulators are

N. Topham (✉)
School of Informatics, The University of Edinburgh, 10 Crichton Street, Edinburgh EH8 9AB, UK
e-mail: npt@staffmail.ed.ac.uk

R. Leupers, O. Temam (eds.), *Processor and System-on-Chip Simulation*,
DOI 10.1007/978-1-4419-6175-4_9,

very slow. By compiling an instrumented version of the source code of a target application to a host machine, it is possible to achieve higher speeds, due to the elision of detailed modeling of low-level target features. Such *compiled simulation* techniques can therefore be many orders of magnitude faster than low-level *interpretive simulators*, although with reduced observability. A fundamental limitation of compiled simulation is that it can be used only in situations where the application code is available in source form, and does not change dynamically. Programs which require an operating system, or which are shrink-wrapped, cannot benefit from compiled simulation. The approach explored in this chapter is based on the concept of *dynamic binary translation* (DBT). This approach begins with interpretive simulation in order to obtain initial profile information. This is interspersed with profile-guided just-in-time translation (JIT) of target code to host code, with optional performance instrumentation, similar to the approach taken by many virtual machine environments. The net result is an extremely high-speed simulator, capable of adapting to the changing dynamic behavior of the application.

Our work focuses on how dynamic adaptation can provide simulators with important advantages in speed and performance modeling. First, we show how extremely high-functional simulation rates can be achieved through the use of DBT applied at the basic block level through to entire pages of target binary. Our approach effectively extends the simulator at run-time with dynamically loaded functions corresponding to the translated regions of the target binary. Second, we show how high-speed performance modeling can be achieved using *statistical machine learning*. In essence, the simulator learns and adapts an internal performance model during phases of comparatively slow cycle-accurate simulation. This model then provides very low-cost performance information during phases of high-speed functional simulation.

The research reported in this chapter was carried out using the Edinburgh high-speed simulator (EHS), developed in the Institute for Computing Systems Architecture at Edinburgh University [1]. In the first half of this chapter, we explore how EHS adapts through dynamic binary translation, and in the second half we look at how machine learning can provide high-speed performance models.

9.2 Fast Simulation Through Dynamic Binary Translation

Interpretive simulation of an instruction set involves repeatedly decoding and emulating the semantics of the target instruction set. Much of the work involved at the interpretive level can be avoided completely by translating a sequence of target instructions into host instructions which perform equivalent transformations on the target processor state. The most effective way to create such a host machine sequence is to use dynamic binary translation within a simulation framework that allows target code to be interpreted, when no translation exists, and which then creates binary translations on-demand when *hot* regions of target code are identified.

In the context of EHS, a *translation unit* is a region of target code that is translated to a single host function. These translated functions (TFs) are dynamically linked into the simulator, extending the simulator binary. To create a set of TFs, simulation time is partitioned into *epochs*, where each epoch is defined as the interval between two successive calls to the binary translation process. During each epoch, new translation-units may be discovered; previously seen but not yet translated translation-units may be re-interpreted; translated translation-units may be discarded (e.g., due to self-modifying code); and translated units may be executed. Throughout this process, the simulator continues profiling all interpreted blocks. The end of each simulation epoch is reached when a predefined number of interpreted basic blocks has been executed. Execution profiles for each physical page are built up during the simulation epoch. This involves maintaining a count of the number of times individual basic blocks have been interpreted and maintaining a summary of this per page.

Translated functions are located with minimal overhead using the translation cache (TC), a hash table of translated function pointers indexed by program counter. If the current PC hits in the TC, its associated TF is executed by dereferencing the associated function pointer. In addition to implementing the target code, all state information for the simulated processor, including the PC value, is updated prior to exiting the TF. This provides precise observability of target state at the boundaries between any translation unit or interpreted block.

If the instruction address is not found in the TC, it is looked up in the translation map (TM). This is a complete table of all translated units and is therefore slower to access than the TC. Each TM entry contains the translated unit's entry address and a pointer to its TF. The TC is loaded from the TM rather like a level-1 cache is loaded from a level-2 cache. Any block that is not present in the TM must be interpreted, and in this case the appropriate profiling information is also gathered.

At the end of each simulation epoch, a profiling analysis phase is initiated prior to binary translation. This scans the profiling structures for frequently executed blocks. When larger translation units than a single basic block are being handled, as discussed in Section 9.2.1, the scanning process will search for hot regions encompassing multiple basic blocks. The hot translation units are translated in batches, comprising translation units from the same physical page. The grouping of translation units in this way enables all translations for a physical page to be easily discarded if, for example, a target page is overwritten by self-modifying code or if the page is reused by a different process.

Translation begins by converting each region to a C function which emulates each target instruction by modifying the processor state according to the semantics of the instruction. PC is updated at the end of the block or on encountering an exception. Exceptions, such as a TLB miss, cause an immediate return from the TF to the main simulation loop, where the exception is handled. Any pending interrupts are detected at block boundaries, at which point control will again be transferred back to the main simulation loop.

The C code for all translations in a given page is compiled to a single shared library, using GCC, which is then loaded by the dynamic linker. Several such libraries may be loaded during each translation phase, and during loading the Translation Map is updated with information about each TF, ensuring a hit in the TM next time that TF is encountered.

9.2.1 Identifying Larger Translation Units

Typically, in DBT simulators, the unit of translation is either the target instruction or the basic block. It has been shown that increasing the size of the translation unit delivers a significant increase in simulation rate [2]. The increase in performance is due to two factors. First, large translation units (LTUs) offer greater scope to the compiler for global optimizations across multiple blocks. Second, less time is spent on the overhead of managing the outer simulation loop.

The EHS simulator supports three different types of LTU, in addition to the basic block. An LTU in this context is a group of basic blocks connected by control-flow arcs, which may have one or more entry and exit points. The different translation modes incorporated into the EHS simulator include basic blocks (BB), a group of basic blocks forming a set of strongly connected components (SCC), the set of maximal control-flow graphs (CFG) identified during profiling, and all blocks from within the same physical page (Page).

In contrast to other DBT simulators, the EHS simulator profiles the target program's execution in order to discover hot *paths* rather than to identify hot *blocks*, some of which may be infrequently executed within a given page. The target program is profiled, and the translation units are created on a per physical-page basis. Figure 9.1 shows the different translation unit types and their entry points. Example

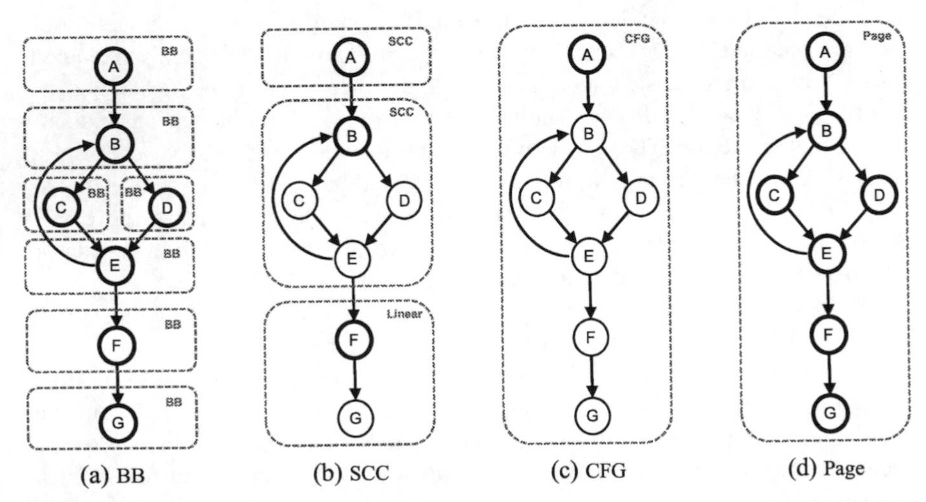

Fig. 9.1 Different types of translation unit. Each figure shows a target program CFG divided into DBT mode translation units. *Dotted lines* indicate the separate translation units while thick-edged circles show the possible entry points

target program CFGs are shown, divided into separate translation units in accordance with their DBT mode. The possible entry points into a translation unit varies with DBT mode. However, entry is always to the address of the first instruction within a basic block.

In BB mode, those basic blocks which are frequently executed at simulation time are identified and scheduled for binary translation. For SCC mode, the block execution path is analyzed to discover SCCs and any attached linear regions. In CFG mode, the block execution path is analyzed to discover hot CFGs. In Page mode, the block execution path is analyzed to discover all hot CFGs within that page, after which those CFGs are then translated as a single unit, creating a single TF.

9.2.2 Persistent Translations

In order to obtain highly efficient translations, the simulator invests a significant effort in finding the most important regions of target code and optimizing their translation to host binary form. It therefore makes sense to re-use these translations from one simulation run to the next, wherever possible. For this reason, the EHS simulator supports *persistent translation*, in which a previous translation for a given hot region will be re-used if the binary signature matches that of a previous invocation.

To observe the benefit of persistent translation, we examined where the simulation time was being spent. To a first approximation, the total simulation time is divided between five main tasks: (1) the main simulation loop calls TFs and interprets instructions; (2) the library loading function loads previously compiled translations from the shared libraries; (3) the page CFG function adds interpreted blocks to the page CFG; (4) the profile analysis function analyses page CFGs and identifies hot translation units; and (5) the dynamic compilation function compiles hot translation units and creates shared libraries. Figure 9.2 shows how much time is spent performing each task for a typical benchmark. On the first run, 80–92% of the time is spent compiling the hot translation-units, with the rest of the time spent in the main simulation loop. On successive runs, which re-use translations from earlier runs, almost 100% of the time is spent in the main simulation loop. Time spent performing the other tasks is all but insignificant. This is the general pattern observed for all benchmarks. Persistence is another form of adaptation, in which the simulator becomes *specialized* for a particular application based on prior translations that have been "learned" during previous runs.

9.2.3 The Performance of DBT

In this section, we present the results from simulating a subset of the EEMBC benchmark suite on the EHS simulator. These results compare the relative performance of the four DBT modes, as well as show the absolute achievable simulation rates on a stated hardware platform. All benchmarks were compiled for the ARC

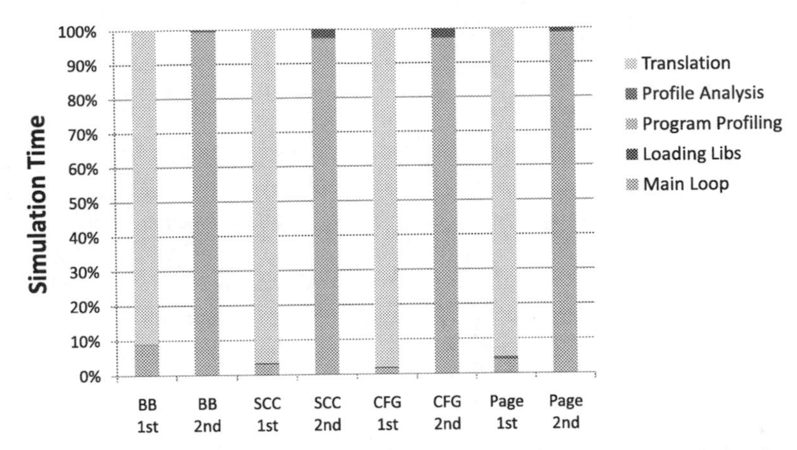

Fig. 9.2 Bitmnp simulation tasks. This figure shows the percentage of total simulation time spent performing each of the five main simulator tasks

700 architecture using `gcc` version 4.2.1 with `-O2` optimization and linked against `uClibc`. The EEMBC lite benchmarks were run for the default number of iterations, and the simulator operated in user-level simulation mode so as to isolate the effects of an underlying operating system.

The simulator itself was compiled, and the translated functions dynamically compiled, with `gcc` version 4.1.2 and `-O3` optimization. All simulations were performed on the workstation detailed in Table 9.1 running Fedora Core 7 under conditions of minimal system load. The simulator was configured to use a simulation epoch of 1000 basic blocks and a low translation threshold to ensure that every translation unit would be translated. The target memory had a physical page size of 8 KB.

Table 9.1 Host configuration

Model	Dell OptiPlex
Processor	1× Intel Core 2 Duo 6700, 2660 MHz
Caches	32KB L1 I/D caches; 4 MB L2 unified cache
FSB frequency	1066 MHz
RAM	2 GB, 800 MHz, DDRII

Figure 9.3a shows the simulator's functional execution rate for each of the four DBT modes. Each benchmark was simulated twice, with each successive simulation loading the translations available on completion of the previous run. Due to the persistence of translations, no target instructions were interpreted on the second simulation run. Figure 9.3b shows the relative increase in simulation speed obtained by the three modes which identify larger translation units. Overall, the use of large translation units yields a 63% speed improvement compared with translations based on individual basic blocks. It is also clear that no single translation mode works best for all benchmarks, indeed each mode is optimal over a different subset of the

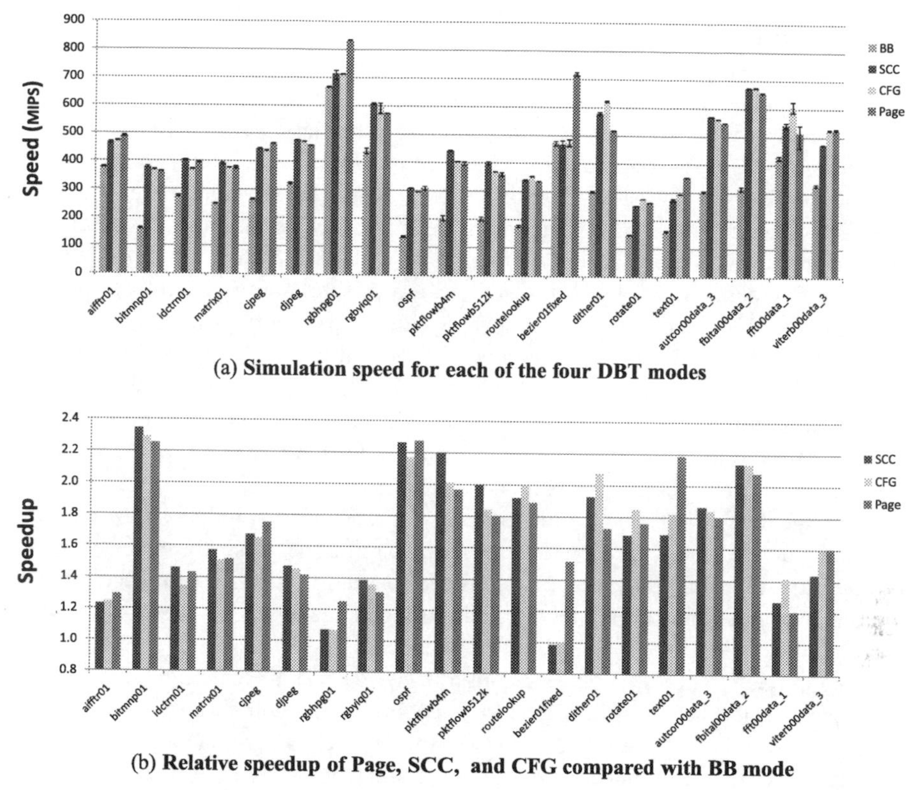

(a) **Simulation speed for each of the four DBT modes**

(b) **Relative speedup of Page, SCC, and CFG compared with BB mode**

Fig. 9.3 Simulation speed results

benchmarks. This suggests the choice of translation mode is another area in which adaptability can play a positive role.

9.2.4 Assessing the Advantages of DBT

The most significant advantage of DBT is its ability to work very effectively with binary representations of target applications. This is in marked contrast with simulators requiring access to C sources [3]. This allows the instructions at a given memory location to change, as they would if code is paged-out when simulating a virtual memory operating system or when self-modifying code is executed. The dynamic insertion and removal of breakpoint instructions, when using a DBT simulator for fast debugging of target code, represents another form of code modification which the EHS is able to handle. Being oblivious of specific API or ABI requirements, a DBT simulator will cope with literally any legal target code. The simulation speed results presented in this section also indicate that DBT-based simulators are able to achieve the highest levels of performance reported in the literature to date.

9.3 Learning-Based Timing Models

There exist huge differences in simulation speed and accuracy between cycle-accurate and instruction-accurate instruction set simulators (ISS). For example, a performance gap of one to two orders of magnitude between Verilog RTL and interpretive simulation and another factor of ten improvement over this for compiled-code simulators is reported in [4]. RTL-to-C translation [5, 6] improves the performance of cycle-accurate simulation, but it still does not match that of instruction-accurate simulation. Cycle-accurate simulators typically operate at 20–100K instructions per second, compared with JIT-translating simulators such as the one described earlier in this chapter which operate typically at several hundred million instructions per second. At this higher speed, however, ISS are not able to provide cycle counts, but only present the user with high-level statistical information such as instruction counts and, possibly, certain micro-architectural events such as the number of cache accesses or predicted branches.

To bridge the gap between instruction- and cycle-accurate simulations, we extend a fast, instruction-accurate ISS with a machine learning based performance estimator. Our goal is to fully automate the construction and use the performance model while at the same time provide the user with a high-speed cycle-approximate simulator.

Our approach to performance estimation is based on statistical regression where training data is used to construct a performance model. Using high-level information from a fast simulator, we apply the performance model to compute a performance estimate for a new program (or program region). Under the assumption that performing the additional computations needed for the performance model evaluation is faster than cycle-accurate simulation, the resulting *cycle-approximate* simulation will benefit from the higher speed of the instruction-accurate simulator while providing the user with performance estimates close enough to the actual cycle count.

Our performance prediction scheme relies on a *hybrid* simulator that can switch between functional and cycle-accurate simulation modes at the boundaries between instructions, basic blocks, or larger program regions. We exploit this feature for the *online* construction of a performance model and allow for *incremental model updates* if it turns out that workload patterns change over time and the model cannot make further predictions with sufficient confidence. Performance is then estimated and accumulated across whole program regions. While this scheme has much in common with the sampling simulation methodology presented in chapter 10, it improves on this in several ways. First, performance models are application-specific, and past behavior of a program can be incorporated in the performance model through incremental updates. Second, the training data sets can be carried forward between simulation runs, and subsequent simulations can benefit from previously tuned performance models.

Before we describe our performance estimation methodology in detail, we introduce the background of statistical regression analysis and, in particular, multiple linear regression.

9.3.1 Statistical Regression

Regression analysis is a statistical method to examine the relationship between a dependent variable y and n independent variables $x = (x_1, \ldots, x_N)$. This relationship is modeled as a function f with $y = f(x)$. The function f chosen depends considerably on the relationship between input vector x and output y. Many different forms of regression analysis exist to compute f, including many variants of linear and non-linear relationships.

If the relationship between y and x is linear, we use the linear regression model shown below, in which $\beta = (\beta_0, \ldots, \beta_N)$ is a matrix of weights.

$$y = \beta_0 + \sum_{i=1}^{N} \beta_i x_i \tag{9.1}$$

The chosen input variables x may only represent a subset of the total factors affecting y, and other "hidden" variables may be present. In this case, equation above can only be an approximation of y. The difference between y and a predicted $\hat{y}$ is the error term $\epsilon = y - \hat{y}$ called the *residual*.

Given m training points $(y_1, x_{1,1}, \ldots, x_{1,N})$ to $(y_m, x_{m,1}, \ldots, x_{m,N})$, we can extend our original design matrix (Eq. 9.1) and construct the following equation system:

$$\begin{aligned} y_1 &= \beta_0 + \beta_1 x_{1,1} + \beta_2 x_{1,2} + \cdots + \beta_N x_{1,N} + \varepsilon_1 \\ \vdots &= \vdots \\ y_m &= \beta_0 + \beta_1 x_{m,1} + \beta_2 x_{m,2} + \cdots + \beta_N x_{m,N} + \varepsilon_m \end{aligned}$$

These equations can be rewritten as $y = X\beta + \varepsilon$, where β represents the vector of regression coefficients or the *weights matrix* and X the *model matrix*.

The parameters of the model β should be selected such that ϵ is minimized, i.e., the regression function produces predictions as close to the actual value as possible. A common method of choosing these parameters is the *least-squares method* that minimizes the sum of the squares of the prediction errors SSE:

$$\text{SSE} = \sum_{i=1}^{m} \left(y_i - \beta_0 - \sum_{j=1}^{N} \beta_j x_{i,j} \right)^2 \tag{9.2}$$

The computed values represent estimates of the regression coefficients β, thus the calculation of a prediction $\hat{y}$ at point x only involves the application of Eq. (9.1). The weights matrix is then computed as $\beta = (X^T X)^{-1} X^T y$.

Unfortunately, this is an expensive operation ($O(m^3)$) due to the calculation of the inverse of the variance matrix $X^T X$. Dependent upon the size of the training data, this can severely hinder performance, especially if the weights matrix must be updated frequently due to newly added training points. A light-weight incremental method with lower computational complexity is discussed in Section 9.3.2.

Unlike their analytical counterparts, statistical models are suitably adapted to dealing with missing or noisy data and can provide additional information on the "quality" of statistical estimations. We use this to compute *confidence intervals* to describe the "reliability" of the estimations. How likely the interval is to contain the predicted value is determined by the *confidence level*. Increasing the desired confidence level will widen the confidence interval. For example, for linear regression, the confidence interval for a given confidence level α can be obtained by

$$x\hat{\beta} \pm \left(S\sqrt{x^{\mathrm{T}}(X^{\mathrm{T}}X)^{-1}x}\, t(n-p;\frac{\alpha}{2})\right) \tag{9.3}$$

where x is the point to be estimated, $\hat{\beta}$ the calculated parameter estimates, S the estimated variance, and $t(n-p;\frac{\alpha}{2})$ the value at $\frac{\alpha}{2}$ of the cumulative distribution function for the t-distribution.

9.3.2 On-Line Performance Model Construction and Incremental Model Updates

Our online scheme for performance model construction [7] extends existing off-line schemes such as [8] in several ways. The main difference is in the use of a hybrid simulator that can switch between instruction- and cycle-accurate simulation modes at any point in time. We exploit this feature to perform performance estimation at a finer granularity, i.e., simulation epochs rather than the whole program, and to continuously update the performance model with new training data if required.

The on-line method alternates phases of fast instruction-accurate and slower, but detailed cycle-accurate simulation. During cycle-accurate simulation phases, a performance model is constructed that is subsequently used for performance estimation when the simulator operates in instruction-accurate mode. We employ a continuous learning algorithm in an effort to ensure the accuracy of the predictions being made. By allowing the training data to be updated when confidence decreases, each prediction made will use the most recent and relevant training information. It also allows predictions to be performed over a much smaller time slice ensuring that any error that does occur affect only a smaller section of the code than the off-line method presented in [8]. The combined learning and prediction within the same simulation will also assist in customizing the training data for the current execution, further reducing any error margins.

The diagram in Fig. 9.4 displays the operation of the simulator in continuous learning mode. As shown in the diagram, the simulator initially begins in cycle-accurate mode, with new training points being added to the predictor. Once enough points have been collected such that the initial training matrix can be calculated, the simulation remains in cycle-accurate mode performing cycle count predictions once every time slice. At each prediction, the accuracy and confidence is checked.

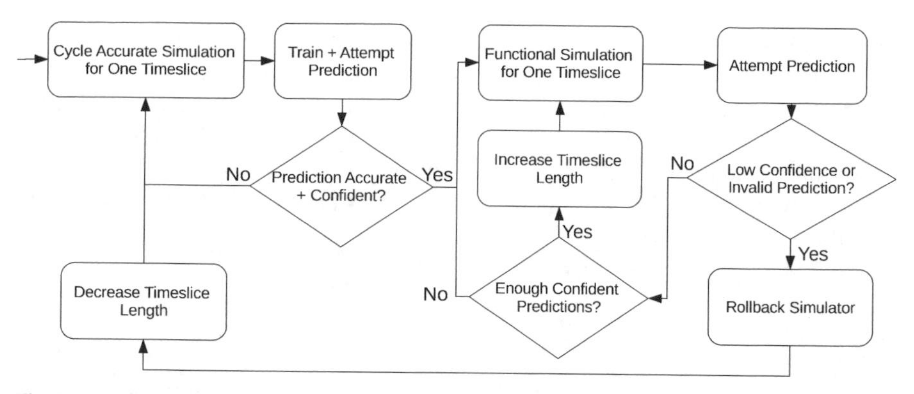

Fig. 9.4 The sequential operation of the instruction set simulator in continuous learning mode

If the prediction is too inaccurate, or the confidence interval of the prediction is too high, then the last time slice is added to the training data and the prediction matrix is updated.

Once the error and confidence interval drop below a pre-programed threshold, the simulation switches to instruction-accurate mode and relies purely upon the confidence interval of each prediction. This is monitored at every prediction, and should it become too high the simulator will fall back into cycle-accurate mode in one of two methods described in the next section. The simulator continues to run in cycle-accurate mode updating the training data once per time slice and attempting further predictions. Again, once the error ratio and the confidence of the prediction drop below their thresholds, the simulator returns to instruction-accurate mode and continues as above.

Once a prespecified number of confident predictions occur, the *time slice* over which the predictions are made is increased in an attempt to gain a further performance gain. Once the confidence interval widens too far, the time slice is reduced again to allow training across a finer grain.

9.3.2.1 Incremental Linear Regression

The standard linear regression algorithm is not well suited for situations where training data points are added incrementally. The calculation of the weights matrix is a computationally expensive operation involving the inversion of a potentially large matrix. This is to calculate the variance matrix for the training data. On occasions when updates to the training matrix are necessary, this calculation involves the recalculation of the weights matrix using the entire training data set. Alternatively, research has been performed into an incremental version to calculate the weights matrix [9].

The alternative linear regression algorithm still requires the initial calculation of the weights matrix as before; however, it optimizes the process of adding new data points, updating the weights matrix once it has been initially calculated. The process involves initially updating the variance matrix $(X^{\mathrm{T}}X)^{-1}$ of the data to include the new training points. The variance matrix will hereafter be referred to as A^{-1}. Computation of the updated variance matrix is optimized in two ways depending on the number of training points added: one method for a single update to the matrix that does not involve the calculation of the inverse of a matrix, and another method for multiple training points added to the training set.

For a single added training point, the updated variance matrix $\hat{A}^{-1}$ is calculated as

$$\hat{A}^{-1} = A^{-1} - \frac{A^{-1}\hat{X}^{\mathrm{T}}\hat{X}A^{-1}}{1 + \hat{X}A^{-1}\hat{X}^{\mathrm{T}}} \tag{9.4}$$

with $\hat{X}$ corresponding to the row of the newly added training point. Given A^{-1} has already been calculated during the initial *fitting* of the weights matrix, there is no requirement for recalculation, allowing for further optimization of this algorithm.

In the case of multiple samples being added to the training data, adding each training point individually using Eq. (9.4) would result in more computation than necessary, hence the alternative equation shown below can be used:

$$\hat{A}^{-1} = A^{-1} - A^{-1}\hat{X}^{\mathrm{T}}(\hat{X}A^{-1}\hat{X}^{\mathrm{T}} + 1)^{-1}\hat{X}A^{-1} \tag{9.5}$$

This method still calculates the inverse of a matrix to update the variance matrix; however, the size of the matrix being inverted $(\hat{X}A^{-1}\hat{X}^{\mathrm{T}}+1)^{-1}$ is much smaller and less computationally expensive than if it were to be performed across the entire data set. Once the updated variance matrix has been calculated, the weights matrix for the regression function is updated to $\hat{\beta} = \beta - \hat{A}^{-1}\hat{X}^{\mathrm{T}}(\hat{X}\beta - \hat{y})$, where $\hat{y}$ represents the column vector of the output variable for the newly added training points. Future predictions use the updated $\hat{\beta}$ as the weights matrix for Eq. (9.1).

9.3.2.2 On-Line Feature Selection

The purpose of feature selection is to determine a subset of the most relevant features of the training data. A number of possible feature selection methods are available to help choose which features affect the cycle count the most. For a continuous learning simulator, it is necessary to have an on-line feature selector as the relevant features are likely to be different for each benchmark.

The calculation of relevant features is performed every time the training data is updated, potentially requiring the recalculation of the training matrix. If the list of relevant features does not change then the new training point can update the matrix using the standard incremental linear regression, however; a, change in relevant features requires a complete recalculation of the training matrix. For this reason, the threshold for deciding which features are relevant must be set such that feature selection changes are infrequent, while ensuring that all of the relevant features are kept. In addition to the stability of chosen feature sets, the feature selection algorithm is required to be efficient as possible, because its repeated execution contributes to the overall time for simulation. For these reasons, the *sum of non-zero values* within a feature is chosen, allowing for an easily determined threshold to be used. Sums can be maintained between training updates, removing the need for repeated passes through the data set.

9.3.2.3 Dealing with a Less Confident Prediction

When a less-confident prediction occurs that is below the threshold, the situation can be handled by two distinct methods: a fast *patch* scheme or a conservative *rollback and replay* scheme.

Patch. The simulator accepts a less-confident prediction, but falls back to cycle-accurate mode and updates the training matrix based on the next few time slices.

Rollback–replay. The simulator does not accept a less-confident prediction, and the simulation is rolled back to a previously known state, i.e., just before the last time slice began, and re-runs the simulation in cycle-accurate mode, updating the training matrix as necessary.

The implementation of the rollback–replay scheme is based on the existing *checkpointing* facility of the simulator that has been modified to store the simulation state in memory rather than external storage. In addition, we have implemented a *copy-on-write* mechanism to further reduce the runtime overhead associated with the regular snapshots of the simulation state.

9.3.3 Prediction Accuracy

In this section, we discuss the accuracy of the performance estimation methodology introduced in this chapter. Table 9.2 summarizes the seven benchmark suites that we have used in our experiments, resulting in a total of 293 applications.

Table 9.2 Overview of the benchmark suites used in our experiments

Benchmark suite	Description
DSPstone	Small DSP kernels
UTDSP	Small DSP kernels and applications
SWEET WCET	Worst case execution time benchmarks
MediaBench	Multimedia applications
EEMBC	Automotive, consumer, digital, entertainment, networking, office automation and telecom applications
Pointer-intensive benchmarks [10]	Non-trivial pointer-intensive applications
StreamIt benchmarks	Cryptography, software radio, audio processing

In our experiments, simulations were performed in sequence, with the initial baseline cycle-accurate simulation, followed by our new simulation method. Figure 9.5 illustrates accuracy that can be expected from our statistical performance estimation scheme. Estimated cycle counts (for a time slice length of 100) are compared to the actual, observed cycle counts, and both sets of values are plotted in the scatter graph in Fig. 9.5a. For programs varying by several orders of magnitude in total execution time, our scheme is highly accurate and the estimations come close to the actual values as can be seen by proximity of data points near or on the ideal straight line. In Fig. 9.5b the distribution of the estimation error is shown. The vast majority of the estimations are centered closely around the 0% error mark, and virtually all predictions fall into the $\pm 2\sigma$ interval corresponding to an error margin of $<11\%$.

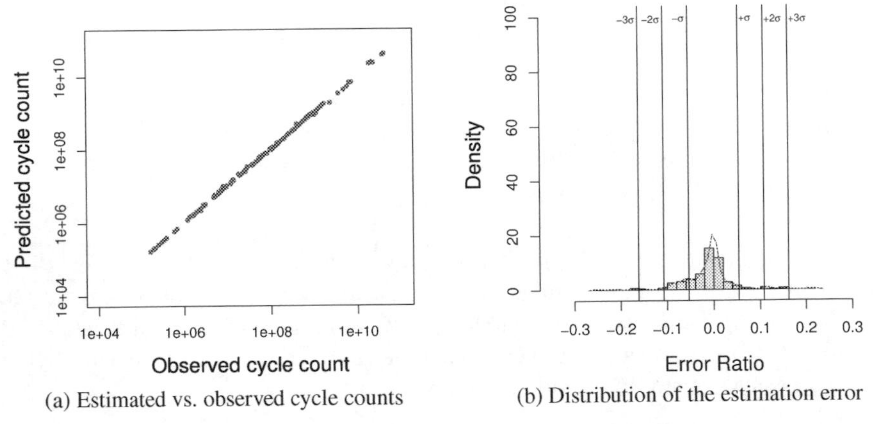

(a) Estimated vs. observed cycle counts (b) Distribution of the estimation error

Fig. 9.5 Accuracy of the estimated cycle counts (for an initial time slice length = 100)

Our performance estimation scheme displayed improvements of $\approx$50% speedup with an mean average error of 2.36% and a maximum error of, at best, 19.97%. Simulations proved that a shorter original time slice is effective at allowing shorter benchmarks to receive the performance benefits without adversely affecting longer benchmarks. They also proved that having a variable prediction time slice posed additional performance benefits without severely affecting error margins; however, this cannot increase too quickly, as this prevents the predictor from receiving adequate training data and as such negates any performance benefits. There is also little to no benefit in rerunning "less-confident" prediction blocks in cycle-accurate mode to ensure accuracy.

Small applications with less than $\approx 10^6$ instructions do not benefit from our scheme as they spend their entire execution in cycle-accurate training mode. However, this is not critical as applications of this size are simulated within less than 1 s even using the slowest available mode of simulation.

Sum of non-zero values feature selection and incremental linear regression have low computational complexity and are very suitable for *online* or *just-in-time* performance modeling. In fact, profiling of the extended simulator has revealed that only 2–3% of the overall execution time are spent for updates of the performance model and predictions and, hence, are making their contribution negligible.

9.4 Conclusions

In this chapter, we have shown how dynamic adaptation, through a combination of JIT binary translation and on-line learning of performance models, enables high-speed cycle-approximate processor simulation. We have given an overview of how these adaptation techniques are deployed in the Edinburgh high-speed simulator. We leverage the precise state observability of EHS to provide a golden reference model for SoC designs, now in working silicon, and the performance models enable rapid design-space exploration for our next-generation SoC research.

The techniques described in this chapter can be extended to include storing the prediction model between simulations. This would allow the simulator to start making predictions from the start of execution and, hence, removing the initial training period. Furthermore, we are currently extending our performance estimation methodology to work with the faster, but even more high-level JIT-compiled simulation mode offered by the simulator used in this chapter. At the same time, we consider the inclusion of other metrics such as power, energy, and thermal distribution in our framework.

References

1. Topham, N., Jones, D.: High speed CPU simulation using JIT binary translation. In: *Proceedings of the 4rd Annual Workshop on Modeling, Benchmarking and Simulation: MoBS'07*, (2007).
2. Jones, D., Topham, N.: High speed CPU simulation using LTU dynamic binary translation. In: *Proceedings of the HiPEAC Conference*. LNCS 5409, Springer, Berlin pp. 50–64, (2009).
3. Gao, L., Kraemer, S., Leupers, R., Ascheid, G., Meyr, H.: A fast and generic hybrid simulation approach using C virtual machine. In: *Proceedings of the 2007 International Conference on Compilers, Architecture and Synthesis for Embedded Systems*, pp. 3–12, Salzburg, Austria (2007).
4. Weber, S.J., Moskewicz, M.W., Gries, M., Sauer, C., Keutzer, K.: Fast cycle-accurate simulation and instruction set generation for constraint-based descriptions of programmable architectures. In *Proceedings of CODES+ISSS'04*, pp. 18–23, Stockholm, Sweden, (2004).
5. Snyder, W., Wasson, P., Galbi, D.: Verilator, `http://www.veripool.com/verilator.html`, (2007).
6. ARC International, ARC VTOC Tool, `http://www.arc.com/software/simulation/vtoc.html`, (2007).
7. Powell, D.C., Franke, B.: Using continuous statistical machine learning to enable high-speed performance prediction in hybrid instruction-/cycle-accurate instruction set simulators. In *CODES+ISSS '09: Proceedings of the 7th IEEE/ACM International Conference on Hardware/software Codesign and System Synthesis*. ACM, New York, NY, pp. 315–324, (2009).
8. Franke, B.: Fast cycle-approximate instruction set simulation. In *Proceedings of the Workshop on Software and Compilers for Embedded Systems (SCOPES'08)* pp. 68–78, Munich, Germany, (2008).
9. Orr, M.J.: Introduction to radial basis function networks. In *Technical Report*, Centre for Cognitive Science, University of Edinburgh, (1996).
10. Austin, T.M.: Pointer-intensive benchmark suite, `http://www.cs.wisc.edu/austin/ptr-dist.html` (2007).

Chapter 10
Representative Sampling Using SimPoint

Greg Hamerly, Erez Perelman, Timothy Sherwood, and Brad Calder

Abstract SimPoint is a technique used to pick what parts of the program's execution to simulate in order to have a complete picture of execution. SimPoint uses data clustering algorithms from machine learning to automatically find repetitive (similar) patterns in a program's execution, and it chooses one sample to represent each unique repetitive behavior. Each sample is then simulated and weighted appropriately, and then together the results from these samples represent an accurate picture of the complete execution of the program.

10.1 Introduction

Nearly all industry-standard benchmarks require the execution of a *suite* of programs. For example, the SPEC CPU2000 benchmark suite consists of 26 different programs, requiring the execution of a combined total of approximately 6 trillion instructions. Still worse, architecture researchers need to simulate each benchmark over a variety of different architectural configurations and design options to find the set of features that provide an appropriate trade-off between performance, complexity, area, and power. The same program binary, with the exact same input, may be run hundreds or thousands of times to examine how, for example, the effectiveness of a given architecture changes with its cache size. Researchers need techniques which can reduce the number of machine-months required to estimate the impact of an architectural modification without introducing an unacceptable amount of error or excessive simulator complexity. We present a method, distributed as a software package called SimPoint, which can meet this need by uncovering and exploiting the structured way in which individual programs change behavior over time.

A program's behavior often changes as it executes. These changes are not random, but rather are often structured as sequences of a small number of recurring

B. Calder (✉)
Microsoft, 2831 134th Ave NE, Bellevue, WA 98005, USA
e-mail: bcalder@microsoft.com

R. Leupers, O. Temam (eds.), *Processor and System-on-Chip Simulation*,
DOI 10.1007/978-1-4419-6175-4_10, © Springer Science+Business Media, LLC 2010

behaviors, which we term *phases*. Identifying this repetitive and structured behavior can be of great benefit, since it means we only need to sample each unique behavior once to create a complete representation of the program's execution. This is the underlying philosophy of SimPoint [2, 5, 6, 10, 13, 14]. SimPoint intelligently chooses a very small set of samples from an executed program called *simulation points* that, when simulated and weighted appropriately, provide an accurate picture of the complete execution of the program. Simulating in detail only these carefully chosen simulation points can save hours of simulation time over a random sampling of the program, while still providing the accuracy needed to make reliable decisions based on the outcome of the cycle level simulation.

Before we developed SimPoint, architecture researchers would often simulate SPEC programs for 300 million instructions from the start of execution, or fast forward 1 billion instructions to try to get past the initialization part of the program. These ad hoc techniques can result in very high error rates. In some of the cases presented later in this chapter, errors as high as 3,736% have been observed. In contrast, SimPoint achieves very low error rates (2% average error, 8% maximum error) and on average reduces simulation time by a factor of 1,500x, compared to simply simulating the whole program. This approach is now widely used both by researchers in the architecture community and by companies such as Intel [9]. Since then, other sampling techniques have been proposed, such as SMARTS, see Chapter 11, or dynamic sampling as implemented within COTSON, see Chapter 4. This chapter describes the core ideas behind SimPoint, including how repetitive phase behaviors can be automatically found in programs with machine learning and how knowledge of these behaviors can improve simulation methodology.

10.2 Methodology

While the techniques presented here have been shown to be effective across many architectures and benchmark suites, to ground our discussion of the concepts we use data from a real set of experiments. Specifically, we present data for the complete set of SPEC CPU2000 programs for multiple inputs using the Alpha binaries from the SimpleScalar website. All of the frequency vector profiles and simulation results were gathered using SimpleScalar [3]. To generate our baseline results, we executed all programs from start to completion using SimpleScalar, gathering the hardware metrics of interest. The baseline microarchitecture model is detailed in Table 10.1.

To examine the accuracy of our approach we provide results in terms of cycles per instruction (CPI) prediction error. CPI is a commonly used performance metric in the architecture community because assuming the architecture changes proposed do not effect the number of instructions committed, then CPI is proportional to execution time. The CPI prediction error is the percent difference between CPI predicted using only simulation points chosen by SimPoint and the baseline (true) CPI of the complete execution of the program.

Table 10.1 Baseline simulation model

I Cache	16k 2-way set-associative, 32 byte blocks, 1 cycle latency
D Cache	16k 4-way set-associative, 32 byte blocks, 2 cycle latency
L2 Cache	1Meg 4-way set-associative, 32 byte blocks, 20 cycle latency
Main Memory	150 cycle latency
Branch Pred	hybrid - 8-bit gshare w/ 8k 2-bit predictors + a 8k bimodal predictor
O-O-O Issue	out-of-order issue of up to 8 operations per cycle, 128 entry re-order buffer
Mem Disambig	load/store queue, loads may execute when all prior store addresses are known
Registers	32 integer, 32 floating point
Func Units	8-integer ALU, 4-load/store units, 2-FP adders, 2-integer MULT/DIV, 2-FP MULT/DIV
Virtual Mem	8K byte pages, 30 cycle fixed TLB miss latency after earlier-issued instructions complete

10.3 Defining Phase Behavior

Since phases are the way we describe recurring behaviors of a program executing over time, we begin by describing phase analysis with a demonstration of the time-varying behavior [12] of two programs from the SPEC 2000 benchmark suite, `gcc` and `gzip`. To characterize the behavior of these programs we have simulated their complete execution from start to finish. Each program executes many billions of instructions, and gathering these results took several machine-months of simulation time. The behavior of each program is shown in the top graphs of Figs. 10.1 and 10.2. The top graph in each figure shows how the CPI rate changes for these two programs over time. Each point on the graph represents the average CPI taken over a window (or interval) of 10 million executed instructions. These graphs show that programs are fairly complex, changing behaviors frequently over the course of their executions.

Note that not only do the behaviors of the programs change over time, they change on the largest of timescales, and even at a large scale one can find repeating behaviors. Programs may have stable behavior for billions of instructions and then change suddenly. In addition to CPI, we have found for the SPEC 95 and 2000 programs that the behavior of *all* of the architecture metrics (branch prediction, cache misses, etc.) tend to change in unison, though not necessarily in the same direction [12, 14]. These corresponding changes are due to underlying changes in program execution behaviors.

The fundamental methodology used in this work is the ability to automatically identify these underlying program changes *without relying on architectural metrics*. To ground our discussion in a common vocabulary, the following is a list of definitions to describe program behavior and its automated classification.

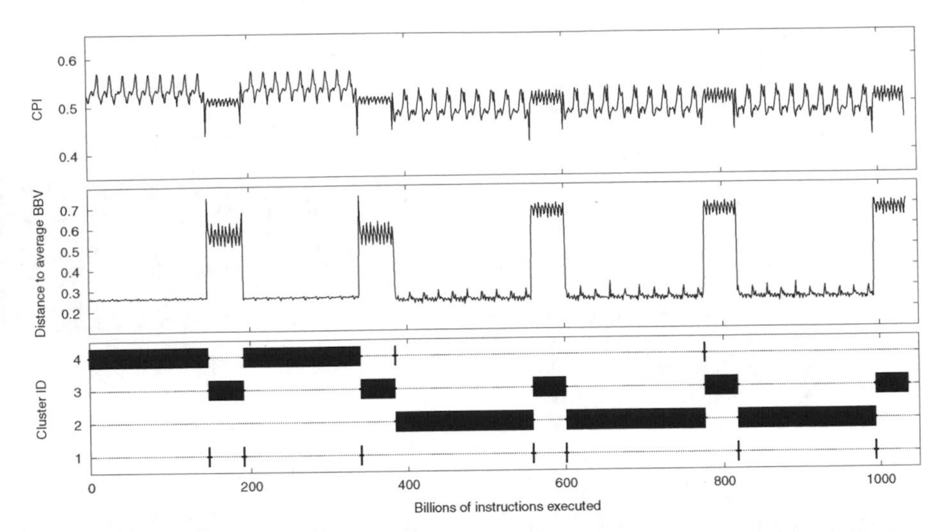

Fig. 10.1 These plots show the relationship between measured performance (CPI) and code usage for the program `gzip-graphic` and SimPoint's ability to capture phase information by only looking at what code is being executed. The *horizontal axis* represents execution of the program over time. Each plotted point represents one 10-million instruction interval. The *top plot* shows the CPI for the executing program. The *middle plot* shows the code usage distance of each interval to the average code usage of the entire program (using basic block vectors as explained in Section 10.4). Lower distances indicate higher similarity. The *bottom plot* shows how SimPoint classifies each interval into one of four phases. The phase transitions correspond to changes in the CPI in the top graph, though SimPoint does not use metrics like CPI to classify intervals

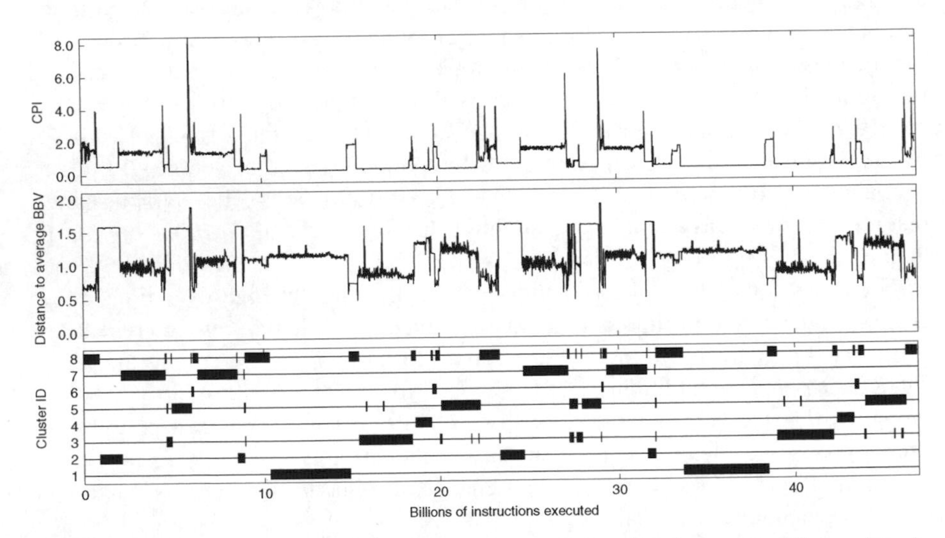

Fig. 10.2 These plots are analogous to Fig. 10.1, but for the benchmark `gcc-166`. This benchmark exhibits more complicated behaviors than `gzip-graphic`, and hence SimPoint uses more phases to represent its behavior

- Interval—To perform our analysis we break a program's execution up into non-overlapping intervals of execution. An interval is a section of contiguous execution (a time slice) of a program's execution. For example, the first 100 million executed instructions would belong to the first interval, the next 100 million to the second interval. In the work presented here all intervals are chosen to be the same length, as measured in the number of instructions committed within an interval. This is usually 1, 10, or 100 million instructions, as used in [10].
- Similarity—A similarity metric measures the similarity in behavior between two intervals of a program's execution and is specific to the representation of those intervals.
- Phase—A set of intervals within a program's execution that all have similar behavior, *regardless* of temporal adjacency. A phase may be made up of intervals which are disjoint in time; we would call this a phase with repeating behavior. A "well-formed" phase should have intervals with similar behavior across various architecture metrics (e.g., CPI, cache misses, branch misprediction). In this chapter we consider the terms "cluster" and "phase" to be equivalent.
- Phase classification—Using machine learning to group intervals from a program/input pair into phases (clusters) with similar behavior.

10.4 The Strong Correlation Between Code and Performance

In this section we describe how we identify phase behavior in an architecture-independent fashion.

10.4.1 Using an Architecture-Independent Metric for Phase Classification

To find program phases, we need a metric of similarity between different parts of a program's execution. In creating this metric it is advantageous to not rely on hardware-based statistics such as cache miss rates or performance (i.e., CPI), since using these would tie the phases to statistics that change depending on the architecture configuration. If such statistics were used, the phases would need to be re-analyzed every time there was a change to some architectural parameter (either statically if the size of the cache changed or dynamically if some policy changes adaptively). This is not acceptable, since our goal is to find a set of samples that can be used across an architecture design space exploration, where many of these parameters may change. To address this, we need a metric that is *independent* of any particular hardware-based statistic, but still relates to the fundamental changes in behavior like those shown in the top graphs of Figs. 10.1 and 10.2.

An effective way to design such a metric is to base it on the behavior of a program in terms of the code that is executed over time. We have shown that there is a very strong correlation [5] between the set of paths executed in a program and the time-varying architectural behavior observed. The intuition behind this is that the executed code determines the behavior of the program. Thus it is possible to find the phases in programs using *only* a metric related to how the code is being exercised (i.e., both what code is touched and how often). The central idea behind SimPoint is that it can find the phase behaviors (as illustrated in the top graphs of Figs. 10.1 and 10.2) by examining only the frequency with which the code parts (e.g., basic blocks) execute over time.

10.4.2 Basic Block Vector

The basic block vector (BBV) [13] is a structure designed to concisely capture information about how a program is changing behavior over time. A basic block is a section of code (i.e., a contiguous set of instructions) that executes from start to finish with one entry and one exit. The metric we use for comparing two time intervals in a program is based on the differences in the execution frequencies for each basic block executed during those two intervals. The intuition behind this is that the behavior of the program at a given time is directly related to the code it is executing during that interval, and basic block vectors provide us with this information.

A program, when run for any interval of time, will execute each basic block a certain number of times. This information provides a code signature for that interval of execution and shows where the application is spending its time in the code. Knowing the basic block distributions for two different intervals gives two separate signatures which we can compare to find out how similar the intervals are. If the signatures are similar, then the two intervals spend about the same amount of time in the same code, and the performance of those two intervals should be similar.

We represent a basic block vector as a one-dimensional array, with one element in the array for each static basic block in the program. Each interval in an executed program is represented by one BBV, and at the beginning of each interval, its corresponding BBV has all zeros. During each interval, we count the number of times each basic block has been entered and record that number into the corresponding element in the vector. This number is weighted by the number of instructions in the basic block, since we want every individual instruction to have the same influence. Therefore, each element in the array is the count of how many times its corresponding basic block has been entered during an interval of execution, multiplied by the number of instructions in that basic block. For example, if the 50th basic block has two instructions and is executed 15 times in an interval, then $bbv[50] = 30$ for that interval. At the end of an interval's execution, we normalize the BBV to sum to 1.

We call the vectors used to guide phase analysis *frequency vectors*, of which basic block vectors are one type. Frequency vectors can represent basic blocks, branch edges, or any other type of program-related structure which provides a representative summary of a program's behavior for each interval of execution. We recently examined frequency vector structures other than basic block vectors for the purpose of phase classification. We have looked at frequency vectors for data, loops, procedures, register usage, instruction mix, and memory behavior [6]. We found that using register usage vectors, which simply counts for a given interval the number of times each register is defined and used, provides similar accuracy to using basic block vectors. In addition, using only loop and procedure branch execution frequencies performs almost as well as using the full basic block information. We also found, for SPEC 2000 programs, that augmenting BBVs by including both code and data access patterns into the vectors did not improve classification over just using code [6].

10.4.3 Basic Block Vector Difference

In order to find patterns in a program we must first have some way of comparing the similarity of two basic block vectors. The operation should take two basic block vectors and return a single number corresponding to how similar (or different) they are.

There are several ways of measuring the similarity of two vectors, such as taking the dot product between the vectors, finding the Euclidean (2-norm) distance of the connecting vector, or Manhattan (1-norm) distance of the connecting vector. The Euclidean distance has been shown to be effective for comparing basic block vectors once they have been subjected to random projection to reduce their dimensionality [10, 14], and this is what we use for the SimPoint analysis. In later work we show that the Manhattan distance, which can be more efficiently implemented in hardware, can also be effective if an on-the-fly phase analysis (e.g., predicting phases during computation), is desired [7, 15].

10.4.4 Correlation Between Code Signatures and Performance

For a detailed study showing that there is a strong correlation between executed code and real performance, please see [5]. The top two graphs of Fig. 10.2 give one illustration of this correlation by showing the time-varying CPI and BBV distance graphs next to each other for `gcc-166`. The top graph plots the CPI for each interval executed showing how the program's CPI varies over time. Similarly, the BBV distance graph plots for each interval the Manhattan distance of the BBV (code signature) for that interval from the whole program's target vector. The whole program's target vector is a BBV that comes from viewing the whole program as a single interval. The same information is also provided for `gzip` in the top two graphs of Fig. 10.1. These graphs show that changes in CPI have corresponding changes in code signatures, which is one indication of strong phase behavior for these applications.

These graphs show a strong correlation between code changes and CPI changes even for complex programs like `gcc`. The graphs for `gzip` show that phase behavior can be found even if the intervals' CPIs have small variance. This brings up an important point about classifying intervals based on code similarity rather than based on similarity of CPI or some other hardware metric. Assume we have two intervals with *different code signatures* but they have very *similar CPIs* because both of their working sets fit completely in the cache. During a design space exploration search, as the cache size changes, their CPIs may differ dramatically if one of them no longer fits into the cache. This is why it is important to perform the phase analysis by comparing the code signatures independent of the underlying architecture. We have found that the BBV code signatures correctly identify differences like these, which cannot be seen by looking at just the CPI.

10.5 Automatically Finding Phase Behavior

In this section we describe the algorithms used to automatically detect patterns using the frequency vectors described in the previous section.

10.5.1 Using Clustering for Phase Classification

A primary goal of SimPoint is to have an automated way of extracting phase information from programs. Data clustering algorithms have been shown to be very effective at breaking the complete execution of a program into phases that have similar frequency vectors [14]. Because the frequency vectors correlate to the overall performance of the program, grouping intervals based on their frequency vectors produces phases that are similar not only in the distribution of program structures used but also in every other architecture metric measured, including overall performance.

The goal of clustering is to divide a set of points into clusters such that points within each cluster are similar to one another (by some metric), and points in different clusters are different from one another. In this way, if we have a good metric to compare intervals to one another, a "phase" is really just a "cluster" of intervals.

SimPoint uses k-means [8], an efficient and well-known clustering algorithm, to split program intervals into phases. One drawback of the k-means algorithm is that it requires the number of clusters k as an input to the algorithm, but we do not know beforehand what value is appropriate. To address this, we run the algorithm for several values of k and then use a penalized likelihood score to guide our final choice for k. Our goal is to choose a clustering with a small number of clusters which still models the program behavior well.

The following steps summarize the SimPoint phase clustering algorithm at a high level.

1. Profile the program and divide its execution into contiguous intervals of fixed or varying length. For each interval, collect a frequency vector tracking the program's use of some program structure (e.g., basic blocks, branch edges).

2. Reduce the dimensionality of the frequency vector data to a much smaller number of dimensions using random linear projection.
3. Run k-means multiple times on the projected data with several values of k in the range from 1 to K (a user-prescribed maximum number of phases). Each run produces a clustering, which is a partition of the data into k different phases/clusters.
4. Score each clustering with different k by the Bayesian information criterion (BIC) [11]. A high BIC score indicates that the clustering is a good fit to the data.
5. Choose the clustering with a small k such that its BIC score is nearly as good as the best observed. The chosen clustering is the final grouping of intervals into phases.

This algorithm has several important parameters: interval length, projected dimension, the maximum number of clusters K, how the BIC is to be used to select the best clustering, etc. Each must be tuned to create accurate and representative simulation points using SimPoint.

10.5.1.1 Random Projection and Sampling of Intervals

Many clustering algorithms suffer from the so-called curse of dimensionality, which refers to the fact that finding an optimal clustering becomes intractable as the number of dimensions increases. For basic block vectors, the number of dimensions is the number of executed basic blocks in the program, which ranges from 2,756 to 102,038 for the SPEC benchmark suite, and could easily grow into the millions for very large programs. We wish to reduce the number of dimensions k-means must analyze, while preserving the structure of the data.

SimPoint uses random linear projection [4] to create a low-dimensional space into which it projects the original basic block vectors. Consider a set of n basic block vectors with d dimensions, represented as an $n \times d$ matrix called X. To obtain the projected version X' having d' dimensions, create a projection matrix P of size $d \times d'$. Each entry in P is chosen uniformly randomly from $[-1, 1]$. Then calculate $X' = X \times P$ using matrix multiplication. Dasgupta [4] shows that random linear projection as a preprocessing step before clustering has several desirable theoretical properties, including preserving the relative structure of the data. We found in earlier work [14] that projecting to 15 dimensions is low enough to be computationally tractable, but sufficiently high to discover the different phases of execution with clustering. Figure 10.3 shows a projection of `gzip` down to two dimensions. The projection clearly retains the program's clustered structure, even when using such a low-dimensional representation.

Some benchmark programs run for such a long time that the number of intervals (i.e., basic block vectors) that SimPoint must analyze becomes prohibitively large. In such cases, k-means may run too slowly or may not run at all due to memory restrictions. It is reasonable to randomly choose a sample of the intervals for k-means to cluster, which speeds up the analysis tremendously. SimPoint has the functionality to automatically choose a sample of a specified size, run k-means on

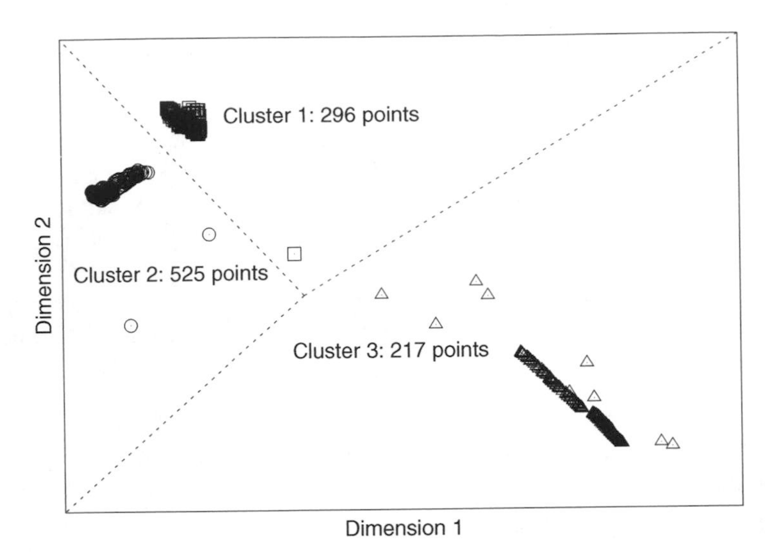

Fig. 10.3 This plot shows a two-dimensional projection of the basic block vectors for the program `gzip`, having 1038 total intervals, and clustered into three clusters with k-means. The lines show divisions between these three primary clusters. Note that SimPoint normally operates in more than two dimensions, but this illustrates the fact that that program behavior does form natural groups that can be found through data clustering even at low dimensions

this sample of intervals, and then assign phase ids to and select simulation points from the *entire* set of intervals (including those left out of the k-means sample). This has been found to significantly reduce the running time of SimPoint, while still maintaining its accuracy.

10.5.1.2 Bayesian Information Criterion

To compare the different clusterings formed for different k, we use the Bayesian information criterion, or BIC [11], as a measure of the "goodness of fit" of a clustering to a data set. The BIC is an approximation of the probability of the clustering, given the data that has been clustered. Thus, the larger the BIC score, the higher the probability that the clustering being scored is a "good fit" to the data being clustered. The BIC formulation we use is appropriate for clustering with k-means; however, other formulations of the BIC could also be used for other clustering models.

There are two parts of the BIC: the likelihood and the penalty. The likelihood is a statistical measure of how well the clustering predicts the observed data. However, the likelihood tends to increase without bound as more clusters are added. Therefore, the second term is a penalty that offsets the likelihood growth based on the number of clusters.

For a set of basic block vectors, SimPoint calculates the BIC score for each k-means clustering, varying k from 1 to K. It then chooses the clustering that achieves a BIC score that is close to the highest BIC score seen.

10.5.2 Clusters and Phase Behavior

The bottom plots in Figs. 10.1 and 10.2 show the results of running our phase-finding clustering algorithm on `gzip` and `gcc`. These results use an interval length of 10 million instructions and the maximum number of phases (K) is set to 10. The horizontal axis corresponds to the execution of the program (in billions of instructions), and each interval is classified to belong to one of the clusters (labeled on the vertical axis).

For `gzip`, the program's execution is partitioned into four clusters. Looking at the middle plot for comparison, the cluster behavior captured by our algorithm lines up quite closely with the behavior of the program. Clusters 2 and 4 represent the large sections of execution that are similar to one another. Cluster 3 captures the smaller phase that lies in between these larger phases. Cluster 1 represents the phase transitions between the three dominant phases. The intervals in cluster 1 are grouped into the same phase because they execute a similar combination of code, which happens to be part of the code behavior in either cluster 2 or 4 and part of code executed in cluster 3. These transition points in cluster 1 also correspond to the same intervals that have large spikes in CPI seen in the top graph (these spikes are due to increased cache misses for those regions).

The bottom plot of Fig. 10.2 shows how `gcc` is partitioned into eight clusters. Comparing this to the middle and top plots in the same figure, we see that even the more complicated behavior of `gcc` is captured well by SimPoint. The dominant behaviors in the top two graphs can be seen grouped together in phases 1, 3, 5, and 7.

10.6 Choosing Simulation Points from the Phase Classification

After the phase classification algorithm has done its job, intervals with similar code usage will be grouped together into the same phases (clusters). Then from each phase, SimPoint chooses one representative interval that will be simulated in detail to represent the behavior of the whole phase. Therefore, by simulating *only* one representative interval per phase, we can extrapolate and capture the behavior of the entire program.

To choose a representative for a cluster, SimPoint picks the interval that is closest (by Euclidean distance) to the cluster's k-means center. The center can be viewed as a pseudo-interval which behaves most like the average behavior of the entire phase. Most likely there is no interval that exactly matches the center, so SimPoint chooses the closest interval. The selected interval is called a *simulation point* for that phase [10, 14]. We can then perform detailed simulation on the set of simulation points.

As part of its output SimPoint also gives a weight for each simulation point. Each weight is a fraction of the total number of instructions represented by the simulation point's cluster divided by the number of instructions in the program. The weight for a phase represents the percentage of execution spent in that phase (cluster) and the percentage of execution the simulation point from that phase should represent when estimating the overall program's execution. With the weights and the detailed simulation results of each simulation point, we can compute a weighted average for the architecture metric of interest (CPI, cache miss rate, etc.) for the entire program's execution.

These simulation points are chosen once for a program/input combination because they are chosen based only on how the code is executed and not based on architecture metrics. Therefore, they only need to be calculated once for a binary/input combination and can be used repeatedly across all of the runs for an architecture design space exploration.

10.6.1 Simulation Time, K, and Interval Granularity

The number of simulation points that SimPoint chooses has a direct effect on the simulation time that will be required for those points. The maximum number of clusters, K, along with the interval length, represents the maximum amount of simulation time that may be needed. When fixed length intervals are used, ($K \times$ interval length) is a limit on the number of simulated instructions.

SimPoint allows users to trade-off simulation time with accuracy. Researchers in architecture tend to want to keep simulation time to below a fixed number of instructions (e.g., 300 million) for a run. If this is a goal, we find that an interval length of 10 million instructions with $K = 30$ provides very good accuracy with reasonable simulation time (220 million instructions on average) for SPEC programs. If even more accuracy is desired, then decreasing the interval length to 1 million and setting $K = 300$ performs well for the SPEC 2000 programs, as does setting $K = \sqrt{n}$ (where n is the number of clustered intervals). Empirically we discovered that as the interval granularity becomes finer, the number of phases discovered increases at a sub-linear rate. The upper bound defined by this square-root heuristic works well for the SPEC benchmarks.

The length of the interval chosen by users of SimPoint depends on their simulation infrastructure and how much they want to deal with warmup. Warmup is the process of initializing the simulator's state (caches, branch predictor, etc.) at the start of a simulation point so that it is the same as if we simulated from the beginning of the program to that point. For many programs, using a long interval length (e.g., more than 100 million instructions) will make warmup unnecessary. This is the approach used by Intel's PinPoint for simulation [9]. They simulate intervals of length 300–500 million instructions so they do not have to worry about implementing warmup in their simulation infrastructure. With such long intervals the architecture structures are warmed up sufficiently during the beginning of the interval's execution to provide accurate simulation results. In comparison, short interval lengths can be used, but this requires having an approach for warming up the architecture state. One way to do this is with an architecture checkpoint, which stores the potential contents of the major architecture components at the start of the simulation point [1]. This can significantly reduce warmup time, since warmup consists of just reading the checkpoint from a file and using it to initialize the architecture structures.

10.6.2 Accuracy of SimPoint

We now show the accuracy of using SimPoint for the complete SPEC 2000 benchmark suite and their reference inputs. Figure 10.4 shows the simulation accuracy results using SimPoint (and prior methods) for the SPEC 2000 programs when compared to the complete execution of the programs. For these results we use an interval length of 100 million instructions and limit the number of simulation points to no more than 10. With the above parameters SimPoint finds four phases for `gzip`, and eight for `gcc`. As described above, one simulation point is chosen for each cluster, so this means that a total of 400 million instructions were simulated for `gzip`. The results show that this results in only a 4% error in performance estimation for `gzip`. We measure the relative prediction error on a performance metric like CPI or IPC for a benchmark as

$$\text{Error} = \frac{|\text{True value} - \text{SimPoint estimate}|}{\text{True value}}$$

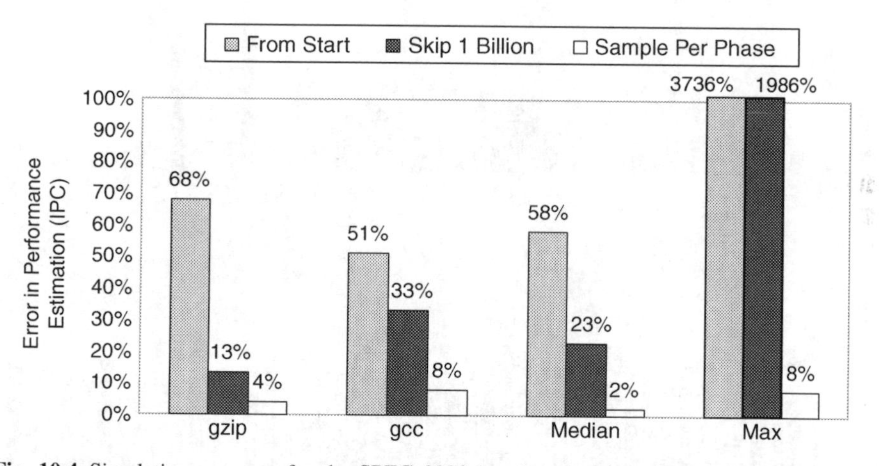

Fig. 10.4 Simulation accuracy for the SPEC 2000 benchmark suite when performing detailed simulation for several hundred million instructions compared to simulating the entire execution of the program. Results are shown for simulating from the start of the program's execution, for fast-forwarding 1 billion instructions before simulating, and for using SimPoint to choose at most ten 100-million-instruction intervals to simulate. The results are shown as percent error of predicted IPC, which is how much the estimated IPC using SimPoint is different from the complete execution of the program. IPC is the inverse of CPI. The median and maximum results are for the complete SPEC 2000 benchmarks

For these results, we compare this estimated IPC using SimPoint to the baseline IPC. IPC (instructions per cycle) is the inverse of CPI and often used instead of CPI when describing performance. The baseline was gathered from spending months of simulation time to simulate the entire execution of each SPEC program. The results in Fig. 10.4 compare SimPoint to how architecture researchers used to choose where to simulate before SimPoint. The first technique was to just simulate the first N million instructions of a benchmark's execution. The second technique was to blindly skip the first billion instructions of execution to get past the initialization of the program's execution and then simulate for N million instructions. The results show that simulating from the start of execution, for the exact same number of instructions as simulated with SimPoint, results in a median error of 58%. If instead, we fast forwarded for 1 billion instructions and then simulate for the same number of instructions as chosen by SimPoint, we see a median 23% IPC error. When using SimPoint to create multiple simulation points we have a median IPC error of 2%. Note that the maximum error seen for the prior techniques is significant for the SPEC programs, but it is very reasonable (only 8%) for SimPoint.

Figure 10.5 shows the accuracy of SimPoint for all of the SPEC 2000 programs, along with the number of simulation points chosen. These experiments use $K = 30$ and an interval size of 10 million instructions per interval. The results show that SimPoint chose 22 simulation points on average.

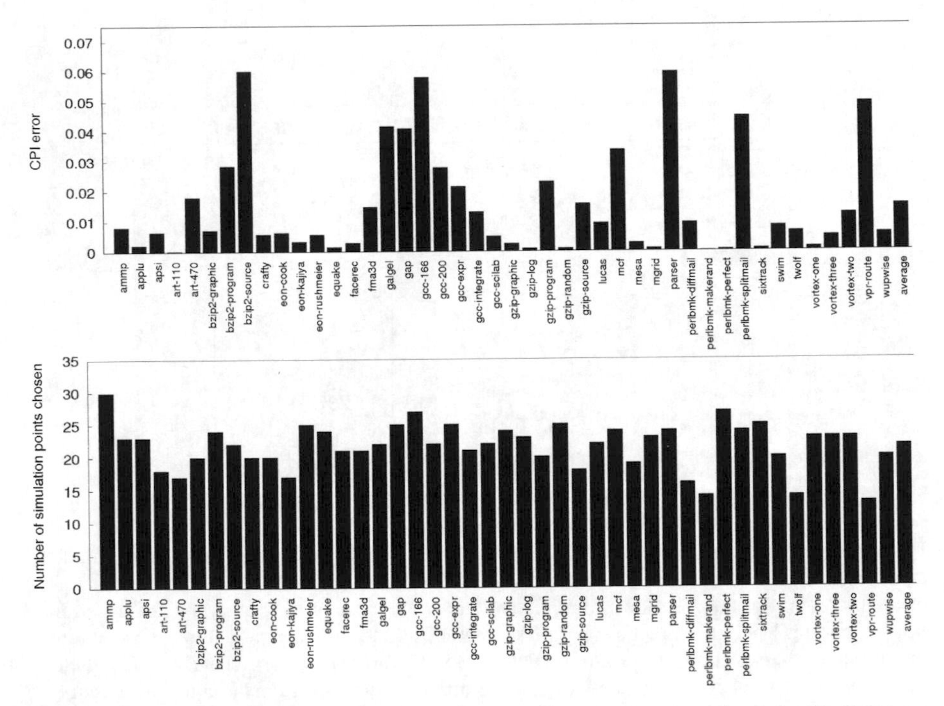

Fig. 10.5 These *plots* show the CPI error and number of simulation points chosen by SimPoint on the SPEC CPU 2000 suite of benchmarks

In terms of CPI, SimPoint is able to achieve a 1.5% CPI error rate averaged across all SPEC 2000 benchmarks, with a maximum error of around 6%. These results require an average simulation time of about 220 million instructions per program. These error rates are sufficiently low to make design decisions, and the simulation time is small enough to do large-scale design space explorations.

10.6.3 Relative Error During Design Space Exploration

The absolute error of a program/input run on one hardware configuration is not as important as tracking the changes in metrics across different architecture configurations. There is a lot of discussion and research about getting lower simulation error rates. But what often is not discussed is that a low error rate for a single configuration is not as important as achieving the same relative error rates across the design space search and having them all biased in the same direction.

We now examine how SimPoint tracks the relative change in hardware metrics across several different architecture configurations. To examine the independence of the simulation points from the underlying architecture, we used the simulation points for the SimPoint algorithm with an interval length of 1 million instructions and the maximum K set to 300. For the program/input runs examined, we performed full program simulations while varying the memory hierarchy, and for every run we used the same set of simulation points when calculating the SimPoint estimates. We varied the configurations and the latencies of the L1 and L2 caches as described by [10].

Figure 10.6 shows the results across 19 different architecture configurations for `gcc-166`. The left y-axis represents the performance in instructions per cycle (IPC) and the x-axis represents different memory configurations from the baseline architecture. The right y-axis shows the miss rates for the data cache and unified L2 cache, and the L2 miss rate is a local miss rate. For each metric, two lines are shown: "True" for the true metric from the *complete* detailed simulation and the "SP" for the estimated metric using our simulation points. For the results, the configurations on the x-axis are sorted by the IPC of the full run.

This figure shows that the simulation points, which are chosen by only looking at code usage, can be used across different architecture configurations to make accurate architecture design trade-off decisions and comparisons. The simulation points are able to track the relative changes in performance metrics between configurations. This means we are able to make the same decision between two architectures, in terms of which one is better, using SimPoint's estimate instead of complete simulation of the program. One interesting observation is that although the simulation results from SimPoint have a bias in its predictions, this bias is consistent across the different configurations for a given program/input. We observed this to be true for both IPC and cache miss rates. We believe one reason for the direction of the bias is that SimPoint chooses the most representative interval from each phase, and

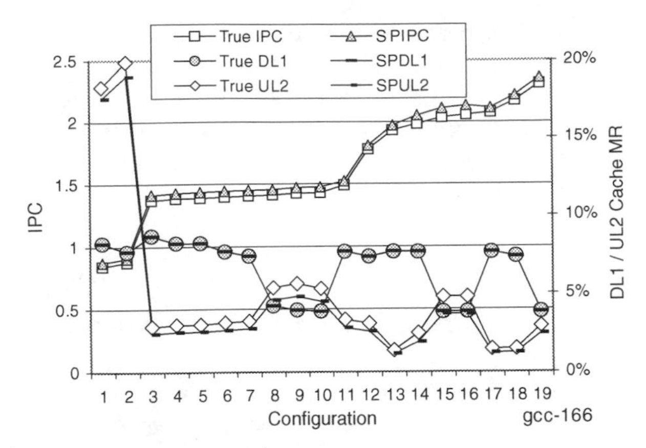

Fig. 10.6 This plot shows the true and estimated IPC and cache miss rates for 19 different architecture configurations for the program gcc. The left y-axis is for the IPC and the right y-axis is for the cache miss rates for the L1 data cache and unified L2 cache. Results are shown for the complete execution of the configuration and when using SimPoint

intervals that represent phase change boundaries are less likely to be fully represented across the chosen simulation points.

10.7 Summary

The main idea behind SimPoint is the realization that programs typically only exhibit a few unique behaviors which are interleaved with one another through time. By finding these behaviors and then determining the relative importance of each one, we can maintain both a high level picture of the program's execution and at the same time quantify the cycle level interaction between the application and the architecture.

The methods described in this chapter are distributed as part of SimPoint [10, 14]. SimPoint automates the process of picking simulation points using an off-line phase classification algorithm based on k-means clustering, which significantly reduces the amount of simulation time required. Selecting and simulating only a handful of *intelligently* picked sections of the full program provide an accurate picture of the complete execution of a program, which gives a highly accurate estimate of performance. The SimPoint software can be downloaded at:

http://www.cse.ucsd.edu/users/calder/simpoint/

For the industry-standard SPEC programs, SimPoint has less than a 6% error rate (2% on average) for the results in this chapter and is 1,500 times faster on average than performing simulation for the complete program's execution. Because of this time savings and accuracy, our approach is currently used by architecture researchers and industry companies (e.g., [9] at Intel) to guide their architecture design exploration.

References

1. Van Biesbrouck, M., Eeckhout, L., Calder, B.: "Efficient sampling startup for uniprocessor and simultaneous multithreading simulation." In: *International Conference on High Performance Embedded Architectures and Compilers*, November (2005).
2. Van Biesbrouck, M., Sherwood, T., Calder, B.: A co-phase matrix to guide simultaneous multithreading simulation. In: *IEEE International Symposium on Performance Analysis of Systems and Software*, March (2004).
3. Burger, D.C., Austin, T.M.: The SimpleScalar tool set, version 2.0. *Technical Report CS-TR-97-1342, University of Wisconsin, Madison*, June (1997).
4. Dasgupta, S.: "Experiments with random projection." In: *Uncertainty in Artificial Intelligence: Proceedings of the Sixteenth Conference (UAI-2000)*, pp. 143–151, Stanford University, June 30–July 3, (2000).
5. Lau, J., Sampson, J., Perelman, E., Hamerly, G., Calder, B.: "The strong correlation between code signatures and performance." In: *IEEE International Symposium on Performance Analysis of Systems and Software*, pp. 236–247, Austin, TX, March 20–22, (2005).
6. Lau, J., Schoenmackers, S., Calder, B.: "Structures for phase classification." In: *IEEE International Symposium on Performance Analysis of Systems and Software*, pp. 57–67, Austin, TX, March 10–12, (2004).
7. Lau, J., Schoenmackers, S., Calder, B.: "Transition phase classification and prediction." In: *11th International Symposium on High Performance Computer Architecture*, pp. 278–289, San Francisco, CA, February 12–16, (2005).
8. MacQueen, J.: Some methods for classification and analysis of multivariate observations. In: LeCam, L.M., Neyman, J. (eds.) *Proceedings of the Fifth Berkeley Symposium on Mathematical Statistics and Probability*, vol 1, pp. 281–297, University of California Press. Berkeley, CA, (1967).
9. Patil, H., Cohn, R., Charney, M., Kapoor, R., Sun, A., Karunanidhi, A.: "Pinpointing representative portions of large Intel Itanium programs with dynamic instrumentation." In: *International Symposium on Microarchitecture*, pp. 81–92, Portland, OR, December 4–8, (2004).
10. Perelman, E., Hamerly, G., Calder, B.: "Picking statistically valid and early simulation points." In: *International Conference on Parallel Architectures and Compilation Techniques*, pp. 244–255, New Orleans, LA, September 27–October 1, (2003).
11. Schwarz, G.: Estimating the dimension of a model. *The Ann Statis*, **6**(2), 461–464, (1978).
12. Sherwood, T., Calder, B.: Time varying behavior of programs. In: *Technical Report UCSD-CS99-630, UC San Diego*, August (1999).
13. Sherwood, T., Perelman, E., Calder, B.: "Basic block distribution analysis to find periodic behavior and simulation points in applications." In: *International Conference on Parallel Architectures and Compilation Techniques*, pp. 3–14, Barcelona, Spain, September 8–12, (2001).
14. Sherwood, T., Perelman, E., Hamerly, G., Calder, B.: "Automatically characterizing large scale program behavior." In: *10th International Conference on Architectural Support for Programming*, pp. 45–57, San Jose, CA, October 5–9, (2002).
15. Sherwood, T., Sair, S., Calder, B.: "Phase tracking and prediction." In: *30th Annual International Symposium on Computer Architecture*, pp. 336–347, San Diego, CA, June 9–11, (2003).

Chapter 11
Statistical Sampling

Babak Falsafi

Abstract Cycle-accurate full-system multiprocessor simulation is impractical because it is often six or more orders of magnitude slower than real hardware. Statistical sampling of microarchitecture simulation reduces overall cycle-accurate simulation for complete program runs by four orders of magnitude and allows for embarrassingly parallel simulation of independent windows of cycle-accurate simulation. Unlike conventional schemes measuring a single or a few simulation windows, statistical sampling aggregates results over hundreds of short measurements to provide a rigorous confidence in estimates.

11.1 Introduction

Cycle-accurate simulation often takes four or more orders of magnitude slowdown as compared to real hardware. Full-system simulation is up to a million times slower than real hardware because it requires simulating many CPUs, peripherals, and other system components that can slow down simulation by another factor of 10–100. Finally, multiprocessor simulation exacerbates this slowdown because it often increases demand on simulation linearly with an increase in the number of processors in the target system. This speed difference leads to intractable turnaround times if simulating the complete execution of computer benchmarks, in particular, for full-system multiprocessor workloads. These benchmarks are often longer than their uniprocessors counterparts to compensate for non-deterministic thread scheduling and perturbation effects from the OS and I/O that can lead to significant short-term performance variation.

Statistical sampling makes cycle-accurate full-system multiprocessor simulation tractable. Application execution is often highly repetitive because of repetitive algorithmic control flow resulting in homogeneity in application performance. Applying statistical sampling theory to cycle-accurate simulation enables exploiting this

B. Falsafi (✉)
EPFL IC ISIM PARSA, Station 14, 1015 Lausanne, Switzerland
e-mail: babak.falsafi@epfl.ch

R. Leupers, O. Temam (eds.), *Processor and System-on-Chip Simulation*,
DOI 10.1007/978-1-4419-6175-4_11, © Springer Science+Business Media, LLC 2010

homogeneity to eliminate redundant measurement and minimizing the total amount of cycle-accurate simulation.

Statistical sampling offers a key advantage over conventional approaches to measurement using cycle-accurate simulation. Unlike conventional approaches performance measurement through a single or a few long windows of cycle-accurate simulation, [7], statistical sampling prescribes a large number of minimal windows of simulation. Therefore, statistical sampling uses theory to determine the optimal total length of cycle-accurate simulation given a confidence in performance estimates, guaranteeing that the results are representative of the full application execution. Therefore, sampling is not only desirable from a turnaround perspective but also from a measurement accuracy and confidence perspective.

Breaking down simulation into a large number of small measurements enables a number of optimizations. Besides over ten-thousand fold reduction in total simulation turnaround, sampling allows for embarrassingly (e.g., >500-way) parallel simulation because measurements often occur over distant windows of execution and are as such independent. Such parallelism is "free," and complementary to any form of parallelism a given cycle-accurate simulator may exploit.

Moreover, randomized sampling allows for online and flexible throttling of simulation turnaround based on the desired error and confidence in measurement. With sampling, researchers can prune the space at earlier design stages quicker with higher error tolerance, tuning the error to lower thresholds at later design stages at the cost of higher turnaround time.

Finally, variability in performance estimates is often higher within an application's execution on a given target platform than across platforms. "Matched-pair" comparison is another statistical tool that optimizes simulation turnaround for a speedup comparison between two designs. Such a study only requires simulating enough windows of execution to capture the variability across the two designs, reducing overall simulation turnaround by up to an order of magnitude.

Sampling simulation, however, in general poses two key challenges. In the following, we summarize these challenges. In this and in the next chapter, we provide solutions with varying trade-offs to these challenges. Sampling requires fine-grain measurement of performance estimates (e.g., over windows of 1,000–100,000 instructions). Conventional performance metrics such as instructions per cycle (IPC), while applicable when simulating uniprocessors, do not accurately gauge forward progress in multiprocessor execution [2] due to synchronization in parallel programs and spinning (e.g., processors that spin while waiting to acquire a lock will exhibit high IPC while not actually contributing to an application's progress in execution).

Second, to avoid cycle-accurate simulation of the entire execution, sampling requires that the machine state prior to each measurement corresponds accurately to the target state as if the machine had seen the entire execution. The full potential of accurate sampling lies in rapidly constructing correct initial state for the large number of fine-grain performance measurements, which we call the *bias problem*. Bias is inherently a difficult problem because the target state, in general, may be timing dependent (e.g., asynchronous OS activity affecting the cache hierarchy or non-determinism in a parallel program's critical path due to synchronization).

In this chapter, we will first present the theory behind sampling and our SMARTS framework. Next, we present the challenges in sampling full-system multiprocessor workloads. Then, we present a checkpoint-based approach to accelerate sampling. Then, we present sampling optimization techniques. Finally, we present results of applying to apply SMARTS to both user-level uniprocessor workload and full-system multiprocessor workload simulation.

11.2 Statistical Sampling in SMARTS

We propose the sampling microarchitecture simulation (SMARTS) [14] framework for simulation sampling. The field of inferential statistics offers well-defined procedures to quantify and to ensure the quality of sample-derived estimates. In this section, we present the basic background on statistical sampling and how it applies to simulation in the SMARTS framework.

Statistical sampling attempts to estimate a given cumulative property of a *population* by measuring only a *sample*, a subset of the population [12]. By examining an appropriately selected sample, one can infer the nature of the property over the whole population in terms of total, mean, and proportion. The theory of sampling is concerned with choosing a minimal but representative sample to achieve a quantifiable accuracy and precision in the estimate. The theory does not presume a normally distributed population. Our goal is to apply this theory to (1) identify a minimal but representative sample from the population for cycle-accurate simulation and (2) establish a confidence level for the error on sample estimates.

Sampling can be defined in terms of the following parameters. The *sample size*, n, is the number of measurements that will be taken, where each measurement in our context corresponds to one continuous window of cycle-accurate simulation. Sampling attempts at minimizing the variability in the measured performance metric of interest (e.g., cycles per instruction.) The required sample size is proportional to the square of the performance metric's (x) coefficient of variation V_x, for example, for 95% confidence (2.0 standard deviations) of $\pm$ 5% error:

$$n \geq \left(\frac{2.0}{0.05} V_x \right)^2 . \tag{11.1}$$

Each measurement is called a *sampling unit*. The sampling unit size, U, greatly affects the selection of n. A larger U averages out short-term variation in performance, reducing V_x, and therefore reducing n. There are, however, large-scale variations in performance which may result in diminishing returns in reducing V_x and cannot be effectively captured with a larger U. Therefore, the best combination of n and U can be determined with (1) knowledge of V_x, which we can obtain with preliminary samples and (2) the strategy to reduce sampling bias (described below).

11.3 Sampling Metrics

In SMARTS, the goal of sampling is to estimate a performance metric of interest with a minimal amount of cycle-accurate simulation. In cycle-accurate simulation, typically measured performance metrics are execution time in cycles or average cycles per instruction (CPI). In general, the former can be derived from the latter and a total instruction count. Therefore, when sampling one measures the coefficient of variation in CPI, namely V_{CPI}. Other metrics such as average energy/power or vulnerability (used in reliability studies) can also be measured but estimating peak or maximum values for a performance metric is challenging because sampling in nature skips cycle-accurate simulation of the entire execution.

The required sample size to estimate CPI at a given confidence is directly proportional to the square of the population's coefficient of variation, $n \propto V_{\text{CPI}}^2$. A benchmark with a small V_{CPI} implies a greater opportunity for accelerated simulation because fewer instructions from the benchmark need to be simulated with cycle-accurate models. A benchmark's instruction stream can be divided into a population using different values of U. Figure 11.1 from [15] plots V_{CPI} as a function of U on an eight-way out-of-order superscalar processor with user-level uniprocessor SPEC benchmarks. V_{CPI} decreases with increasing U because short-term CPI variations within a window of U instructions are hidden by averaging over the sampling unit. The V_{CPI} curves for all benchmarks share the same general shape, with a steep negative slope for U less than 1,000, leveling off thereafter.

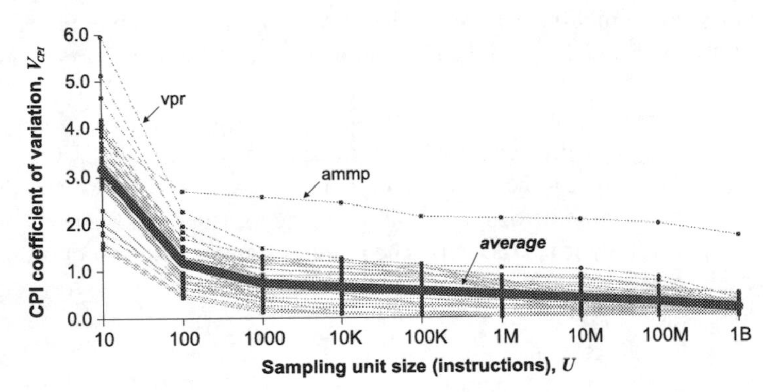

Fig. 11.1 Coefficient of variation of CPI for a simulated eight-way out-of-order superscalar processor as a function of sampling unit size [15]

Unfortunately, aggregate CPI (or IPC) does not accurately measure forward progress in multiprocessor workloads [2]. In multiprocessor benchmarks, contention to enter a critical section often results in spinning. As such, while aggregate CPI may be quite low (because spinning often proceeds at higher than average program speeds), the actual forward progress on the critical path of execution has much higher CPI. Therefore, in general, sampling parallel program execution while measuring CPI requires spin or critical path prediction. In the next chapter, we present techniques to allow accurate measurement of forward progress for generalized parallel workloads.

In the specific case of transactional servers workloads, one often reports performance in terms of *transactions per second*. Unfortunately, transactions are too long for a simulator to execute, and their completion rate has a high coefficient of variation. Figure 11.2 from [13] plots the coefficient of variation for transaction throughput, $V_{\text{TRANSACTION}}$, for several server benchmarks measured on a real four-core multiprocessor system. Points on the plot correspond to sampling units of a logarithmically increasing number of transactions (labeled every factor of 10), with the mean time to complete those transactions indicated on the x-axis. Because transaction completions are bursty and transactions take seconds to execute, measuring $V_{\text{TRANSACTION}}$ as a sampling parameter would require years of simulation time on a full-system multiprocessor simulator.

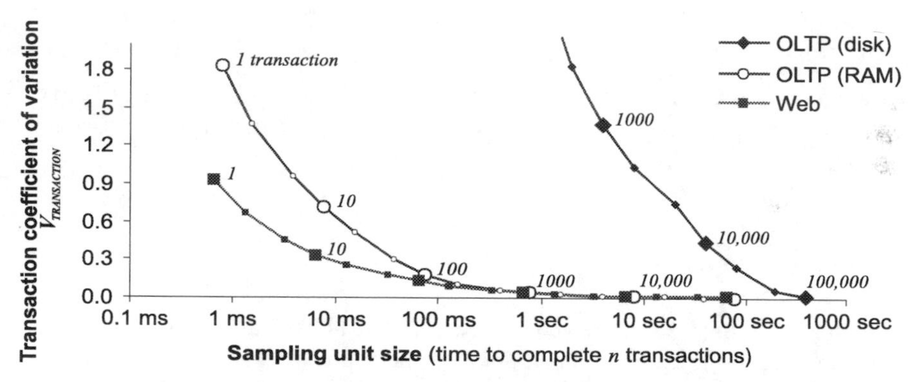

Fig. 11.2 Coefficient of variation of transactions on a real four-processor system. Sample size is quadratically related to coefficient of variation. For example, measurements of 10 transactions each for OLTP (RAM) yield a $V_{\text{TRANSACTION}}$ of 0.7. Thus, from eq. (11.1) to achieve a 95% chance of ±5% error requires about 800 100-ms measurements [13]

Fortunately, Hankins et al. [8] made the observation that the amount of work a database server performs to complete a transaction often does not vary greatly, and the average time user-mode instructions take to complete is linearly proportional to transaction throughput. We have found that this result applies across generalized transactional workloads from database to web servers. This relationship holds because applications yield to the operating system when they are not making forward progress (e.g., the OS idle loop or spin loops in OS locking primitives).

The linear relationship between user-instruction throughput and transaction throughput allows us to sample *user instructions per cycle* (U-IPC) to assess transaction throughput. Sampling U-IPC is advantageous because the variance of U-IPC is lower than that of transaction throughput at vastly smaller measurement sizes. Thus, we can simulate shorter portions of the benchmarks for each measurement while achieving the same confidence. Figure 11.3 plots the coefficient of

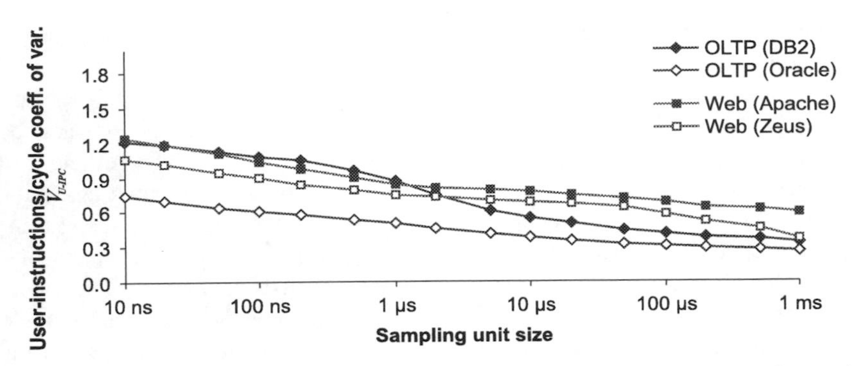

Fig. 11.3 Coefficient of variation of user-level IPC (instructions per cycle) measured on a simulated 16-processor system. The sampling unit size for a particular $V_{\text{U-IPC}}$ is three orders of magnitude smaller than for the same $V_{\text{TRANSACTION}}$. Thus, sampling U-IPC requires 1,000 times less than total simulation [13]

variation for U-IPC across a range of measurement sizes. Using U-IPC as the target metric saves three orders of magnitude in simulation time versus using transaction throughput.

11.4 Sampling Bias

Sampling *bias* refers to errors introduced into the individual measurements that make up a sample (e.g., by the measurement methodology) and are not accounted for by statistical confidence calculations.

The cause of bias in simulation sampling is inaccurate initial target state at the beginning of a sampling unit. The target state in simulation includes both the architectural state (i.e., registers, memory, and I/O) and microarchitectural state (e.g., pipeline queues, cache hierarchies, and interconnection network). A bias-free sample requires that the target state at the start of a measurement corresponds that generated if the entire execution were simulated using cycle-accurate simulation. Sampling is only effective if one can rapidly reproduce this initial state to speed up simulation in-between measurements while introducing only a small bias.

There are a number of common bias sources in simulation. The common architectural source of bias is in multiprocessor programs where synchronization creates non-determinism in execution paths. Approximate target models in-between measurements to accelerate simulation may perturb timing in critical sections of parallel programs, thereby leading to diverse execution paths and entirely different sample populations.

The most common cause of bias in microarchitectural state is the cold-start effect of unwarmed structures, referred to as the *warming* problem. The simplest form of simulation sampling would use instruction emulation to generate only architectural (i.e., program) state in-between measurements to speed up simulation. Such sampling, for instance, would result in empty caches and incorrectly low performance estimates and empty interconnect networks that produce overly optimistic performance.

11.5 Architectural State Bias

Architectural state bias is the result of generating architectural (i.e., register, memory, and I/O) state while not modeling timing accurately. In this chapter, we present results on user-level uniprocessor simulation sampling where such a bias does not occur, and full-system multiprocessor simulation of commercial server workloads which are highly tuned parallel workloads whose execution is not affected by the speed of critical sections. Techniques to mitigate this source of bias for generalized workloads are discussed in the subsequent chapter.

In multiprocessor benchmarks, instruction interleaving across processors varies over multiple benchmark runs and can cause changes in the dynamic instruction stream as races (e.g., for locks) resolve differently. On real platforms, designers often account for such non-determinism by running a benchmark repetitively to exercise multiple instruction interleavings. Simulation, however, is often deterministic and because of its prohibitively low speeds precludes execution over long windows of (real) time. Therefore, simulating the execution of generalized multiprocessor benchmarks on target hardware to reach multiple interleavings is difficult.

Fortunately, most of today's commercial multiprocessor server workloads consist of *throughput applications* for which one can build instruction interleavings efficiently. These include online transaction processing, decision support queries, and web serving. These applications consist of long (or unbounded) sequences of randomly arriving transactions, queries, or server requests (for simplicity, in the rest of this chapter we will refer server requests as transactions). In these applications, transactions arrive randomly, so that a single run covers the range of possible instruction interleavings. Recent results indicate that one can draw a sample by selecting measurement locations over a time window that has proven reliable on real hardware (e.g., ~30 s [1]). Figure 11.4 from [13] illustrates sampling a throughput benchmark.

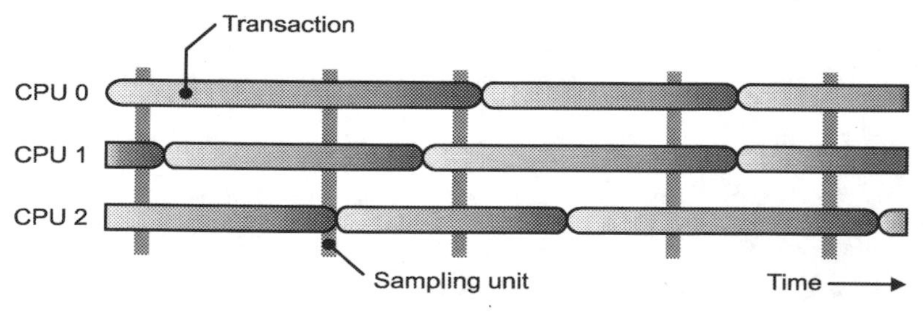

Fig. 11.4 Sampling a throughput application. Each sampling unit measures performance of a random interleaving of transactions. The sample's population is then defined as the runtime that effectively covers all reachable transaction interleavings [13]

11.6 Microarchitectural State Bias

Microarchitectural state bias arises from inaccuracies in the microarchitectural state update (e.g., pipeline state, cache hierarchies, branch predictor and prefetcher tables, and the interconnection network) in-between measurements.

The conventional approach to ameliorating microarchitectural state bias is to "warm" the state through cycle-accurate simulation prior to each measurement. We refer to this conventional approach as *detailed warming*. While detailed warming is effective for structures that only have short-term history (e.g., pipeline queues or network queues) whose warming requirements can be bounded, warming for generalized target state and across benchmarks is unbounded [15]. Unbounded detailed warming, however, offsets two of the primary benefits from sampling, namely minimizing the required amount of cycle-accurate simulation and having a confidence on the measured error.

There are a number of techniques in the literature to address warming and timing-dependent updates to microarchitectural state including techniques to analytical model timing and microarchitectural state updates in-between measurements all the way to techniques to estimate dynamically the required amount of detailed warming. In the following chapter, we will present dynamic techniques to bound warming. In this chapter, we present the SMARTS approach to warming.

11.7 Warming in SMARTS

In SMARTS, we make the observation that microarchitectural state can be divided into two groups: (1) short-history state belonging to components with a history and warmup requirement (e.g., queues) and (2) long-history state which can be updated with approximate models but whose warmup is in general unbounded. We propose detailed warming for the first group with an analytically bounded window. For the second group, we propose *functional warming*, that continuously warms microarchitectural state with very long history with approximate timing models that are dramatically faster than cycle-accurate simulation (Fig. 11.5).

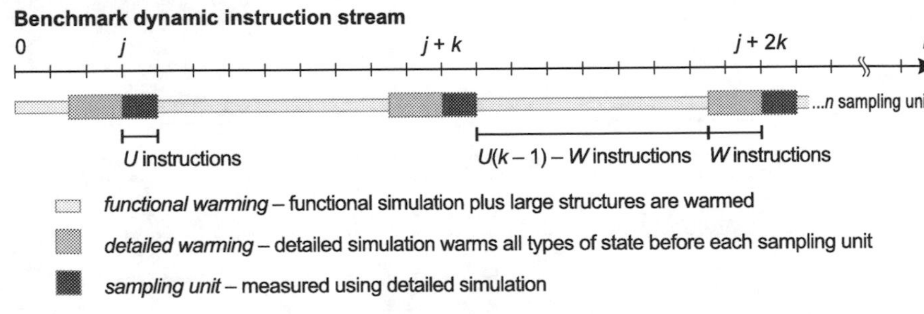

Fig. 11.5 Systematic sampling as performed in SMARTS. Two modes of simulation are used: functional simulation and detailed simulation. The need for determined warmup for large structures, such as caches, is eliminated by performing continuous functional warming [15]

A caveat to the functional warming approach is that it may not always be able to accurately reproduce the correct microarchitectural state if correct warming requires exact knowledge of detailed execution. Moreover, timing-dependant behavior (e.g., operating system scheduling activity) requires timer approximation. If functional warming simulates instructions in order, it also may not accurately reflect the artifacts of out-of-order and speculative event ordering. Cain et al. [4] have suggested that out-of-order and speculative ordering has minimal impact on CPI and other performance metrics. Wunderlich et al. [15] corroborate these results and present an analysis of the residual biases after functional warming. We believe functional warming is the most cost-effective approach to achieve accurate CPI estimation with simulation sampling.

11.8 Checkpointed Warming

Although functional warming enables accurate performance estimation, it limits SMARTS's speed, occupying more than 99% of simulation runtime. Functional warming dominates simulation time because the entire benchmark's execution must be functionally simulated, even though only a tiny fraction of the execution is simulated using detailed microarchitecture timing models.

The second shortcoming of the original SMARTS framework is that functional warming requires simulation time proportional to benchmark length rather than sample size. As a result, the overall runtime of a SMARTS experiment remains constant even when the measured sample size is reduced (e.g., by relaxing an experiment's statistical confidence requirements). Moreover, functional warming time will increase with the advent new benchmark suites that lengthen benchmarks to scale with hardware performance improvement.

We developed *checkpointed warming* [13] as an alternative to functional warming to reduce simulation turnaround time without sacrificing accuracy. In checkpointed warming, a *live-point* stores the necessary data to reconstruct warm state for a simulation sampling execution window. Although modern computer architecture simulators frequently provide checkpoint creation and loading capabilities [11], current checkpoint implementations: (1) do not provide complete microarchitectural model state and (2) cannot scale to the required checkpoint library size (~10,000 checkpoints per benchmark) because of multi-terabyte storage requirements.

The key challenge in checkpointed warming is generating "reusable" live-points that allow for simulating a range of microarchitectural configurations of interest. Because the majority of microarchitectural state can be reconstructed dynamically with limited simulation, live-points must only handle the functionally warmed structures (e.g., branch predictors and caches). For these structures, researchers can often place limits on the configurations of interest (e.g., through trace-based studies). Moreover, there are ways to store "super-set" information for caches [10] and branch predictors [3] to allow for simulating multiple configurations with one live-point.

We also make the observation that checkpointed microarchitectural state can be minimal because the state needed is that for a limited execution window in a sampling unit and not the entire microarchitectural state up to an given execution point. As such, live-points can reduce checkpoint storage by three orders of magnitude over conventional checkpointing schemes.

Checkpointed warming, however, can remain a bottleneck in studies where libraries are often not reusable and must be recreated. Chapter 4 targets techniques to trade-off accuracy and confidence for accelerated warming. FPGA technologies (e.g., ProtoFlex [5]) can also help reduce library generation time by orders of magnitude by implementing simple instrumentation (e.g., cache simulation) in hardware.

11.9 Sampling Optimizations

There are a number of ways in which turnaround can be optimized using checkpointed sampling.

11.9.1 Parallel Simulation

Sampling using checkpoints leads to embarrassing parallel simulation. A typical sample may have 500–1,000 measurements, each of which can proceed independently. Each point can be simulated independently. Live-point independence allows us to parallelize simulation over many host machines, reducing the overall latency to obtain results.

11.9.2 Online Results

With live-points, each complete library forms an unbiased uniform sample of a workload. Given such a library, a randomly selected subset within the library also forms a uniform sample. Simulating live-points in a random order therefore allows one to stop simulation after an arbitrary number of live-points for "online results" while having simulated a valid uniform sample (Fig. 11.6). Similarly, one can use this property to provide a continuous update of estimated results as points are processed [9, 13].

11.9.3 Matched-Pair Sample Comparison

A typical computer architecture study involves comparing performance metrics of one design against another (base) design. When comparing designs, designers often need relative rather than absolute performance. Because changes in design under consideration often do not lead to dramatic changes in performance, the variability

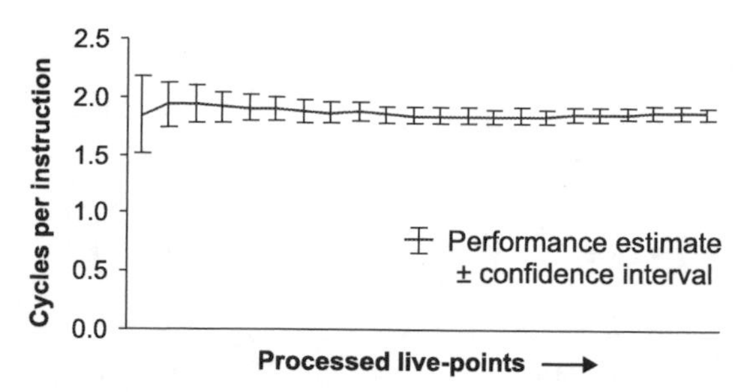

Fig. 11.6 Online results example. Results converge toward their final values, and confidence improves as more live-points are processed [13]

in performance during execution within a given design is often higher than the variability across designs.

Matched-pair sample comparison [6, 13] allows for exploiting this reduced variability in the performance metric. Figure 11.7 depicts how we can measure the same points for both designs and build a confidence interval directly on the delta-performance observed on each point. The reduced variability allows us to achieve the same confidence on relative performance with a smaller sample than required for absolute performance estimates.

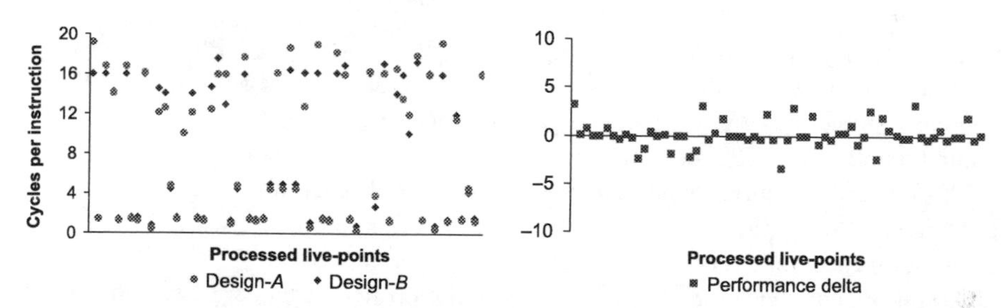

Fig. 11.7 Matched-pair comparison example: performance data for the same sampling unit measured under two microarchitecture designs (*left*) and performance deltas for each sampling unit (*right*). The lower variability of performance deltas reduces sample size by 3.5–150 times [13]

11.10 Sampling Speedup and Cost

We summarize our experiences using the SMARTS methodology to estimate performance of both SPEC CPU2000 and commercial multiprocessor applications in Table 11.1. Our experiments with SPEC CPU2000 are fully described in [15]. Our multiprocessor design experiments model a 16-way distributed-shared-memory multiprocessor loosely based on the HP GS1280 and Compaq Piranha designs and are fully described in [13]. Our commercial applications are scaled to use 10 GB data sets.

Table 11.1 Multiprocessor simulation sampling with Flexus

	SPEC CPU (TurboSmartSim)	OLTP (DB2)	OLTP (Oracle)	Web (Apache)	Web (Zues)
Runtime on real hardware	1.5 h			30 s	
Simulation time of sampling	0.6 CPU years		10–20 CPU years		
Sampling parameters					
Performance metric	Cydes/Instr.		User- instructions/cycle		
Detailed warming	2,000 instr.		100,000 cycles		
Sampling unit size	1,000 instr.		50,000 cycles		
Target confidence interval	99.7% ± 3%		95% ± 5%		
Absolute performance estimate					
Typical sample size	8816[a]	301	100	774	289
Simulation time	91 CPU s[a]	39 CPU h	25 CPU h	140 CPU h	58 CPU h
Live-point library size	12 GB	28 GB	4 GB	74 GB	19 GB
Typical relative performance estimate					
Typical sample size	3511	193	30	91	183
Simulation time	36 CPU s[a]	25 CPU h	7 CPU h	17 CPU h	37 CPU h

[a]Per reference input (mean). There are 45 reference inputs in SPEC CPU2000

Simulation without sampling is simply impractical, requiring CPU-years to provide accurate results with high confidence. In contrast, the SMARTS methodology enables high confidence estimates of performance in minutes for SPEC CPU2000 and with only a few CPU-days for commercial applications. Matched-pair comparison further reduces turnaround time by up to a factor of seven for design comparison experiments. By parallelizing simulations on a compute cluster, we can complete even multiprocessor experiments in just a few hours.

Unfortunately, because in full-system simulation live-points are built on top of a full-system functional emulator's (i.e., Virtutech Simics') checkpointing mechanism, each live-point must include either a complete snapshot of system state or delta images of memory and disk state from another live-point. As such, full-system server simulation (for the commercial applications) uses checkpoints that take from 10 to 200 MB requiring a live-point library storage of as much as 74 GB. However, even on real systems, commercial applications have high disk space requirements (in excess of 25 GB for a 100 warehouse TPC-C installation). Live-point library storage requirements are well within the capabilities of modern high-capacity disks.

11.11 Summary

Statistical sampling provides the key to fast and accurate performance evaluation of computer systems through simulation. The SMARTS experimental methodology provides an avenue to assess system performance with a minimal sample while providing rigorous statistical measures of confidence. Through its sample design and live-points SMARTS enables simulation-based assessment of uniprocessor applications in minutes instead of weeks and throughput-based multiprocessor applications in hours instead of years.

References

1. Alameldeen, A.R., David A. Wood: Variability in architectural simulations of multi-threaded workloads. In: *Proceedings of the International Symposium on High-Performance Computer Architecture*, Anaheim, CA, February (2003).
2. Alameldeen, A.R., Wood, D.A.: IPC considered harmful for multiprocessor workloads. *IEEE Micro* **26**(4), 8–17, (2006).
3. Barr, K.C., Asanovic, K.: Branch trace compression for snapshot-based simulation. *International Symposium on Performance Analysis of Systems and Software* 25–36, Anstin, TX (2006).
4. Cain, H.W., Lepak, K.M., Schwartz, B.A., Lipasti, M.H.: Precise and accurate processor simulation. In: *Workshop on Computer Architecture Evaluation using Commercial Workloads*, Boston, MA, February (2002).
5. Chung, E.S., Papamichael, M., Nurvitadhi, E., Hoe, J.C., Mai, K., Falsafi, B.: ProtoFlex: Towards scalable, full-system multiprocessor simulations using FPGAs. *ACM Transactions on Reconfigurable Technology and Systems*, Article 15, **2**(2), June (2009).
6. Ekman, M., Stenstrom, P.: Enhancing multiprocessor architecture simulation speed using matched-pair comparison. In: *Proceedings of the International Symposium on Performance Analysis of Systems and Software*, Austin, TX, March (2005).
7. Hamerly, G., Perelman, E., Lau, J., Calder, B.: Simpoint 3.0: Faster and more flexible program analysis. *J Instruction-Level Parallelism*, September (2005).
8. Hankins, R., Diep, T., Annavaram, M., Hirano, B., Eri, H., Nueckel, H., Shen, J.P.: Scaling and characterizing database workloads: Bridging the gap between research and practice. In: *Proceedings of the 36th Annual IEEE/ACM International Symposium on Microarchitecture (MICRO 36)*, San Diego, CA, December (2003).
9. Hellerstein, J.M., Haas, P.J., Wang, H.J.: Online aggregation. In: *Proceedings of the International Conference on Management of Data*, Tucson, AZ, May (1997).
10. Hill, M.D., Smith, A.J.: Evaluating associativity in CPU caches. *IEEE Trans Comput* **C-38**(12), 1612–1630 December (1989); **83**, February (1997).
11. Magnusson, P.S., Christensson, M., Eskilson, J., Forsgren, D., Hallberg, G., Hogberg, J., Larsson, F., Moestedt, A., Werner, B.: Simics: A full system simulation platform. *IEEE Comput* **35**(2), 50–58, February (2002).
12. Jain, R.: *The Art of Computer System Performance Analysis: Techniques for Experimental Design, Measurement, Simulation, and Modeling*, Wiley-Interscience, New York, NY, April (1991).
13. Wenisch, T.F., Wunderlich, R.E., Falsafi, B., Hoe, J.C.: Simulation sampling with live-points. In: *Proceedings of the International Symposium on Performance Analysis of Systems and Software*, Austin, TX, June (2006).
14. Wunderlich, R.E., Wenisch, T.F., Falsafi, B., Hoe, J.C.: Accelerating microarchitecture simulation via rigorous statistical sampling. International Symposium on computer Architecture, 84–95, San Diego, CA, (2003).
15. Wunderlich, R.E., Wenisch, T.F., Falsafi, B., Hoe, J.C.: Statistical sampling of microarchitecture simulation. *ACM Transaction on Modeling and Computer Simulation (TOMACS)* **16**(3), 197–224, July (2006).

Chapter 12
Efficient Cache Modeling with Sparse Data

Erik Hagersten, David Eklöv, and David Black-Schaffer

Abstract Obtaining good application performance requires tuning for effective use of the cache hierarchy. However, most tools to analyze cache usage either generate architecture-specific results (e.g., hardware performance counters) or incur prohibitively high overheads for real-world workloads (e.g., trace-based simulations). This chapter reviews several recently introduced techniques that address these issues to efficiently model cache systems and coherent memory hierarchies in an architecturally independent manner. The techniques utilize only sparse, architecturally independent runtime information that can be collected with an overhead of 10–30%. This information is then processed by statistical models to quickly predict cache behavior across a range of architectures. With these approaches, accurate modeling is possible from data sampled with rates as low as 1 in 10^6 memory accesses.

12.1 Introduction

This chapter introduces several techniques for low-overhead architecture-independent analysis of programs to help rapidly identify the shortcomings of the application's usage of a given cache hierarchy. Such analysis is essential for good performance on processors with high penalties for accessing off-chip memory, which includes nearly all commercial processors today. In the case of multiprocessor systems, the situation is further complicated when multiple threads access the same data and their interactions through the coherency protocol. In general such optimization requires that the developer tunes the code for data locality and cache usage. However, these properties are difficult to understand with existing tools and hardware, which both complicates the program analysis and the final evaluation of implemented optimizations. The techniques reviewed here provide insight into the cache behavior of an application executing on real-world data in an architecturally independent manner.

E. Hagersten (✉)
Acumem AB, Döbelnsgatan 17, Uppsala, SE 75237, Sweden
e-mail: eh@it.uu.se

R. Leupers, O. Temam (eds.), *Processor and System-on-Chip Simulation*,
DOI 10.1007/978-1-4419-6175-4_12,

12.2 Modeling Technology Requirements

While any technology must be both accurate and informative to be useful, the requirements for this modeling technology to be *successful* are that it must be portable and efficient. For the technology to be portable, it must be easily usable across a range of commodity platforms. Efficiency requires that the technology has a small enough impact on performance to enable the evaluation of real-world data sets. Without either of these features, the usefulness of the technology for developer optimization would be severely hampered.

The portability requirement has several direct implications. As commercial software is rarely written and compiled to suit a single architecture implementation, the same binary will be expected to execute on multiple generations of architectures manufactured by several different companies, such as the x86 implementations by Intel and AMD. Thus, this technology should be capable of modeling different architectures based on information from a single run on a single architecture. This implies further that the technology should not depend on hardware extensions that are not widely available in order to attain reasonable efficiency. In addition to such architecture independence, the technology should be as unobtrusive as possible, preferably requiring no manual modification of the application for analysis.

Efficiency provides not only convenience but is critical for correctness. When analyzing the behavior of the memory system it is of great importance that the studied application runs a realistic data set, since a study based on a reduced data set may lead to incorrect conclusions. We consider a slowdown of more than 100%, e.g., requiring twice as much time to perform the analysis as running the application natively, prohibitively high.

12.3 Choice of Overall Modeling Technique

To put these modeling techniques in perspective, it is worthwhile to compare them to a range of existing approaches. The most traditional, and accurate, methods of analyzing cache behavior are based on *full address traces* of the program. A full address traces includes the address and type of all memory accesses performed by the application (and possibly the system) during its execution. Such full address traces can be run through cache simulators or analyzed directly to evaluate the memory behavior of the application to any level of desired accuracy.

Full address traces are typically gathered by running the application in a simulation environment, instrumenting the application and recording the trace at runtime, or using hardware extensions to assist in the runtime capture of memory accesses. Each of these approaches has significant efficiency or portability implications. Acquiring full application traces, whether through architectural simulation [4], functional simulation [9], or instrumentation [8] incurs overheads well in excess of the goals for the techniques presented here. To reduce this overhead, hardware extensions have been proposed [11], but no standard has been adopted to make their use portable. Further, a full address trace, regardless of the efficiency with which it is captured, is typically quite large for a program of any significant size. Storing this information to disk, as well as post-processing it, therefore incurs a significant overhead. Sampling techniques have been proposed to limit the simulation activity to only study some subtraces (Chapter 11).

An alternative to capturing the full address trace of an application is to use existing hardware performance monitors to capture information during the execution of an application [12]. While this method entails little performance overhead, the data tends to be in the form of coarse statistics such as overall cache miss rates across all processes. This aggregate data does not provide the insight or flexibility required to model varying architectures.

Instead of capturing and analyzing the dynamic behavior of an application, one can also statically analyze the program [6]. While this has the benefit of not impacting runtime performance, it is inherently unable to capture data-dependent execution patterns and must be conservative in its analysis of aliasing and data sharing.

The approach described in this chapter is a compromise between the accuracy of a full address trace and the speed of hardware performance counters. At runtime, a *sparse sample* of architecturally independent access information is collected. The sparse sample data is then fed into a fast statistical model that analyzes the program's behavior on a range of different architectures. The combination of sparse sampling and architecturally independent data results in an approach that is both efficient and portable.

12.4 Sparse Reuse Distance Sampling

To achieve good performance, only sparse runtime information is collected while the application is running. A sample rate determines what fraction of memory accesses should be monitored on average, with the individual memory access to be monitored selected randomly to achieve the target sample rate. For each monitored memory access, the number of subsequent memory accesses is counted until the same data unit is accessed again. The recorded data is the *reuse distance* between two accesses to the same data unit, called a *reuse pair*. Typically, the size of the data unit to monitor is the cache-line size of the architecture later to be modeled. Together these sparse samples constitute an *application fingerprint*, which is a collection of the sampled reuse pair information, recorded as the program executes. Berg et al. showed good results with sample rates in the range of 10^{-6} in a study of SPEC CPU 2000 benchmarks [2].

There are two reasons why reuse distance is an attractive property to measure at runtime. First, the measurement is architecturally independent and measures the locality property of the binary without depending on the design choices of the cache hierarchy of the host system. Second, this information can be captured with very low overhead. As discussed in the appendix to this chapter, monitoring when a piece of data will be touched again can be efficiently implemented with readily available watchpoint mechanisms, and counting the number of intervening memory accesses is cheaply implemented using ubiquitous hardware counters. For long-running applications, capturing such sparse data results in a sampling overhead of between 10 and 30%.

This process is illustrated in Fig. 12.1, which shows a stream of memory accesses performed by an application running on the host computer. The application fingerprint consists of the memory reuse pairs highlighted in the figure. The first access chosen to be monitored is access number 1, which references data unit A. The sampler then has to count how many memory references occur until data unit A is referenced again. If another memory reference (number 5) is selected to be monitored, the sampler must now look for the next access to data unit B as well and count its intervening memory accesses. The sampler saves the information about this reuse pair in the application fingerprint. For this simple example the fingerprint will contain the two reuse distances of length 5 and 3. For the basic modeling discussed in the first part of this chapter, this is all the information needed. The commercial tools by Acumem store even more information and metadata in the fingerprint to allow for a more complete analysis.

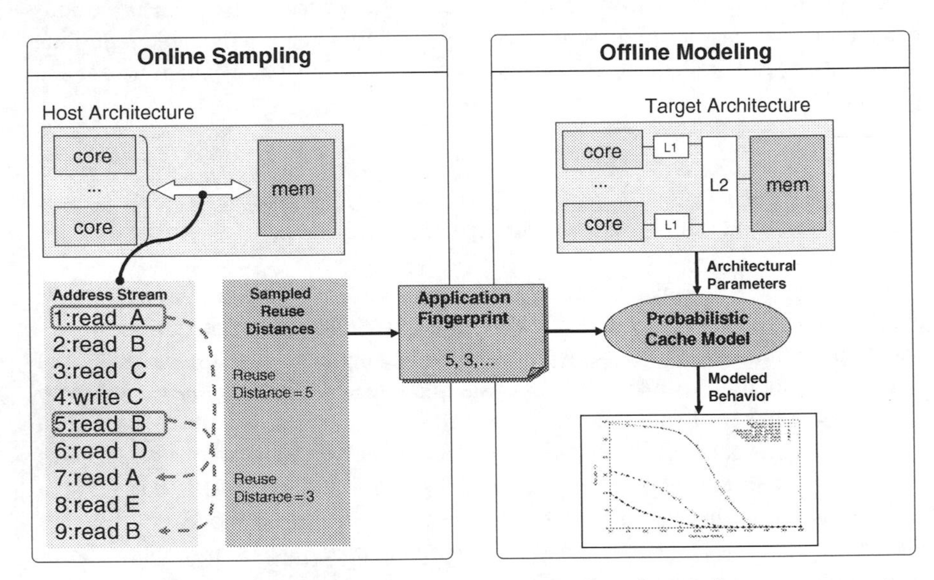

Fig. 12.1 The technology presented here uses online sampling of reuse distances to generate a sparse application fingerprint at runtime. The fingerprint is then fed into a probabilistic cache model along with the parameters of the target architecture

Note that the sparseness of collected information described here is more fine grained than many other sampling techniques (Chapter 11), which require fairly large contiguous traces to be collected.

12.5 Probabilistic Cache Modeling and Sample Windows

The fingerprint of reuse distances captures architecturally independent locality information at runtime with a low overhead. However, in order to interpret the implications of the locality information when running on a target architecture, we need to model the behavior of the application on the target architecture, as described in the next few sections. This requires a transformation of sparse reuse information into the miss ratio on the target architecture. This transformation is done by evaluating a probabilistic cache model that uses the fingerprint of the application and the architectural description of the target architecture to determine the resulting cache miss ratio of the application on the target architecture. The target architecture is described by basic parameters such as cache size, cache-line size, and replacement strategy, and, for multiprocessor memory systems, the model takes into account cache topology, cache sharing, and coherence activity.

Note that the probabilistic algorithm does not require any cache warming before information can be gathered. While traditional time sampling must use long contiguous subtraces to warm cache models and avoid cold misses, we can sample reuse distances in many short bursts, called *sample windows*. Typically, a two-level time-sampling scheme is used when capturing the fingerprint: first, an appropriate number of sample windows are scattered throughout the execution time, either randomly or guided by phase detection, possibly using techniques similar to SMARTS (Chapter 11) or SimPoint (Chapter 10); second, time sampling is used within each sample window to choose the memory references for which reuse distances should be measured. Contrary to SMARTS and SimPoint, each sample window only contains sparse samples and not contiguous subtraces. This sample-based approach is based on the assumption that a subset of samples, captured closely in time, is most often from the same phase of the execution and will therefore be representative of the overall behavior of that phase. With this assumption, varying application phases are transparently handled by applying the statistical model to each sample window independently. However, in any given application fingerprint there are likely to be some sample windows that cross application phase boundaries, as seen in Fig. 12.2. These phase-crossing windows will produce inaccurate estimates from the statistical model, but as long as the proportion of such windows is small, the overall results will not suffer significantly. Our experience indicates that around 200 sample windows with 200 samples per window are required to produce accurate results for complex applications. The number of samples needed reflects the overall application complexity and not the execution time.

This chapter describes three probabilistic models: the random replacement cache model (StatCache), the LRU cache model (StatStack), and the coherent multiprocessor cache model (StatCacheMP). Each of these models use similar sparse fingerprint information and related statistical models to provide efficient, accurate, and portable application analysis in an architecturally independent manner.

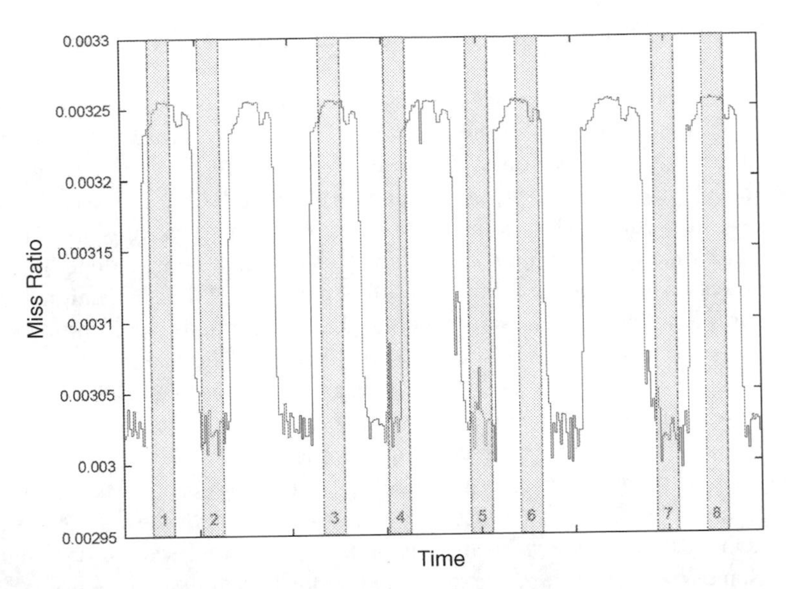

Fig. 12.2 The sample windows are spread throughout the application's execution, which allows different application phases to be handled transparently. As long as only a small number of windows cross phase boundaries, as window 4 does here, the overall modeling will accurately reflect the phase behavior of the application

12.6 StatCache: Modeling Random Replacement Caches

StatCache is a technique for estimating an application's cache miss ratio for a fully associative cache with a random replacement policy [1, 2]. Random replacement caches lend themselves very naturally to probabilistic modeling. For example, consider a cache with L cache lines, such as the one shown in Fig. 12.3. The likelihood of a cache hit depends on the number of replacements performed since the accesses data was last known to be in the cache. On a replacement, a cache line is randomly selected for eviction. The likelihood that a given cache line is evicted is $1/L$, and the likelihood that it is not evicted is $1–1/L$. After R replacements, the likelihood that the cache line is still in the cache is $(1–1/L)^R$. For a sampled reuse pair, we know that the accessed data must be in the cache after the first access in the pair is performed. Either the access is a hit, in which case data was already in the cache, or it is a miss, in which case it is brought into the cache. The miss probability of the second accesses in the pair can therefore be expressed as a function of the number of replacements performed between the two accesses.

Miss probability after R replacements:

$$m(R) = 1 - (1 - 1/L)^R$$

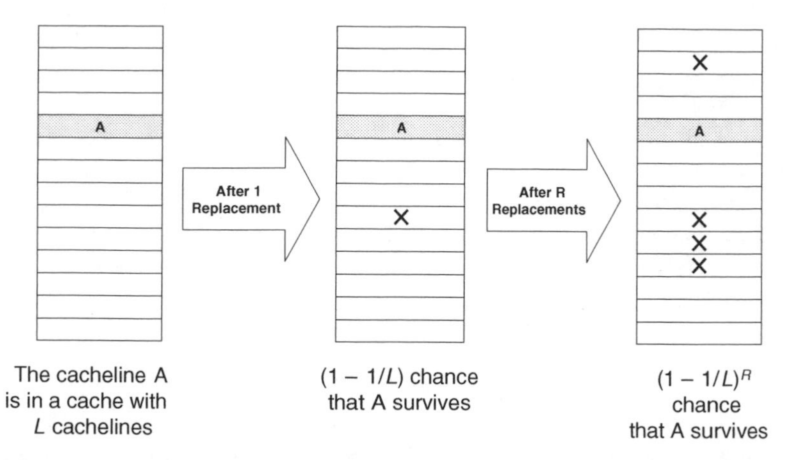

Fig. 12.3 The likelihood of a hit in a fully associative random replacement cache with L cache lines depends on the number of replacements performed since the last time the data was known to be in the cache

Assuming for a while that the miss ratio M is constant in a sampling window and known to us, we can estimate the number of replacements performed between the two accesses in a reuse pair. If the reuse distance of a reuse pair is D, then the number of replacements is $R \approx D \bullet M$. Using the miss probability equation above, we can express the expected miss probability as function of reuse distance.

Expected miss probability:

$$m(D \cdot M) = 1 - (1 - 1/L)^{D \cdot M}$$

The calculated probability of a cache miss for the samples captured in Fig. 12.1 is shown in Fig. 12.4. Memory reference 7 will miss with a probability of $m(5 \cdot M)$, and memory reference 9 will miss by a probability of $m(3 \cdot M)$.

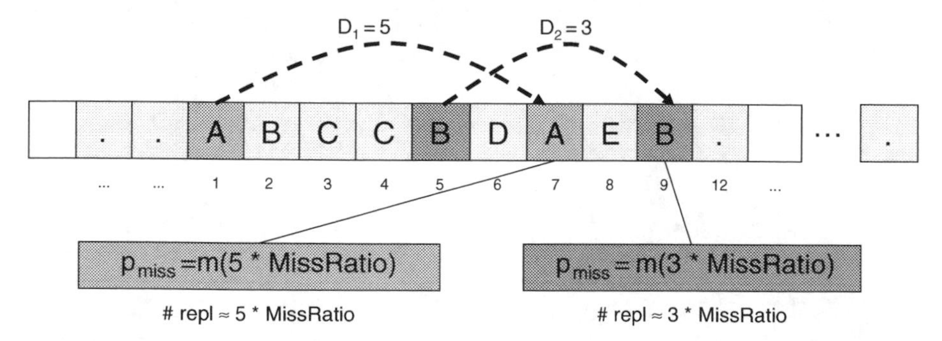

Fig. 12.4 Estimating the expected miss probability for each of the reuse pairs from Fig. 12.1 by calculating the expected number of replacements using the overall miss ratio M

If we capture N samples in a sample window, each with the reuse distance D_i, we can express the expected total number of misses for these samples as the sum of each sample's expected miss probability. We can also calculate the expected number of misses as the miss ratio for the window (M) multiplied by the number of samples in the window (N).

By equating these two ways of expressing the number of misses we obtain the following equation with the miss ratio M as the sole unknown.

Cache model equation:

$$\sum_{i=0}^{N} m(D_i \cdot M) = N \cdot M$$

The above equation can be readily solved for the miss ratio (M) for each sample window using numerical methods. Typically, it takes only a fraction of a second to find solutions for all sample windows. The miss ratio for any cache size can be computed using the same fingerprint by simply changing the value of L in the expected miss probability expression and resolving the cache model equation. The total miss ratio of an application as a function of cache size can then be calculated by averaging the miss ratios of all windows.

The accuracy of the miss ratio computed by StatCache has been evaluated by comparing the results of a traditional trace-driven cache simulator for many of the applications in the SPEC CPU 2000 benchmark suite. Figure 12.5 shows the miss ratio estimated by StatCache next to the miss ratio of the cache simulator for three of the SPEC CPU 2000 applications. More comprehensive comparisons can be found in the StatCache papers [1, 2].

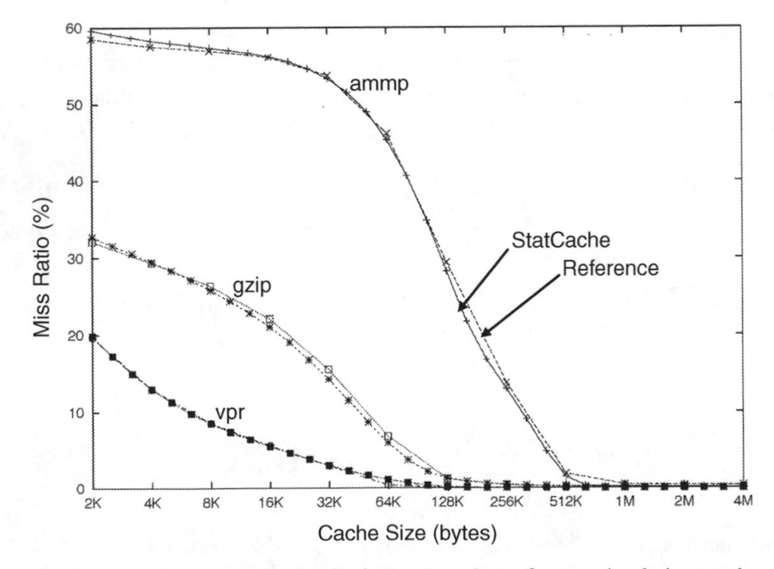

Fig. 12.5 Estimated miss ratios from StatCache compared to reference simulation results

12.7 StatStack: Modeling LRU Replacement Caches

StatStack extends the StatCache model to estimate an application's cache miss ratio for a fully associative cache with an LRU replacement policy, while using the same fingerprint input data. In the previous section it was shown that it is sufficient to estimate the number of replacements between the two accesses in a reuse pair to compute the miss probabilities for a random replacement cache. For LRU caches, we are instead looking to estimate the number of *unique* data objects accessed between the two accesses in a reuse pair, called the *stack distance* [10]. The stack distance directly tells us if the second of the two accesses in a reuse pair results in a cache hit for an LRU cache: if the stack distance is less than the size of the cache it is a hit, otherwise it is a miss. However, when the fingerprint is collected, we only count the number of memory accesses and not the number of unique data objects. The reuse distances in the fingerprint therefore needs to be translated into stack distances to model an LRU cache.

Figure 12.6 summarizes the overall strategy for translating reuse distances to stack distances. In the figure, we see that the stack distances of the reuse pair (*Start*=1, *End*=7) are 3, since there are 3 unique data objects (B, C, and D) accesses between *Start* and *End*. Further, we note that the number of unique data objects is equal to the number of reuse pairs whose first access is performed between *Start* and *End* and whose second access is performed after *End*, which are (4,12), (5,9), and (6,last).

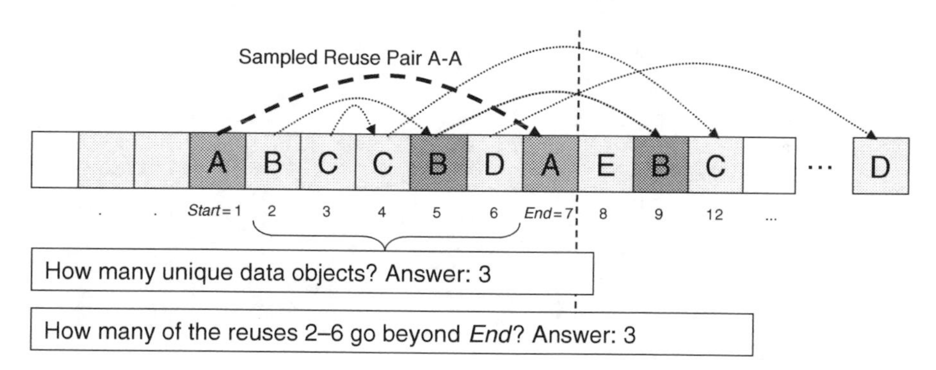

Fig. 12.6 The overall strategy estimating the expected stack distance based on sparse samples with reuse distance information

This relation between stack and reuse distance holds for all reuse pairs. However, a sparse fingerprint does not contain enough reuse information for the above method to be directly applicable. To cope with this lack of information we use an approximation, which relies on the assumption of homogeneous behavior in the sample windows and the representativeness of the fingerprints we collect. We use the distribution of the reuse distances in the fingerprint for a given sample window to approximate the reuse behavior of all the memory accesses in the window. This distribution allows us to estimate the likelihood that second access of a reuse pair whose first access is between *Start* and *End* is performed after *End*. The resulting likelihood indicates whether or not the access should be counted toward the number of unique data objects accessed between *Start* and *End*, which is the desired stack depth.

For example, in Fig. 12.6, the likelihood that the second access of the reuse pair starting with access 4 is performed after *End* is equal to the fraction of reuse pairs with a reuse distances greater than 7–4 = 3 in the reuse distance distribution. Adding up these likelihoods for all reuse pairs starting between *Start* and *End* gives us the expected stack distance of the reuse pair (*Start, End*), which can then be used to compute whether the *End* access results in a cache miss.

Using the above method, we can compute the miss ratio of a sampling window as follows: first, we use the fingerprint to estimate the distribution of reuse distances in the window. Second, we use this distribution to estimate the expected stack distances of all reuse pairs recorded for the window and build an expected stack distance distribution. Finally, to compute the miss ratio for a given cache size, we use the expected stack distance distribution to compute the fraction of reuse pairs with a stack distance greater than or equal to the cache size.

The miss ratios estimated by StatStack have been compared with those of traditional trace-driven cache simulators for all the 28 benchmarks of the SPEC CPU 2006 suite and show remarkably good accuracy. Five such comparisons are shown in Fig. 12.7. The full evaluation and a rigorous error and sensitivity analysis can be found in [5].

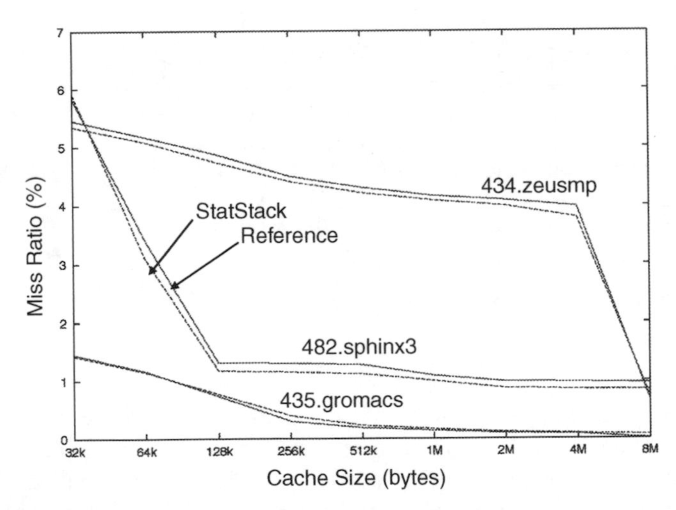

Fig. 12.7 Estimated miss ratios from StatStack compared to reference simulation results

12.8 StatCache-MP: Modeling Multithreaded Cache Interaction

StatCache-MP has extended the techniques of StatCache to also model the coherence activities for multiprocessors with coherent shared memory, such as SMP, NUMA, or multicore processor systems [3]. StatCache-MP can differentiate between many types of misses, such as read capacity misses, write capacity misses, read cold misses, write cold misses, read coherence misses, and write coherence misses. In this section we describe the extended sampling mechanism of StatCacheMP as well as its simple way of modeling coherence activities for MSI coherence protocol.

The probabilistic cache models described so far estimate cache behavior solely for single-threaded execution. In order to also model the inter-thread coherence activities and shared caches, the sparse information gathered in a fingerprint has to be extended in two ways. First, we need to distinguish between read and write accesses. This information can be extracted from the StatCache sample framework if the sampler records the type, read or write, for both the first and the second access of an access pair. Second, while measuring the reuse distance between the two accesses of an access pair, an intervening write access from another thread to the same cache line should be detected, as shown in Fig. 12.8. The identity of all threads with at least one such intervening write access is recorded for each access pair in the fingerprint. Cache misses caused by invalidation are counted as coherence misses only if the cache line otherwise would have remained in the cache at the time of the next local read or write access.

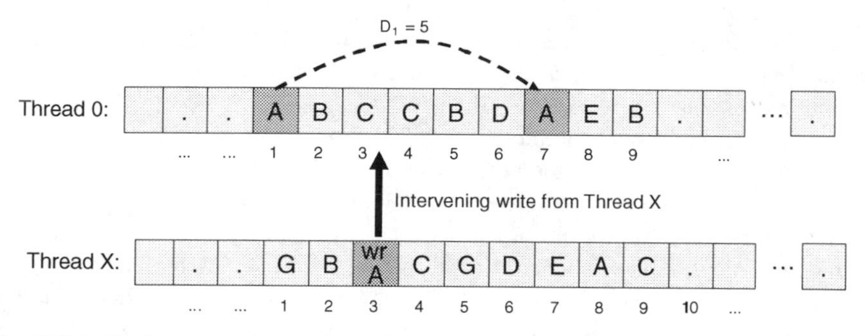

Fig. 12.8 In StatCache-MP intervening write accesses from other threads are detected during a reuse pair, and the identity of the intervening thread is recorded

As with the previously described probabilistic models, StatCacheMP can model several different cache topologies based on a fingerprint collected from one execution. Like StatCache, different cache sizes and cache-line sizes may be modeled, but StatCacheMP can also vary the degree of cache sharing in the cache hierarchy. For example, no coherence activity will be recorded if the intervening Thread X of Fig. 12.8 is modeled to share a cache with Thread 0. This is demonstrated in Fig. 12.9, where miss ratios for the same parallel application are modeled for three

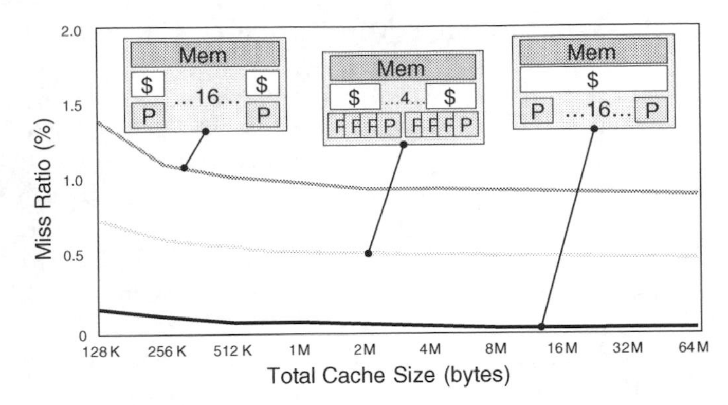

Fig. 12.9 Results of StatCacheMP modeling several different cache topologies from a single application fingerprint of Lu_NC from Splash2 running with 16 threads

different cache topologies using the same fingerprint. Most of the misses experienced for a topology with private caches disappear when all threads are modeled to share a cache.

For StatCacheMP, a change in the modeled architecture's degree of cache sharing requires a small alteration in the capacity miss equation for all the samples from threads sharing a cache. Further details, including an explanation of how cold misses are estimated, can be found in the StatCache-MP paper [3].

While allowing the efficient exploration of a huge design space, StatCache-MP also introduces possible sources of error to the estimations. A different cache configuration could result in an execution order with different coherence misses. This source of error is similar to that of trace driven or functional simulation in that it does not model contention or latency in the correct way and is effectively the result of an architecturally dependent interleaving of the execution. However, we have not found any major performance discrepancies due to this limitation.

StatCache-MP was evaluated using a trace-based evaluation methodology. First, a full memory trace from multithreaded execution was collected. This trace was used to drive the reference simulation and the same trace was sampled to produce the information for a sparse fingerprint. StatCache-MP produces miss-curves similar to Figs. 12.5 and 12.7 for all cache levels in the system, but can also distinguish between several different miss types. Figures 12.10 and 12.11 show a comparison between the reference simulation and the StatCache-MP model for eight applications and two different cache-line sizes. The overall miss ratio, as well as the classification into different miss categories, matches quite well between the two methods.

Note that the simple modeling of StatCache-MP only detects invalidation misses. No upgrade misses are detected. While upgrade misses also could cause latency for an application, especially for stronger memory models, their detection would lead to a more complicated sampling model. Repeated upgrade misses to the same cache line will also most likely appear in concert with invalidation misses, which are detected by the model.

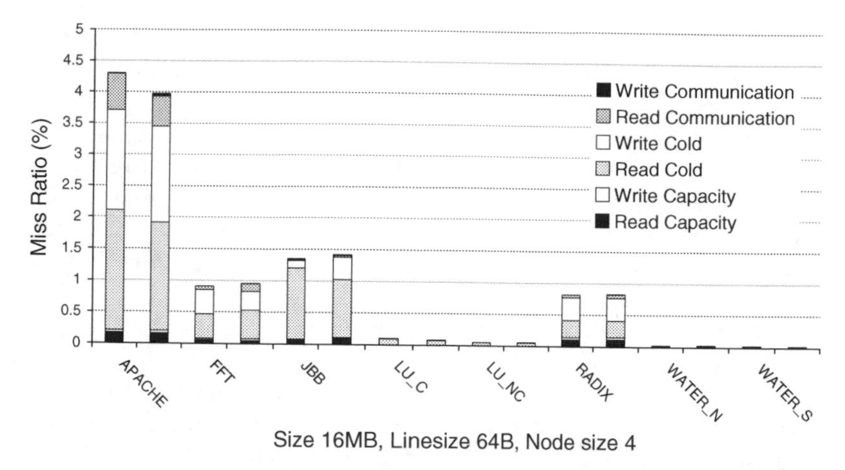

Fig. 12.10 StatCache-MP results (*left*) and reference simulation output (*right*) for four nodes, each with four cores sharing a 4 MB cache with 64 B cache lines. The different categories show very similar results for both methods

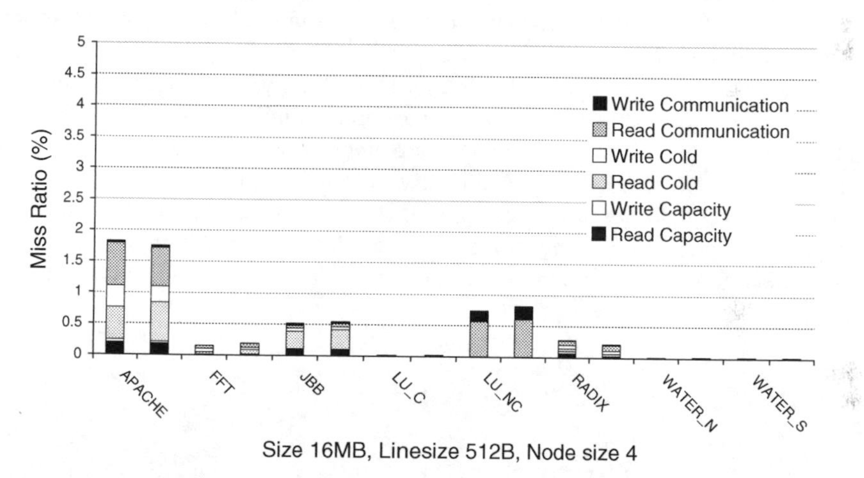

Fig. 12.11 StatCache-MP results (*left*) and reference simulation output (*right*) for 512 B cache lines. It is interesting to note that some applications experience lower cold and capacity misses with a large cache line (FFT), while the communication misses increase for others due to false sharing (LU_NC)

12.9 Acumem: Tools for Code Optimization

The three approaches described in this chapter have been extended and refined by Acumem to build a set of commercial tools targeting developers [7]. The technologies to model memory hierarchies and coherence protocols have been extended with heuristics to identify performance problems commonly found in applications. The Acumem tools list the performance problems on a per-instruction, per-loop, or per-application basis, along with their corresponding source code and performance statistics. This technology has also been used to give profile-guided feedback to compilers and dynamic binary optimization schemes.

As with the techniques discussed in this chapter, the Acumem sampler gathers an application fingerprint at runtime on a host computer. The collected information is sparse but contains significantly more information per sample, for more accurate modeling and detection of performance problems. For example, call stack information as well as some instruction sequences are stored in the fingerprint to enable deeper analysis.

The Acumem tools collect fingerprints from single-threaded and multi-threaded executions in a completely transparent manner. In fact, the fingerprint collector can attach to an already running process with many threads, collect data for a while, and then detach from the process, which will continue its execution undisturbed. The sampling approach is also more sophisticated than that outlined for StatCache or StatStack. The sample rate is dynamically adjusted with the number of samples taken gradually decreasing as the execution progresses. At the same time, some of the earlier captured fingerprint information is discarded for long-running applications to obtain a more representative set of samples. The application fingerprint information is captured in a file that is typically a couple of megabytes, regardless of the execution time.

The fingerprint collection is based on dynamic binary rewriting, which enables the use of unmodified binaries in a language- and compiler-agnostic manner. Typically, long-running applications experience only a 10% execution overhead. In between sample windows the original binary runs at full speed.

Acumem technology extends beyond modeling random caches, LRU caches, and various topologies of coherent caches, to model hardware prefetching, cache efficiency and communication efficiency. These techniques model how much of each

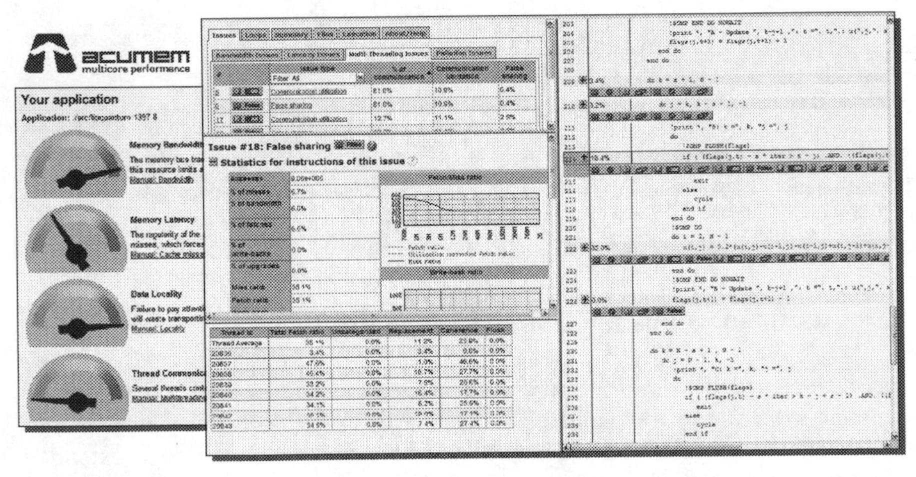

Fig. 12.12 Acumem's tools pinpoint performance problems and provide intuitive feedback to programmers

cache line is actually used before eviction, called cache utilization, as well as the fraction of the data of a write-back cache line that is dirty, called write-back utilization. The StatCacheMP approach has also been refined to model upgrade misses and measure the fraction of a communicated cache line that is actually used by a consumer.

These features enable Acumem tools to provide much richer and more precise feedback to the programmer than the more limiting traditional hardware-counter-based approach. The tools not only identify the exact operations that cause misses in a cache system or those instructions that waste memory bandwidth but also reveal the reason for the misses and suggest possible fixes. Acumem's tools can often estimate the fraction of cache misses and DRAM accesses removed by each proposed fix, which is essential in determining if a fix is worth the effort (Fig. 12.12).

12.10 Appendix: Example of Trap-Based Sampler Implementation

The initial StatCache implementation was based on static code instrumentation [1]. When sampling every 10,000th memory reference, this approach generated a slowdown of 600%. This overhead motivated the development of the more efficient sampling algorithm described in [2] and this appendix, which both samples fewer memory references and enables the use of OS watchpoints. Combined these reduce the sampling overhead to about 30%. While the sample information collected is architecturally independent, the sampling implementation itself is of course highly dependent of the ISA as well as the underlying architecture mechanisms of the architecture and operating system.

As indicated earlier, sampling of reuse distance fingerprint information can be implemented with the following three support mechanisms.

1. *Sample selector.* The sample selector randomly selects memory-reference instructions and starts monitoring the cache-line-sized piece of data that is read or written by that instruction.
2. *Watchpoint mechanism.* The watchpoint mechanism monitors the selected cache-line-sized pieces of memory and informs the sampling runtime system when the application accesses monitored memory.
3. *Memory reference counter.* The sampling runtime system uses the memory reference counter to calculate reuse distances.

The StatCache sampling approach is based on mechanisms available in nonprivileged mode execution in Solaris, but similar mechanisms are available in most operating systems. The runtime system of the tool was built as a module that is linked to the analyzed application, which requires relinking, but no changes to the object code. The initial implementation did not consider library code, but the lack of need to instrument code should make it feasible to also include the impact of such code on cache behavior.

Though sampling is easy in theory, the *sample selector* was the hardest to implement in practice. We found that the results were very sensitive to sample skew—much more so than to, for example, errors in the measured reuse distance. To accurately schedule sampling, the UltraSPARC data cache read operations counter was used to gather samples. This register can generate interrupts on overflow and, therefore, by resetting it to a value of `UINT32_MAX`—sample_distance, it was possible to stop the processor after an arbitrary number of load instructions. On overflow, the process was sent a signal, caught by the runtime system. However, the overflow traps of the counter were not very precise, which meant that the processor could continue to execute instructions after the overflow occurred before stopping.

To address this, we set the counter to stop the processor a number of load instructions before the intended sample position. When the trap was taken, the data cache load counter was read to determine how many instructions remain to execute before the next sample. The runtime system then single-stepped through the code to reach the sample point. Effective addresses were calculated by decoding the instructions and reading the appropriate register values. Implementing single-stepping turned out to be somewhat tricky, as the operating system had problems with such things as handling execution watchpoints in the delay slot of branches with the annul flag set.

The Solaris operating system also provides the needed *watchpoint system*. It uses the page protection system to catch loads and stores to certain memory locations. This watchpoint system was used to set up watchpoints that trapped on the instruction that touched a selected cache-line-sized data object the next time.

A *memory reference counter* is needed to measure reuse distances. Solaris provided the per-process virtualized read-access counter DC_rd. The operating system saved the values on context switches, while hardware gates made sure only user-level read accesses were counted. Unfortunately, the UltraSPARC III used for implementation was not able to count both load and store instructions simultaneously. Therefore we used the data load counter. The post-processing system would then scale the reuse distances using the relative frequencies of dynamic load and store instructions, which were recorded at run time from the samples.

Combined, these techniques enabled the implementation of a relatively unobtrusive, accurate, and low-overhead data reuse sampler. This sampling approach has been further developed for the commercial tools provided by Acumem to enhance the quality of the data collected, improve portability across different architectures, and reduce the execution overhead.

References

1. Berg, E., Hagersten, E.: StatCache: A probabilistic approach to efficient and accurate data locality analysis. In: *Proceedings of the 2004 IEEE International Symposium on Performance Analysis of Systems and Software (ISPASS-2004)*, Austin, TX, USA, March (2004).
2. Berg, E., Hagersten, E.: Fast data-locality profiling of native execution. In: *Proceedings of ACM SIGMETRICS 2005*, Ban, Canada, June (2005).
3. Berg, E., Zeffer, H., Hagersten, E.: A statistical multiprocessor cache model. In: *Proceedings of the 2006 IEEE International Symposium on Performance Analysis of Systems and Software (ISPASS-2006)*, Austin, TX, USA, March (2006).

4. Burger, D., Austin, T.M.: The SimpleScalar tool set, version 2.0. In: *SIGARCH* (1997).
5. Eklöv, D., Hagersten, E.: StatStack: Efficient modeling of LRU caches. In: *Proceedings of the 2010 IEEE International Symposium on Performance Analysis of Systems and Software (ISPASS-2010)*, White Plains, NY, USA, March (2010).
6. Ferdinand, C., Wilhelm, R.: Efficient and precise cache behavior prediction for real-time systems. *J Real-Time Syst* **17**(2–3), (1999).
7. Hagersten, E., Nilsson, M., Vesterlund, M.: Improving cache utilization using Acumem VPE in tools for high performance computing. In: *Proceedings of the 2nd International Workshop on Parallel Tools for High Performance Computing*, HLRS, Stuttgart, Springer, July (2008).
8. Luk, C. et al.: *Pin: Building Customized Program Analysis Tools with Dynamic Instrumentation*. ACM Press, New York, NY, USA, (2005).
9. Magnusson, P. et al.: Simics: A full system simulation platform. *IEEE Comput Soc* **35**(2) 50–58, (2005).
10. Mattson, R.L. et al.: Evaluation techniques for storage hierarchies. *IBM Syst J* **9**(2), (1970).
11. Noordergraaf, L., Zak, R.: SMP system interconnect instrumentation for performance analysis. In: *Proceedings of the 2002 ACM/IEEE Conference on Supercomputing*, Baltimore, Maryland, USA (2002).
12. Intel, Intel VTuen Performance Analyzer, http://software.intel.com/en-us/intel-vtune/, June (2010).

Chapter 13
Statistical Simulation

Lieven Eeckhout and Davy Genbrugge

Abstract The idea of statistical simulation is to measure a number of program execution characteristics from a real program execution through profiling, then generate a synthetic trace from it, and finally simulate that synthetic trace as a proxy for the original program. The important benefit is that the synthetic trace is much shorter compared to a real program trace. Current frameworks report simulation speedups of up to four orders of magnitude compared to detailed simulation with average (absolute) prediction errors in the order of a few percents. This beneficial speed-accuracy trade-off makes statistical simulation a valuable tool in the computer designer's toolbox, especially for early-stage processor design space explorations.

13.1 Introduction and Motivation

Computer system design is an extremely time-consuming, complex process, and simulation has become an essential part of the overall design activity. Simulation is performed at many levels, from circuits to systems, and at different degrees of detail as the design evolves. Consequently, the designer's toolbox holds a number of evaluation tools, often used in combination, that have different complexity, accuracy, and simulation speed properties.

For simulation at the microarchitecture level, detailed models of register transfer activity are typically employed. These simulators track instructions and data on a clock cycle basis and typically provide detailed models for features such as instruction issue mechanisms, caches, load/store queues, and branch predictors, as well as their interactions. For input, microarchitecture simulators take sets of benchmark programs including both standard and company proprietary suites. These benchmarks may each contain hundreds of billions or even trillions of dynamically

L. Eeckhout (✉)
ELIS – Ghent University, Sint-Pietersnieuwstraat 41, B-9000 Gent Belgium
e-mail: lieven.eeckhout@elis.ugent.be

R. Leupers, O. Temam (eds.), *Processor and System-on-Chip Simulation*,
DOI 10.1007/978-1-4419-6175-4_13,

executed instructions, and typical simulators run many orders of magnitude slower than real processors. The result is a relatively long runtime for even a single simulation.

However, processor simulation at such a high level of detail is not always appropriate nor is it called for. For example, early in the design process when the design space is being explored and a high-level microarchitecture is being determined, too much detail is unnecessary. When a processor microarchitecture is initially being defined, a number of basic design decisions need to be made. These decisions involve basic cycle time and instruction per cycle (IPC) trade-offs, cache and predictor sizing trade-offs, and performance/power trade-offs. At this stage of the design process, detailed microarchitecture simulations of specific benchmarks are not feasible for a number of reasons. For one, the detailed simulator, itself, takes considerable time and effort to develop. Second, benchmarks restrict the studied application space being evaluated to those specific programs. To study a fairly broad design space, the number of simulation runs can be quite large. Finally, highly accurate performance estimates are illusory, anyway, given the level of design detail that is actually known.

Similarly, for making system-level design decisions, where a processor (or several processors) may be combined with many other components, a very detailed simulation model is often unjustified and/or impractical. Even though the detailed processor microarchitecture may be known, simulation complexity increases with an increasing number of processors; also, larger benchmark programs are typically required for studying system-level behavior.

In this chapter, we describe statistical simulation which can be used to overcome many of the shortcomings of detailed simulation for those situations where detailed modeling is impractical, or overly time consuming. Statistical simulation measures a well-chosen set of characteristics during program execution, generates a synthetic trace with those characteristics, and simulates the synthetic trace. If the set of characteristics reflects the key properties of the program's behavior, accurate performance/power predictions can be made. The statistically generated synthetic trace is several orders of magnitude smaller than the original program execution; hence, simulation finishes very quickly. The goal of statistical simulation is to reduce design space exploration time; it is not intended to replace detailed simulation but to be a useful complement. Statistical simulation can be used to identify a region of interest in a large design space that can, in turn, be further analyzed through slower but more accurate and more detailed architectural simulations. In addition, statistical simulation requires relatively little new tool development effort. Finally, it provides a simple way of modeling superscalar processors as components in large-scale systems where very high detail is not required or practical.

This chapter is organized as follows. We first discuss the general framework of statistical simulation and make a distinction between statistical simulation applied to single-core versus multi-core processor modeling. We also discuss possible applications for statistical simulation. Subsequently, we present some results to illustrate accuracy and speed. The description in this chapter is tied to our own prior work in this area; however, several other groups have made contributions in statistical simulation and modeling, which we describe toward the end of the chapter.

13.2 Statistical Simulation

13.2.1 Overview of the Paradigm

Statistical simulation consists of three steps as is shown in Fig. 13.1. In the first step, a collection of execution characteristics is measured for a given computer program—this is called statistical profiling. Subsequently, the obtained statistical profile is used to generate a synthetic trace. Finally, this synthetic trace is simulated on a trace-driven statistical simulator. We now discuss these three steps in more detail and make a distinction between single-core and multi-core processor modeling.

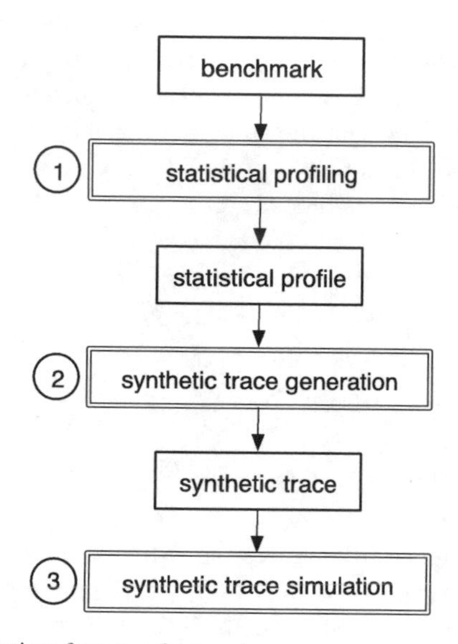

Fig. 13.1 Statistical simulation: framework overview

13.2.2 Statistical Simulation of Single-Core Processors

We start with discussing statistical simulation for single-core processors. This discussion is largely based on the work done by Eeckhout et al. [5].

13.2.2.1 Statistical Profiling

Statistical profiling characterizes the program execution at hand in a statistical way, i.e., the various program characteristics of interest are modeled through distributions and probabilities. Statistical profiling can be done using specialized trace-driven simulation (e.g., as an extension to a functional simulation) as well as through instrumentation tools such as ATOM [25] or Pin [18]. This is done by running the (instrumented) program at hand with a given input; in other words, the statistical profile captures both the program and its input. Collecting the profile is done quickly compared to detailed cycle-level simulation. We make a distinction between characteristics that are independent of versus dependent on the underlying microarchitecture.

Microarchitecture-Independent Characteristics

The core structure of the statistical profile is the statistical flow graph (SFG) which models the program's control flow behavior as a Markov chain. Consider, for example, the following basic block execution sequence AABAABCABC; Figure 13.2 shows the corresponding first-order SFG. Each node in the SFG represents a basic block and its history. This is shown through the node labels A|A, B|A, A|B, etc.: A|B means basic block A given that basic block B is its precedent in the execution. The percentages at the edges represent the transition probabilities between the nodes. For example, there is a 33.3 and 66.6% probability to execute basic block A and C, respectively, after executing basic blocks AB.

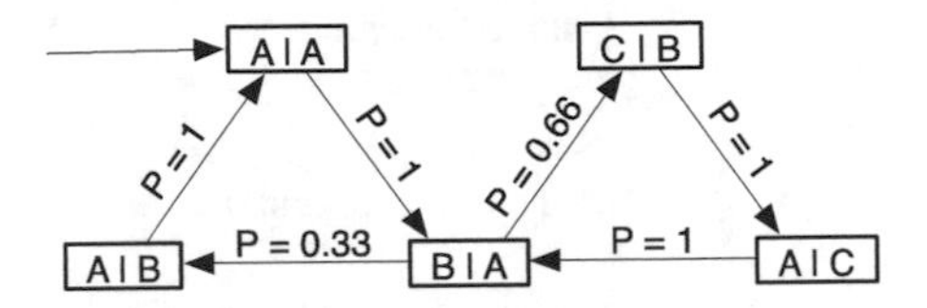

Fig. 13.2 An example first-order statistical flow graph (SFG)

All other program execution characteristics that are included in the statistical profile are collected for each node in the SFG. This implies that the same basic block appearing multiple times in the SFG (with different histories) may have different characteristics. This enables modeling path-dependent program behavior.

For each basic block associated with a node in the SFG we record the instruction type for each instruction. We classify the instruction types in a number of instruction classes according to their semantics, such as load, store, conditional branch, integer ALU, multiply, floating-point ALU. For each instruction, we record the number of input registers. For each input register we also record the dependency distance which is the number of dynamically executed instructions between the production of a register value (register write) and its consumption (register read). We only consider read-after-write (RAW) dependencies since our focus is on out-of-order architectures in which write-after-write (WAW) and write-after-read (WAR) dependencies are dynamically removed through register renaming as long as enough physical registers are available. Although not done so far, this approach could be extended to also include WAW and WAR dependencies to account for a limited number of physical registers. Note that recording the dependency distance requires storing a distribution since multiple dynamic executions of the same static instruction can result in different dependency distances. In theory, this distribution could be very large due to large dependency distances. In practice though, we can limit this distribution because the processor's instruction window is limited anyway.

Note that the characteristics discussed so far are independent of any microarchitecture-specific organization. In other words, these characteristics do not rely on assumptions related to processor issue width, window size, number of ALUs, instruction execution latencies, etc. They are therefore called microarchitecture-independent characteristics.

Microarchitecture-Dependent Characteristics

In addition to the above characteristics we also measure a number of characteristics that are related to locality events, such as cache hit/miss and branch predictability behavior. These characteristics are hard to model in a microarchitecture-independent way (although recent efforts have made significant progress [2], see also Chapter 12). Therefore a pragmatic approach is taken and characteristics for specific branch predictors and specific cache configurations are computed through specialized cache and branch predictor simulation. Note that although this approach requires the simulation of the complete program execution for specific branch predictors and specific cache structures, this does not limit its applicability. Indeed, it is well known that multiple cache configuration can be simulated in parallel using a single-pass algorithm [11].

We collect the following three characteristics per branch: (i) the probability for a taken branch, which will be used to limit the number of taken branches that are fetched per clock cycle during synthetic trace simulation; (ii) the probability for a fetch redirection, which corresponds to a branch target misprediction in conjunction with a correct taken/not-taken prediction for conditional branches—this typically results in (a) bubble(s) in the pipeline; and (iii) the probability for a branch misprediction: a BTB miss for indirect branches and a taken/not-taken misprediction for conditional branches.

The cache characteristics typically consist of the following probabilities: (i) the L1 instruction cache miss rate, (ii) the L2 cache miss rate due to instructions, (iii) the L1 data cache miss rate, (iv) the L2 cache miss rate due to data accesses only, (v) the instruction translation lookaside buffer (I-TLB) miss rate, and (vi) the data translation lookaside buffer (D-TLB) miss rate. Extending to additional levels of caches and TLBs is trivial.

Accurate memory data flow modeling

In our previous work [9], we note that the memory data flow model as described above, i.e., the way how the data cache behavior is modeled, is simplistic in at least three significant ways. First, it assigns hits and misses to loads and stores and does not model delayed hits. A delayed hit, i.e., a hit to an outstanding cache line, is modeled as a cache hit although it should see the latency of the outstanding cache line. Second, it does not model load bypassing and load forwarding, i.e., it is assumed that loads never alias with preceding stores. Third, it does not model cache miss patterns or the number of instructions in the dynamic instruction stream between misses. However, cache miss patterns have an important impact on the available memory-level parallelism. Independent long-latency load misses that are close enough to each other in the dynamic instruction stream to make it into the reorder buffer together overlap their execution, thereby exposing memory-level parallelism (MLP). Given the significant impact of memory-level parallelism on overall performance, a performance model lacking adequate MLP modeling may yield large performance prediction errors. Based on these observations, we proposed three enhancements: (i) cache miss correlation, or measuring cache statistics conditionally dependent on the global cache hit/miss history, for modeling cache miss patterns and memory-level parallelism, (ii) cache line reuse distributions for modeling accesses to outstanding cache lines, and (iii) through-memory read-after-write dependency distributions for modeling load forwarding and bypassing. For more details, we refer to [9].

13.2.2.2 Synthetic Trace Generation

The second step in the statistical simulation methodology is to generate a synthetic trace from the statistical profile. The synthetic trace generator takes as input the statistical profile and produces a synthetic trace that is fed into the statistical simulator. Synthetic trace generation uses random number generation for generating a number in [0,1]; this random number is then used with the inverse cumulative distribution function to determine a program characteristic. The synthetic trace is a linear sequence of synthetic instructions. Each instruction has an instruction type, a number of source operands, an inter-instruction dependency for each register input (which denotes the producer for the given register dependence), I-cache miss info, D-cache miss info (in case of a load), and branch miss info (in case of a branch). The locality miss events are just labels in the synthetic trace describing whether the load is an L1 D-cache hit, L2 hit, or L2 miss and whether the load generates a TLB miss. Similar labels are assigned for the I-cache and branch miss events.

13.2.2.3 Synthetic Trace Simulation

The trace-driven simulation of the synthetic trace is very similar to the trace-driven simulation of real program traces. However, the synthetic trace simulator needs to model neither branch predictors nor caches—this is part of the trade-off that dramatically reduces development and simulation time. However, special actions are needed during synthetic trace simulation for the following cases.

When a branch is mispredicted in an execution-driven simulator, instructions from an incorrect path are fetched and executed. When the branch gets executed, it is determined whether the branch was mispredicted. In case of a misprediction, the instructions down the pipeline need to be squashed. A similar scenario is implemented in the synthetic trace simulator: when a mispredicted branch is fetched, the pipeline is filled with instructions from the synthetic trace as if they were from the incorrect path. When the branch gets executed, the synthetic instructions down the pipeline are squashed and synthetic instructions are re-fetched (and now they are considered from the correct path).

For a load, the latency will be determined by whether this load is an L1 D-cache hit, an L1 D-cache miss, an L2 cache miss, or a D-TLB miss. For example, in case of an L2 miss, the access latency to main memory is assigned.

In case of an I-cache miss, the fetch engine stops fetching for a number of cycles. The number of cycles is determined by whether the instruction causes an L1 I-cache miss, an L2 cache miss, or a D-TLB miss.

13.2.3 Statistical Simulation for Multi-core Processors

The approach to statistical simulation as described above models memory data flow behavior through cache miss statistics, i.e., the statistical profile captures the cache miss rates of the various levels in the cache hierarchy. Although this is sufficient for the statistical simulation of single-core processors, it is inadequate for modeling chip multiprocessors with shared resources in the memory hierarchy, such as shared L2 and/or L3 caches, shared off-chip bandwidth, interconnection network, and main memory. Co-executing programs on a chip multiprocessor affect each other's performance through conflicts in the shared resources, and the level of interaction between co-executing programs is greatly affected by the microarchitecture—the amount of interaction can be very different on one microarchitecture compared to another. Hence, cache miss rates profiled from single-threaded execution are unable to model conflict behavior in shared chip-multiprocessor resources when co-executing multiple programs. We therefore need a different approach for multi-core processors: we need to model memory access behavior in the synthetic traces in a way that is independent of the memory hierarchy so that conflict behavior among co-executing programs can be derived during multi-core simulation of the synthetic traces. We refer the interested reader to [10] for an in-depth description; we now briefly summarize the key ideas in the following sections.

13.2.3.1 Microarchitecture-Independent Statistical Profiling

To enable the simulation of shared caches in a multi-core processor, we add two additional program characteristics in the statistical profile, namely the *cache set profile* and the *per-set LRU stack depth profile*. For doing this, we assume a large shared cache, i.e., the largest cache one may be potentially interested in during design space exploration. The cache access patterns can then be derived for smaller caches from the profile measured for the largest cache. When computing this profile, we determine the cache set that is to be accessed in the largest cache of interest for every memory reference. In addition, we also determine the LRU stack depth in the given set. The maximum LRU stack depth kept track of during profiling is a with a being the associativity of the largest shared cache of interest. The cache set and per-set LRU stack depth profile can be used to estimate cache miss rates for caches that are smaller than the largest shared cache of interest. All accesses to an LRU stack depth of at least a' will be cache misses in an a'-way set-associative cache. Likewise, all accesses to sets s and $s + S/2$ for a cache with S sets will result in accesses to set s for a cache with $S/2$ sets.

When generating a synthetic trace we probabilistically generate a cache set and LRU stack depth accessed for each memory reference based on the above profiles. Simulating the synthetic trace on a CMP with a shared cache then requires that we effectively simulate the shared cache(s). In statistical simulation of a single-core processor on the other hand, caches do not need to be simulated since cache misses are simply flagged as such in the synthetic trace. Statistical simulation of a multi-core processor with a shared cache, on the other hand, requires that the caches are simulated in order to model shared cache conflict behavior.

13.2.3.2 Modeling Time-Varying Program Behavior

A critical issue to the accuracy of statistical simulation for modeling multi-core processor performance is that the synthetic trace has to capture the original program's time-varying phase behavior. The reason is that overall performance is affected by the phase behavior of the co-executing programs: the relative progress of a program is affected by the conflict behavior in the shared resources [26]. For example, extra cache misses induced by cache sharing may slow down a program's execution. This one program running relatively slower may result in different program phases co-executing with the other program(s), which, in turn, may result in different sharing behavior and thus faster or slower relative progress.

To model the time-varying behavior we divide the entire program trace into a number of instruction intervals; an instruction interval is a sequence of consecutive instructions in the dynamic instruction stream. We then collect a statistical profile per instruction interval and generate a synthetic mini-trace. Coalescing these mini-traces yields the overall synthetic trace. The synthetic trace then captures the original trace's time-varying behavior.

13.2.3.3 Multi-threaded Workloads

So far, we were only concerned about multi-program workloads, i.e., workloads that consist of multiple independent programs that co-execute. For multi-program workloads, the only source of interaction is through the shared resources. Multi-threaded workloads add another form of interaction, namely through synchronization and cache coherence. Nussbaum and Smith [22] extend the uniprocessor statistical simulation method to multi-threaded programs running on shared-memory multiprocessor (SMP) systems. To do so, they extended statistical simulation to model synchronization and accesses to shared memory. Hughes and Li [13] more recently proposed the notion of a synchronized statistical flow graph to incorporate inter-thread synchronization.

13.2.4 Applications

Statistical simulation has a number of potential applications.

- *Design space exploration.* An important application for statistical simulation is processor design space exploration. Statistical simulation does not aim at replacing detailed cycle-accurate simulation, primarily because it is less accurate (e.g., it does not model cache accesses along mispredicted paths, it simulates an abstract representation of a real workload). Rather, it aims at providing guidance for decision making early in the design process. Fast decision making is important to reduce the time to market when designing a new microprocessor.
  A single statistical profile and a single synthetic trace can be used to explore a very large part of the design space. The reason is that most of the characteristics are microarchitecture-independent, and the microarchitecture parameters that can be varied include processor width, pipeline depth, number of functional units, buffer sizes, cache size, and associativity—these are the parameters one is most likely interested in exploring in the early stages of the design. Only a limited number of parameters require recomputing the statistical profile (and the synthetic trace), e.g., cache line size, branch predictor.
- *Program behavior characterization.* Another interesting application for statistical simulation is program characterization. When validating the statistical simulation methodology in general and the characteristics included in the statistical profile more in particular, it becomes clear which program characteristics must be included in the profile for attaining good accuracy. That is, this validation process distinguishes program characteristics that influence performance from those that do not.
- *Workload characterization.* The statistical profile can be viewed of as an abstract workload model. In other words, one can compare workloads by comparing their respective statistical profiles. Moreover, one can explore the workload space by varying the various characteristics in the statistical profile (which are hard to vary, if possible, using real workloads) and relate these variations in program behavior to variations in performance. This can lead to valuable insights—Oskin et al. [23] provide such a case study.
- *Large system evaluation.* As discussed so far in this chapter, statistical simulation addresses the time-consuming simulation problem of single-core and multi-core processors. However, for larger systems containing several processors, such as multi-chip servers, clusters of computers and datacenters, simulation time is even a bigger problem. Statistical simulation may be an important and interesting approach for such large systems.

13.3 Evaluation

We now present some evaluation results for statistical simulation in terms of accuracy and simulation speed. We also illustrate the power of statistical simulation through a case study from a practical research study.

13.3.1 Accuracy

Figure 13.3 shows IPC (instructions retired per clock cycles) as obtained through detailed cycle-level simulation as well as through statistical simulation for an aggressive superscalar out-of-order processor. The results are taken from [9] and assume 10 billion instructions taken from the SPEC CPU2000 benchmarks. The synthetic traces generated through statistical simulation count 1 million instructions—a four-order of magnitude reduction in simulation time. These results show that statistical simulation is accurate compared to detailed simulation with IPC prediction errors varying between −2.4 and 3.8% (average absolute prediction error of 1.6%).

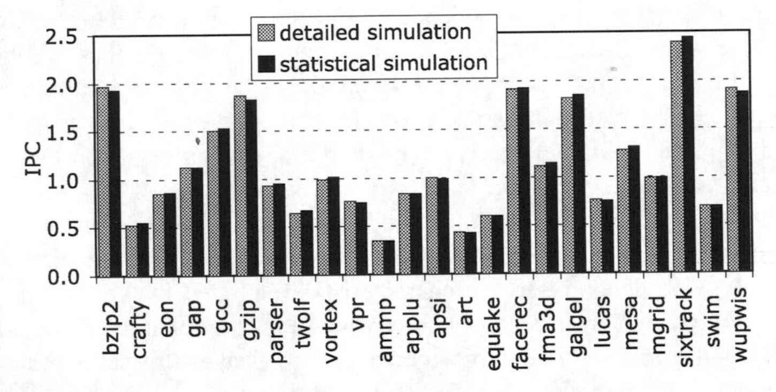

Fig. 13.3 Accuracy for statistical simulation versus detailed simulation for a single-core processor

Figure 13.4 shows similar results for a multi-core processor executing a heterogeneous multi-program workload. The results are taken from [10] and consider four 8-program workloads. The graphs show system throughput (STP) and average normalized turnaround time (ANTT)[1]; STP is a higher-is-better metric and represents performance from a system perspective (STP equals weighted speedup [24]), whereas ANTT is a smaller-is-better metric quantifying performance from a user perspective (ANTT is the reciprocal of the hmean metric [19]). The important observation here is that statistical simulation is able to accurately track detailed simulation. The average prediction error is 4.9 and 5.6% for STP and ANTT, respectively. For example,

[1]STP is a system-oriented metric and ANTT is a user-oriented metric; see [8] for a detailed description on these multi-program performance metrics.

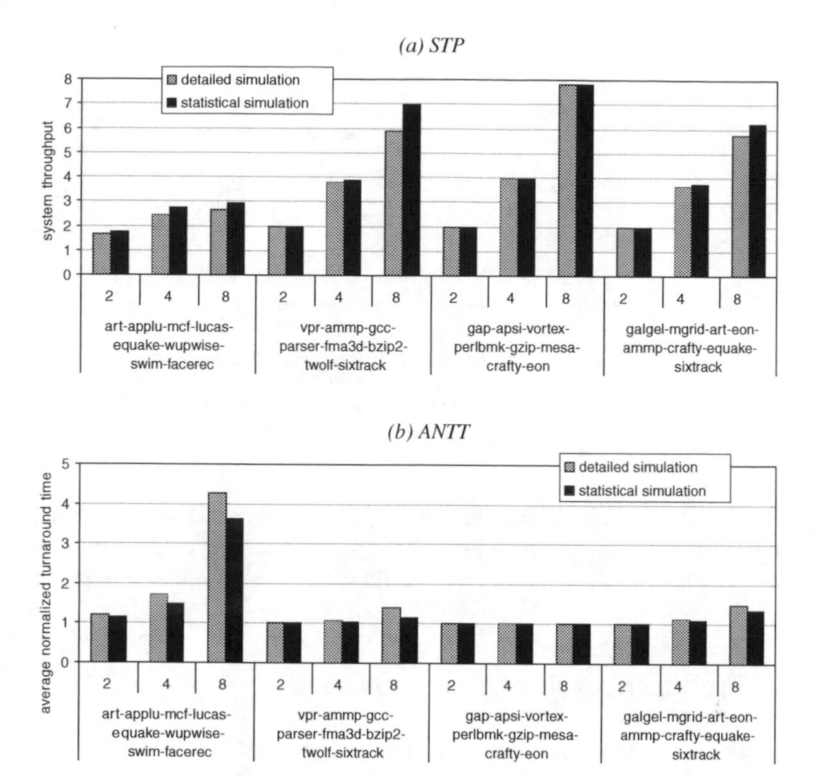

Fig. 13.4 Accuracy for statistical simulation versus detailed simulation in terms of STP and ANTT for a multi-core processor with two, four, and eight cores

it shows that eight cores has a detrimental effect on turnaround time while only marginally improving system throughput for the leftmost workload mix; this suggests that conflict behavior in the shared resources (L2 cache, off-chip bandwidth, and main memory) deteriorates per-program performance. For the other three workload mixes on the other hand, we observe significant benefits in system throughput with only minor effects on job turnaround time.

13.3.2 Simulation Speed

Figure 13.5 shows the average (absolute) IPC prediction error as a function of the length of the synthetic trace. For a single-program workload, the prediction error stays almost flat, i.e., increasing the size of the synthetic trace beyond 1M instructions does not increase prediction accuracy. For multi-program workloads on the other hand, the prediction accuracy is sensitive to the synthetic trace length, and sensitivity increases with the number of programs in the multi-program workload. This can be understood intuitively: the more programs there are in the multi-program workloads, the longer it takes before the shared caches are warmed up and the

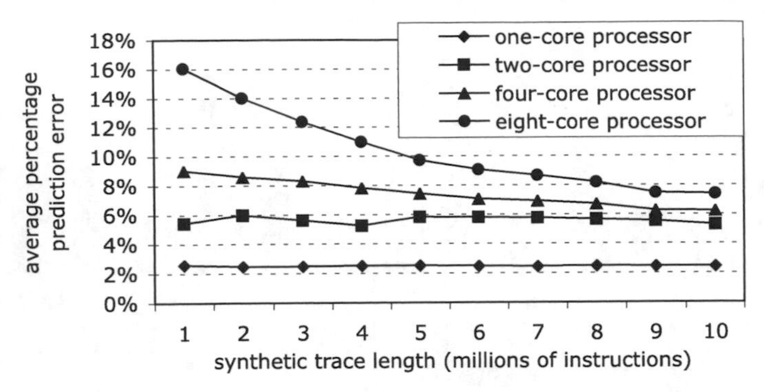

Fig. 13.5 Evaluating simulation speed: average IPC prediction error as a function of synthetic trace length

longer it takes before the conflict behavior is appropriately modeled between the co-executing programs. The results in Fig. 13.5 demonstrate that 10M-instruction synthetic traces yield accurate performance predictions, even for eight-core processors.

13.3.3 Case study: Design Space Exploration

For demonstrating the value of statistical simulation for exploring new architecture paradigms, we now consider a case study in which we evaluate performance of a multi-core processor in combination with 3D stacking [17]—this case study is taken from [10]. We compare the performance of a four-core processor with a 16 MB L2 cache connected to off-chip DRAM memory through a 16-byte wide memory bus against an eight-core processor with integrated on-chip DRAM memory (through 3D stacking) and no L2 cache and a 128-byte wide memory bus. We assume a 150-cycle access time for external memory and a 125-cycle access time for 3D-stacked memory. Figure 13.6 quantifies STP and ANTT for these two design points for four different eight-benchmark mixes. The eight-core processor with 3D stacked memory achieves substantially higher system throughput than the four-core processor with the on-chip L2 cache. This increase in system throughput is offset by a decrease in job turnaround time. The improvement in system throughput and the reduction in turnaround time varies across workload mixes, and statistical simulation can accurately track performance differences between both design alternatives.

13.4 Other Work in Statistical Simulation

Statistical simulation for modeling single-core processor performance has received considerable interest over the past decade. Noonburg and Shen [20] model a program execution as a Markov chain in which the states are determined by the microarchitecture and the transition probabilities by the program. Iyengar et al. [14] use a

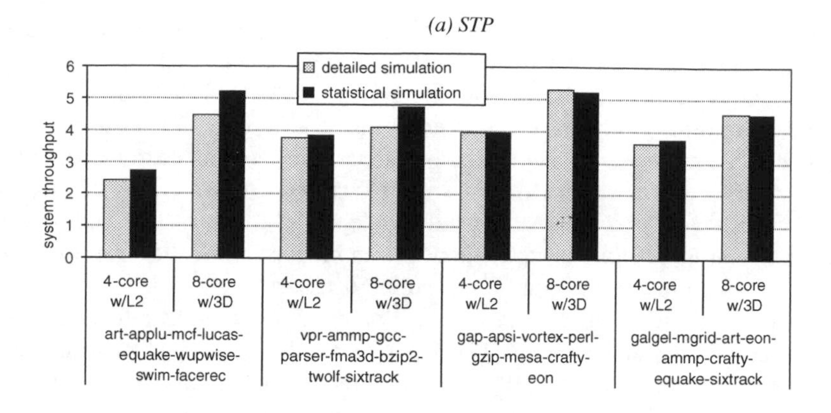

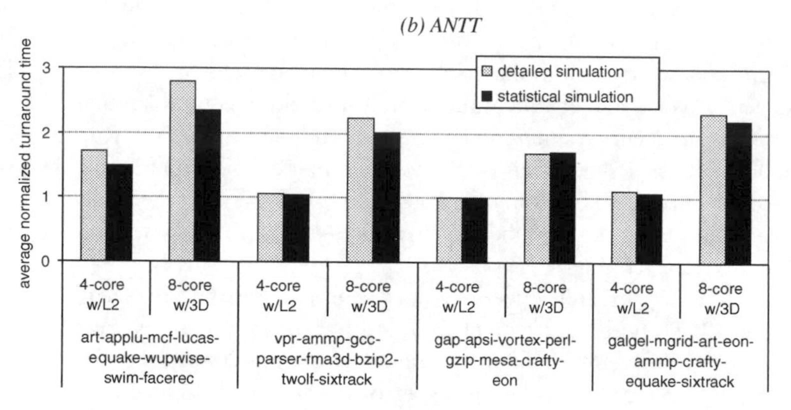

Fig. 13.6 Case study that considers a four-core processor with an on-chip L2 cache and off-chip main memory versus an eight-core processor with 3D stacked memory and no L2 cache

statistical control flow graph to identify representative trace fragments; these trace fragments are then coalesced to form a reduced program trace. The statistical simulation framework considered in this chapter is different in its setup: we generate a synthetic trace based on a statistical profile. The initial models proposed along this line were fairly simple [3, 6, 7]: the entire program characteristics in the statistical profile are typically aggregate metrics, averaged across all instructions in the program execution. Oskin et al. [23] propose the notion of a graph with transition probabilities between the basic blocks while using aggregate statistics. Nussbaum and Smith [21] correlate various program characteristics to the basic block size in order to improve accuracy. Eeckhout et al. [5] propose the statistical flow graph (SFG) which models the control flow in a statistical manner; the various program characteristics are then correlated to the SFG. In our own prior work [9], we further improve the overall accuracy of the statistical simulation framework through accurate memory data flow modeling: we model cache miss correlation, store-load dependencies and delayed hits, and report an average IPC prediction error of 2.3% for a wide superscalar out-of-order processor compared to detailed simulation.

Nussbaum and Smith [22] and Hughes and Li [13] extend the statistical simulation paradigm toward multi-threaded programs. Cache behavior is still modeled based on cache miss rates though; by consequence, they are unable to model shared caches as observed in modern CMPs. In our own work [10], we model memory data flow behavior in a microarchitecture-independent way and in addition we model time-varying behavior which allows for modeling conflict behavior in shared resources as seen in contemporary multi-core processors.

Recent work also focused on generating synthetic benchmarks rather than synthetic traces. Hsieh and Pedram [12] generate a fully functional program from a statistical profile. However, all the characteristics in the statistical profile are microarchitecture-dependent, which makes this technique useless for microprocessor design space explorations. Bell and John [1] generate short synthetic benchmarks using a collection of microarchitecture-independent and microarchitecture-dependent characteristics similar to what is done in statistical simulation. Their goal is performance model validation of high-level architectural simulation against RTL-level cycle-accurate simulation using small but representative synthetic benchmarks. Joshi et al. [15] generate synthetic benchmarks based on microarchitecture-independent characteristics only and leverage that framework to automatically generate stressmarks (i.e., synthetic programs that maximize power consumption, temperature), see [16].

A couple other research efforts model cache performance in a statistical way. Berg et al. [2] propose light-weight profiling to collect a memory reference reuse distribution at low overhead and then estimate cache miss rates for random-replacement caches. Chandra et al. [4] propose performance models to predict the impact of cache sharing on co-scheduled programs. The output provided by the performance model is an estimate of the number of extra cache misses for each thread due to cache sharing. These statistical models are limited to predicting cache effects and do not predict overall performance.

13.5 Summary

The chapter revisited the general concept of statistical simulation. Statistical simulation yields a beneficial speed-accuracy trade-off: simulation speed is enhanced by several orders of magnitude with limited impact on accuracy, which makes it a valuable tool for driving design space explorations during early stages of the processor design cycle.

One promising avenue for future research in statistical simulation is to extend it toward (very) large-scale systems such as clusters of computers, supercomputers, datacenters. Simulating large computer systems is completely infeasible through detailed cycle-accurate simulation; hence, higher level simulation models, such statistical simulation, may be a viable solution in that space. Another avenue for future work is to generate synthetic workloads that are ISA independent. In particular, generating synthetic workloads in a high-level programming language (such as C) would make them easily portable across systems and would even enable compiler research next to architecture research.

Acknowledgments This research is partially funded by projects G.0232.06, G.0255.08, and G.0179.10 by the Fund for Scientific Research–Flanders (Belgium) (FWO Vlaanderen), as well as the UGent-BOF projects 01J14407 and 01Z04109.

References

1. Bell, Jr., R., John, L.K.: "Improved automatic testcase synthesis for performance model validation." In: *Proceedings of the 19th ACM International Conference on Supercomputing (ICS)*, Cambridge, MA, USA, pp. 111–120, June 20–22 (2005).
2. Berg, E., Hagersten, E.: "Fast data-locality profiling of native execution." In: *Proceedings of the International Conference on Measurements and Modeling of Computer Systems (SIGMETRICS)*, Banff, Alberta, Canada, pp. 169–180, June 6–10 (2005).
3. Carl, R., Smith, J.E.: Modeling superscalar processors via statistical simulation. In: *Workshop on Performance Analysis and its Impact on Design (PAID), held in conjunction with the 25th Annual International Symposium on Computer Architecture* (ISCA), Barcelona, Spain, June 28, (1998).
4. Chandra, D., Guo, F., Kim, S., Solihin, Y.: "Predicting inter-thread cache contention on a chip-multiprocessor architecture." In: *Proceedings of the Eleventh International Symposium on High Performance Computer Architecture (HPCA)*, San Francisco, CA, USA, pp. 340–351, Feb 12–16 (2005).
5. Eeckhout, L., Bell Jr., R.H., Stougie, B., De Bosschere, K., John, L.K.: "Control flow modeling in statistical simulation for accurate and efficient processor design studies." In: *Proceedings of the 31st Annual International Symposium on Computer Architecture (ISCA)*, München, Germany, pp. 350–361, June 19–23 (2004).
6. Eeckhout, L., De Bosschere, K.: "Hybrid analytical-statistical modeling for efficiently exploring architecture and workload design spaces." In: *Proceedings of the 2001 International Conference on Parallel Architectures and Compilation Techniques (PACT)*, Barcelona, Spain, pp. 25–34, Sept 10–12 (2001)
7. Eeckhout, L., De Bosschere, K., Neefs, H.: Performance analysis through synthetic trace generation. In: *The IEEE International Symposium on Performance Analysis of Systems and Software (ISPASS)*, Austin, TX, USA, pp. 1–6, Apr 24–25 (2000).
8. Eyerman, S., Eeckhout, L.: System-level performance metrics for multi-program workloads. *IEEE Micro* **28**(3), 42–53 (2008).
9. Genbrugge, D., Eeckhout, L.: Memory data flow modeling in statistical simulation for the efficient exploration of microprocessor design spaces. *IEEE Trans Comp* **57**(10), 41–54 (2007).
10. Genbrugge, D., Eeckhout, L.: Chip multiprocessor design space exploration through statistical simulation. *IEEE Trans Comp* **58**(12), 1668–1681 (2009).
11. Hill, M.D., Smith, A.J.: Evaluating associativity in CPU caches. *IEEE Trans Comp* **38**(12), 1612–1630 (1989).
12. Hsieh, C., Pedram, M.: Micro-processor power estimation using profile-driven program synthesis. *IEEE Trans Comp Aid Des Inte Circuits Syst* **17**(11), 1080–1089 (1998).
13. Hughes, C., Li, T.: "Accelerating multi-core processor design space evaluation using automatic multi-threaded workload synthesis." In: *Proceedings of the IEEE International Symposium on Workload Characterization (IISWC)*, pp. 163–172 (2008).

14. Iyengar, V.S., Trevillyan, L.H., Bose, P.: "Representative traces for processor models with infinite cache." In: *Proceedings of the Second International Symposium on High-Performance Computer Architecture (HPCA)*, San Jose, CA, USA, pp. 62–73, Feb 3–7 (1996).
15. Joshi, A.M., Eeckhout, L., Bell, Jr., R., John, L.K.: Distilling the essence of proprietary workloads into miniature benchmarks. *ACM Trans Archi Code Opti (TACO)* **5**(2), Article No. 10 (2008).
16. Joshi, A.M., Eeckhout, L., John, L.K., Isen, C.: "Automated microprocessor stressmark generation." In: *Proceedings of the International Symposium on High-Performance Computer Architecture (HPCA)*, Salt Lake City, UT, USA, pp. 229–239, Feb 16–20 (2008).
17. Kgil, T., D'Souza, S., Saidi, A., N, B., Dreslinski, R., Reinhardt, S., Flautner, K., Mudge, T.: PicoServer: Using 3D stacking technology to enable a compact energy efficient chip multiprocessor. In: *Proceedings of the Twelfth International Conference on Architectural Support for Programming Languages and Operating Systems (ASPLOS)*, San Jose, CA, USA, pp. 117–128, Oct 21–25 (2006).
18. Luk, C.K., Cohn, R., Muth, R., Patil, H., Klauser, A., Lowney, G., Wallace, S., Reddi, V.J., Hazelwood, K.: "Pin: Building customized program analysis tools with dynamic instrumentation." In: *Proceedings of the ACM SIGPLAN Conference on Programming Languages Design and Implementation (PLDI)*, Chicago, IL, USA, pp. 190–200, June 12–25 (2005).
19. Luo, K., Gummaraju, J., Franklin, M.: "Balancing throughput and fairness in SMT processors." In: *Proceedings of the IEEE International Symposium on Performance Analysis of Systems and Software (ISPASS)*, Tucson, AZ, USA, pp. 164–171, Nov 4–6 (2001).
20. Noonburg, D.B., Shen, J.P.: "A framework for statistical modeling of superscalar processor performance." In: *Proceedings of the Third International Symposium on High-Performance Computer Architecture (HPCA)*, San Antonio, TX, USA, pp. 298–309, Feb 1–5 (1997).
21. Nussbaum, S., Smith, J.E.: "Modeling superscalar processors via statistical simulation. In: *Proceedings of the 2001 International Conference on Parallel Architectures and Compilation Techniques (PACT)*," Barcelona, Spain, pp. 15–24, Sept 10–12 (2001).
22. Nussbaum, S., Smith, J.E.: "Statistical simulation of symmetric multiprocessor systems." In: *Proceedings of the 35th Annual Simulation Symposium 2002*, Washington, DC, USA, pp. 89–97, April 16–22 (2002).
23. Oskin, M., Chong, F.T., Farrens, M.: "HLS: Combining statistical and symbolic simulation to guide microprocessor design." In: *Proceedings of the 27th Annual International Symposium on Computer Architecture (ISCA)*, Vancouver, British Columbia, Canada, pp. 71–82, June 10–14 (2000).
24. Snavely, A., Tullsen, D.M.: "Symbiotic jobscheduling for simultaneous multithreading processor." In: *Proceedings of the International Conference on Architectural Support for Programming Languages and Operating Systems (ASPLOS)*, Cambridge, MA, USA, pp. 234–244, Nov 12–15 (2000).
25. Srivastava, A., Eustace, A.: ATOM: "A system for building customized program analysis tools." In: *Technical Report 94/2, Western Research Lab, Compaq* (1994).
26. Van Biesbrouck, M., Sherwood, T., Calder, B.: "A co-phase matrix to guide simultaneous multithreading simulation." In: *Proceedings of the International Symposium on Performance Analysis of Systems and Software (ISPASS)*, Austin, TX, USA, pp. 45–56, March 10–12 (2004).

Part III
Impact of Silicon Technology

Chapter 14
Memory Modeling with CACTI

Naveen Muralimanohar, Jung Ho Ahn, and Norman P. Jouppi

Abstract Modern systems consist of a hierarchy of memory arrays made of different storage elements such as SRAM, DRAM, FLASH, etc. The organization of a memory array significantly impacts its delay, power, bandwidth, and area parameters which in turn impacts overall system performance and cost. When evaluating a new architecture or exploring new memory designs, it is crucial to get an early estimate of memory access time and power for given input parameters. For studies related to memory hierarchy design, it is also necessary to optimize memory arrays so that they meet specific delay, area, and power constraints. This chapter presents an integrated analytical tool called CACTI that models power, delay, area, and cycle time (bandwidth) of all the components in a modern memory system. As all parameters are calculated based on the same technology and circuit parameters, they are mutually consistent, and the analytical models employed in CACTI have been verified to be within 12% of high fidelity SPICE models. This chapter explains the fundamental building blocks of memory arrays along with their analytical delay, power, and area models and details tradeoffs that exist when designing them.

14.1 Introduction

In computer architecture and design, the classic performance equation is given by the product of three terms,

$$CPUtime = InstructionCount \times CyclesPerInstruction \times ClockCycleTime$$

InstructionCount for a program is easy to obtain once there is a working compiler for an architecture. The public availability of cycle-accurate simulators allows *CyclesPerInstruction* to be quantified with some effort, and most of the content of

N.P. Jouppi (✉)
Exascale Computing Lab, Hewlett-Packard Company, 1501 Page Mill Road, MS 1181 Palo Alto, CA 94304, USA
e-mail: norm.jouppi@hp.com

R. Leupers, O. Temam (eds.), *Processor and System-on-Chip Simulation*,
DOI 10.1007/978-1-4419-6175-4_14, © Springer Science+Business Media, LLC 2010

computer architecture courses involves ways to improve this term. Unfortunately it is relatively difficult to get a definitive number for the clock cycle time of a proposed machine organization or microarchitecture without actually performing a detailed design. As the size and complexity of designs increase, the accurate modeling of the clock cycle time of proposed design is becoming increasingly important.

Furthermore, note that the three performance equation terms are not independent of each other. For example, a complex instruction set can decrease the instruction count but have the side effect of increasing the cycles per instruction more than enough to offset the advantage of decreased instruction path length. Similarly, many changes to machine organization or microarchitecture can have very significant impacts on clock cycle time. Thus studying only one term in isolation can paint an incomplete picture and lead to dramatically suboptimal designs overall.

Along with performance, other considerations have recently become of key importance. The power of computer systems has become a key limiter of not only handheld platforms but also of servers in large datacenters. And the area of designs is the most important factor in determining the fabrication cost of a chip. Ideally we would like to be able to simultaneously and consistently quantify the clock cycle time, power, and area of entire proposed designs.

Of the components of modern computer systems, memory is arguably the most important. The area of modern chip multiprocessors is often dominated by the total area of their caches. Memory certainly dominates the system when the silicon area of the main memory is taken into account. Memory in modern computer systems can range from register files, buffers, and various levels and types of caches to main-memory DRAMs. By modeling all these memories accurately we can model a significant portion of a proposed machine and quantitatively evaluate new architectural ideas impacting the memory hierarchy. The CACTI tool has grown to support modeling this entire range of memory hierarchy components over the past 15 years.

14.2 Application Space of CACTI

CACTI is a versatile tool capable of modeling any storage unit in memory hierarchy from small register files to large DRAM banks. The following sections present modeling basics of CACTI and the feature set of the tool, starting with basic RAM structures.

14.2.1 RAM

CACTI's RAM model is a simple form of a memory array that models register files or other on-chip buffers. Figure 14.1a shows the logical structure of a RAM organization. It consists of an array of static-random access memory (SRAM) cells with a centralized decoder and logic circuits to store and retrieve data from cells. Each memory cell is equipped with access transistors, which are controlled by the decoder

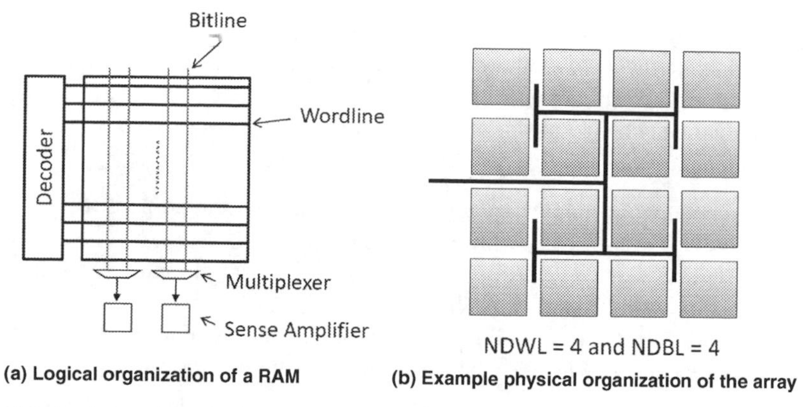

Fig. 14.1 Logical and physical organization of the RAM

output and provide access to the cell to perform read or write operations. To reduce the area overhead of the decoder, an output from the decoder typically activates an entire row of cells connected using a wire known as a wordline. In a typical access, an address request to the RAM is first provided as input to the decoder, which then activates the appropriate wordline. The contents of an entire row of cells are placed on bitlines, which are written to the cells for a write operation or delivered to the output for a read operation. While the data placed on the bitlines can be used to directly drive logic for small arrays, the access latency of such a model can be very high even for moderately sized arrays due to large bitline capacitances. To reduce the access time, sense amplifiers are employed to detect a small variation in bitline signal and amplify them to the logic swing. The use of sense amplifiers also limits the voltage swing in bitlines leading to energy savings.

However, the monolithic model shown in the Fig. 14.1a does not scale with the array size. Due to low-swing signaling employed in bitlines and silicon area overhead associated with repeaters, wordlines and bitlines cannot be repeated at regular intervals, causing their delay to increase quadratically with the size of the array. Furthermore, the throughput of a RAM is a function of wordline and bitline delay and the bandwidth of a single array deteriorates quickly as the array size grows. To reduce this quadratic delay impact, a large array is typically divided into multiple sub-arrays employing hierarchical decoders. The first-level decoder (also referred to as a pre-decoder, not shown in the figure) uses a subset of the address bits to identify the appropriate sub-array. The second-level row decoder takes the remaining address bits to activate the appropriate row of cells. The sub-arrays are connected through an interconnect fabric to transfer addresses and data within the RAM. In order to reduce design complexity, an interconnect with easily predictable timing characteristics is essential. CACTI models a balanced H-tree network as shown in the Fig. 14.1b to connect various sub-arrays. In addition to providing a uniform access time, an H-tree also provide pipelined access without using complex switching circuits.

The number of sub-arrays in a RAM and the height and width of sub-arrays play a key role in determining the power, area, and timing characteristics. For example, reducing the sub-array count would enable tighter packing of cells leading to increased area efficiency (cell area over total area), but would result in increased delay due to longer wordlines and bitlines. A large sub-array count would give better delay characteristics but result in increased silicon area consumption. CACTI uses three fundamental variables to define an array organization.

- NDWL—Number of column partitions in the array i.e., the number of segments a single logical wordline is partitioned into.
- NDBL—Number of row partitions in the array i.e., the number of segments a single logical bitline is partitioned into.
- NSPD—Number of sets stored in each row of a sub-array. For the given NDWL and NDBL values, this decides the aspect ratio of the sub-array.

The tool takes in the following major parameters as input: size, block size (also known as line size), number of ports, technology generation, and number of independent banks (that do not share address and data lines). It then performs a detailed design space exploration that considers variations of NDWL, NDBL, and NSPD along with different circuit components with varying power/delay characteristics. As output, it produces access and cycle time, leakage and dynamic power, and area of the RAM organization that best matches area, delay, and power constraints input to the tool.

14.2.2 Cache

A cache structure is similar to a RAM, but includes an additional tag array to determine a hit or miss status of a block. Depending on the delay model of cache and the type of interconnect employed for address and data transfers, caches are classified into uniform cache access (UCA) and non-uniform cache access (NUCA). The following sections discuss CACTI's modeling of UCA and NUCA caches.

14.2.2.1 Uniform Cache Access

Figure 14.2 shows the logical structure of a uniform cache access (UCA) organization. Similar to a RAM, the address request to the cache is first provided as input to the decoder, which then activates a wordline in the data array and the tag array. The contents of an entire row in both data and tag arrays are placed on the bitlines and sent to sense amplifiers. The multiple tags read out of the tag array are then compared against the input address to detect if one of the ways of the set contains the requested data. The comparator logic drives the multiplexer that finally forward at most one of the ways read out of the data array back to the controller. In practice, the tag and data arrays are too large to be efficiently implemented as single large structures. Hence, similar to a RAM, CACTI partitions each cache storage array into *sub-arrays* to reduce wordline and bitline delays.

The energy to access the data array of a cache is typically greater than its tag array. As cache size increases, the difference in access energy can be significant.

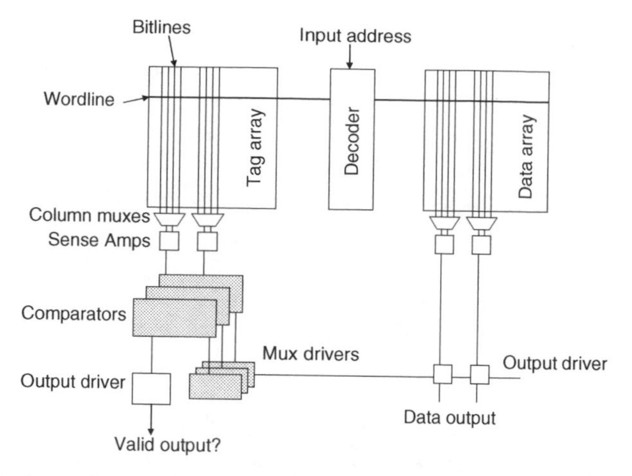

Fig. 14.2 Logical organization of a cache

While accessing both tag and data array in parallel reduces the net access time of a cache, such a design incurs high power overhead and is less efficient in situations where the cache miss rate is high. To better match a cache design to its miss rate, CACTI models three different accessing modes for UCA caches: sequential mode, fast mode, and semi-parallel (or normal) mode. In the sequential mode, access to the data array happens only after the tag array finishes its lookup and signals the data array that the access is a hit. Moreover, only the data corresponding to the matching tag is read out of the array. For cache misses, the tag array directly sends the miss signal to the controller and the data array never gets accessed leading to power savings. However, the sequential accessing can lead to increased access time for hits. In the fast mode, both tag and data array accesses happen in parallel and the entire set is read out of data array. After the tag match is over, on a hit, the relevant block from the set is sent to the cache controller and the rest of the blocks go unused. If the tag array signals a miss, then the entire set read out of the data array gets wasted. This design trades off power to minimize access time. In a semi-parallel access, similar to the fast mode, both tag and data array lookups happen in parallel. However, unlike the fast mode, after the lookup, the data array waits for the tag to send the hit signal and the block number before sending out the data. On a hit, only the relevant block gets sent through the H-tree. In addition to the basic input arguments discussed earlier, to model a cache, the tool also takes cache associativity and access mode as input parameters and outputs both tag and data array organization along with their power, delay, and area values.

14.2.2.2 NUCA

In a uniform cache access (UCA) model the access time of a cache is determined by the delay to access the sub-array farthest from the cache controller. For large caches, the disparity in access time between the closest and farthest sub-arrays can be very high. A more scalable approach is to replace the H-tree bus with a packet-switched on-chip grid network as shown in Fig. 14.3. The latency for a bank is determined by the delay to route the request and response between the bank that contains the data and the cache controller [9]. Since the access time of such a cache is a function of bank location, it is referred to as non-uniform cache access. CACTI builds upon this model and adopts the following algorithm to identify the optimal NUCA organization [11, 12].

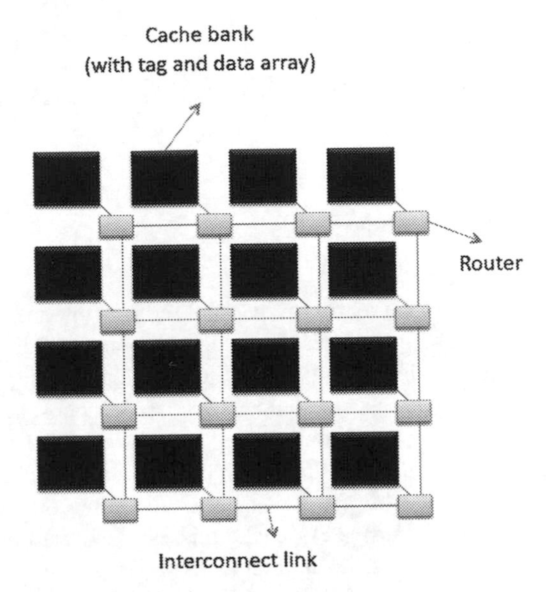

Fig. 14.3 A NUCA cache with 16 banks

The tool first iterates over a number of bank organizations: the cache is partitioned into 2^N banks (where N varies from 1 to 12); for each N, the banks are organized in a grid with 2^M rows (where M varies from 0 to N). For each bank organization, CACTI's UCA model is employed to determine the optimal sub-array partitioning for the cache within each bank. Each bank is associated with a router. The average delay for a cache access is computed by estimating the number of network hops to each bank, the wire delay encountered on each hop, and the cache access delay within each bank. The user has the flexibility to specify the router traversal delay and the operating frequency of the network (which defaults to 5 GHz). However, based on the process technology and the router model, the tool will calculate the maximum possible network frequency [14]. If the assumed frequency is greater than the maximum possible value, the tool will downgrade the network frequency to the maximum value.

In a NUCA cache, more partitions lead to smaller delays (and power) within each bank, but greater delays (and power) on the network (because of the constant overheads associated with each router and decoder). Hence, the above design space exploration is required to estimate the cache partition that yields optimal delay or power. While modeling a NUCA cache for a multicore processor, the bandwidth of a cache is another important factor and it is critical to have a sufficient number of banks to meet the requirements of multiple cores. CACTI's design space exploration also includes the effect of contention in the network. This contention model itself has two major components: router contention and bank contention. If the cache is partitioned into many banks, there are more routers/links on the network and the probability of two packets conflicting at a router decreases. Thus, a many-banked cache is more capable of meeting the bandwidth demands of a many-core system. Further, certain aspects of the cache access within a bank cannot be easily pipelined. The longest such delay within the cache access (typically the bitline and sense-amp delays) represents the cycle time of the bank—it is the minimum delay between successive accesses to that bank. A many-banked cache has relatively small banks and a relatively low cycle time, allowing it to support higher throughput and lower wait-times once a request is delivered to the bank. Both of these two components (lower contention at routers and lower contention at banks) tend to favor a many-banked system. This aspect is also included in estimating the average access time for a given cache configuration.

To improve the search space of the NUCA model, CACTI also explores different router types and wire types for the links between adjacent routers. The wires are modeled as low-swing differential wires as well as global wires with different repeater configurations to yield many points in the power/delay/area spectrum. The sizes of buffers and the number of virtual channels within a router have a major influence on router power consumption as well as router contention under heavy load. By varying the number of virtual channels per physical channel and the number of buffers per virtual channel, CACTI achieves different points on the router power-delay trade-off curve.

The contention values for each considered NUCA cache organization are empirically estimated for typical workloads and incorporated into CACTI as lookup tables. While modeling a NUCA cache, in addition to UCA parameters, the tool takes as the parameters, number of cores and the position of the cache in the hierarchy (level 2 or level 3). These parameters are then used to estimate contention in the NUCA network for various cache organizations.

14.2.3 Dynamic Random Access Memory (DRAM)

DRAM is a high-density memory technology with each cell consisting of just an access transistor and a storage capacitor. CACTI models two different types of DRAM memories: logic process-based DRAM (LP-DRAM) and traditional commodity DRAM. The key difference between them lies in the fabrication process technology employed. Embedded DRAM is made using the same technology used to manufacture SRAM or processor logic and has superior latency properties over commodity DRAM. On the other hand, commodity DRAM is heavily optimized for density and has better retention time (more charge per cell) compared to LP-DRAM.

At the organization level, CACTI's DRAM array model is similar to a RAM array. However, the low area of DRAM cells (compared to SRAM) requires adjustments to peripheral circuit components. Furthermore, unlike SRAM, DRAM loses its data due to charge leakage and requires a mechanism to restore its content at regular intervals. CACTI includes the following features to model DRAM memories.

14.2.3.1 Cell

The area of a typical commodity DRAM cell is 1/20th of an SRAM cell and 1/3rd of an LP-DRAM cell. Hence, for a given die area, a DRAM array packs 7–20 times more cells compared to an SRAM array. This not only makes a DRAM array denser but also places additional constraints on peripheral circuits. For example, decoder drivers and sense amplifiers employed for a DRAM array should be downsized to match the cell area. This in turn increases the delay of wordlines and sensing logic. CACTI's design space exploration takes into account the effect of downsizing while finding the optimal DRAM array organization.

14.2.3.2 Destructive Readouts and Writeback

After reading a DRAM cell, the cell loses its content due to charge redistribution between the cell and the capacitive bitline, and there is a need for data to be written back into the cell. In addition to this, after a writeback the bitlines need to be restored to their precharged value. These writeback and restore operations increase the random cycle time of a DRAM array. CACTI include these operations when calculating delay and energy values.

14.2.3.3 Refresh

Even in the absence of any read or write operations, DRAM cells continue to leak charge and requires periodic refreshing. The refresh period is a function of amount of charge stored in a cell and varies between commodity DRAM and LP-DRAM. This is because the LP-DRAM cells employ transistors with oxides that are much thinner than those of commodity DRAM cells. CACTI considers the overhead of refreshing while modeling DRAM power.

14.2.3.4 Operational Mode

While modeling DRAM using CACTI, a user is provided with an option to either model it as a traditional main memory that follows JEDEC standard [8] (with models for ACTIVATE, READ, WRITE, and PRECHARGE commands) or as a SRAM-like model with just READ and WRITE commands. In the latter case, activate and precharge operations happen along with a read or write without any explicit request for them. CACTI automatically adjusts its timing and power calculations for an array based on its access model.

14.3 Modeling Access Timing, Area, and Power

CACTI models the delay, power, and area of eight major components of a cache: decoder, wordline, bitline, sense-amp, comparator, multiplexer, output driver, and inter-bank wires. To model NUCA caches, CACTI also models router components such as arbiters, flipflops, and crossbars. These components are connected using buffers and drivers. The sizing of transistors in various cache components and drivers are calculated using the method of logical effort [18].

14.3.1 Delay

The delay of a logic or wire is governed by its RC time constant (where, R is resistance, C is capacitance). To calculate the delay, a circuit is first transformed into its equivalent RC tree. The resistance and capacitance values of gates and wires are calculated from transistor sizes (estimated using logical effort) and ITRS projections [7]. The time constant of a path in an RC tree is equal to the sum of the delay through each segment, which in turn is defined as the product of resistance and the downstream capacitance. For example, consider an interconnect fragment shown in the Fig. 14.4 consisting of repeaters and a branch. Its equivalent RC tree is shown in Fig. 14.5. The time constant of the path from input to output 1 is given by

$$\tau = R_1(C_1 + C_{\text{wire}} + C_{21} + C_{22}) + \frac{RC}{2} + R(C_{21} + C22) + R_2 C_{22} \qquad (14.1)$$

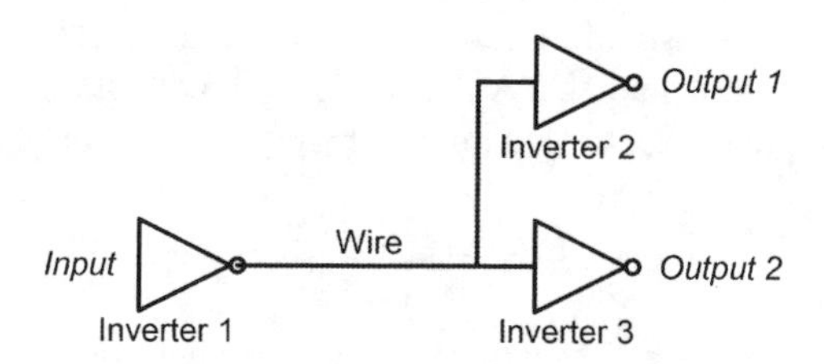

Fig. 14.4 Interconnect fragment with a branch

In the above equation, R_1 and C_1 are the resistance and capacitance of the inverter 1. R_2, C_{21}, and C_{22} are the resistance, the input capacitance, and the output capacitance of the inverter 2. R and C correspond to the effective resistance and capacitance of the connecting wire segment. To include the effect of slow rising signals at the input on delay, the timing model proposed by Horowitz is employed [6]. Equation (14.2) calculates the final delay of a circuit.

$$\text{delay}_r = t_f \sqrt{[\log \frac{v_{\text{th}}}{V_{\text{dd}}}]^2 + 2t_{\text{rise}} b(1 - \frac{v_{\text{th}}}{V_{\text{dd}}})/t_f} \qquad (14.2)$$

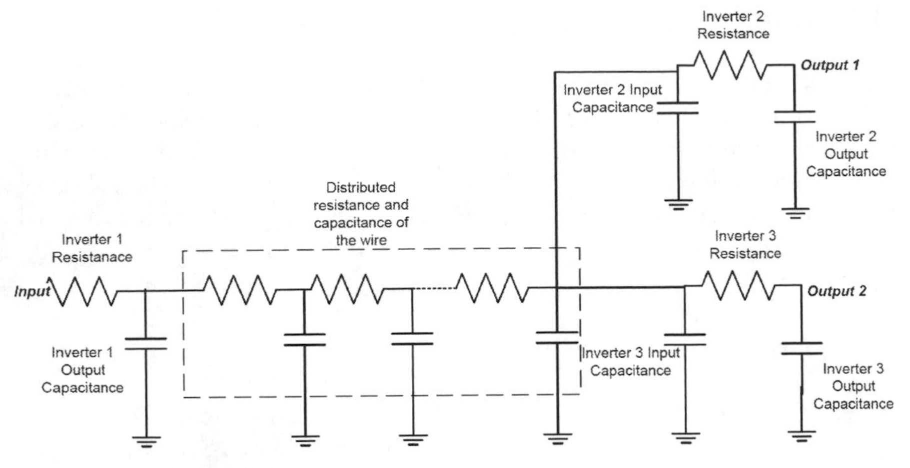

Fig. 14.5 RC tree of the fragment

where t_f is the time constant of the tree, v_{th} is the threshold voltage of the transistor, t_{rise} is the rise time of the input signal, and b is the fraction of the input swing in which the output changes.

14.3.2 Power

The energy consumed by a logic or wire is primarily due to charging and discharging of its parasitic capacitances. In addition to this, a small fraction of power gets wasted due to leakage current in transistors. While the contribution of leakage to the total power is typically small, for deep sub-micron technologies, leakage power contribution can be significant [2]. CACTI models both dynamic and leakage power of all the cache components. The dynamic energy consumed by a gate is given by

$$E_{\text{dyn}} = 0.5\ CV^2$$

The multiplicative factor of 0.5 in the equation assumes consecutive charging and discharging cycles for each gate. Energy is consumed only during the charging cycle of a gate when its output goes from zero to the operating voltage. C in the equation refers to the parasitic capacitance (which includes gate and drain capacitances of transistors) of the gate.

There are multiple sources of leakage current in a transistor: sub-threshold leakage, gate-induced drain current, current due to reverse biased pn junction, punch through current, and gate tunneling current [1]. CACTI models the most dominant component of leakage which is drain-to-source leakage. This component of leakage in a gate depends on two factors: the state of a transistor (leakage current is factored in only when a transistor is in an off-state) and stacking level. This means, in a complementary MOS gate, sub-threshold leakage is accounted either in pull-up transistors or in pull-down transistors. As leakage current in PMOS is different from NMOS, CACTI considers the average of leakage due to pull-up and pull-down transistors. Another factor that affects the sub-threshold current is stacking of transistors. For example, the leakage current in NMOS transistors stacked in a NAND gate is less than the leakage current of a similarly sized NMOS in an inverter. CACTI uses the model proposed by Narendra et al. [13] to determine the reduction in leakage due to stacking. Leakage power of a transistor is given by

$$P_{\text{leak}} = T_{\text{width}} I_{\text{leak}} V \times S_{\text{adj}}$$

where T_{width} refers to the width of the transistor and I_{leak} is the leakage current for a given technology. S_{adj} is the adjustment factor that models the effect of stacking of transistors in a gate.

14.3.3 Area

The area of a memory structure is primarily dominated by the area of memory cells, peripheral logic, and connecting wires. The area of a cell depends on the type of the storage medium and the number of ports in an array. CACTI follows ITRS projections to model area of a single ported cell. For multi-ported cells, the area of additional access transistors required to support multiple ports are added to the single-ported cell area.

To model area of peripheral circuits, complex logic is first decomposed into simple gates. CACTI assumes a standard cell layout for gates as shown in Fig. 14.6. For a given process technology, the area of a standard cell can be determined by a set of process-specific variables shown in Table 14.1. Logic gates with stacked transistors

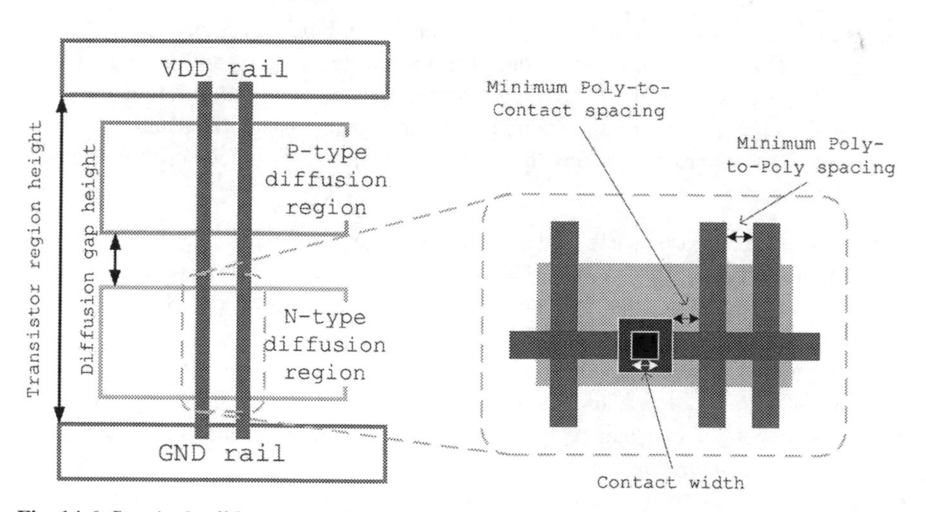

Fig. 14.6 Standard cell layout template

Table 14.1 Technology parameters required by the gate area model

$H_{\text{n-diff}}$	Maximum height of n diffusion of a transistor
$H_{\text{p-diff}}$	Maximum height of p diffusion of a transistor
$H_{\text{gap-bet-same-diffs}}$	Minimum gap between diffusions of the same type
$H_{\text{gap-bet-opp-diffs}}$	Minimum gap between n and p diffusions
$H_{\text{power-rail}}$	Height of V_{DD} power rail
W_{p}	Minimum width of poly
$S_{\text{p-p}}$	Minimum poly-to-poly spacing
W_{c}	Contact width
$S_{\text{p-c}}$	Minimum poly-to-contact spacing

typically share a common diffusion and do not incur any area overhead due to metal contacts. This is also true for large transistors that are folded to fit into the standard cell template. CACTI's area model considers the reduction in area due to folding or stacking of transistors in a gate [19].

The contribution of wires to the total memory area depends on the type of the wire employed for address/data interconnects. A plain wire (wires with no repeaters) or unequalized differential low-swing wires can be laid on top of memory arrays and do not require silicon area. However, wires with repeaters or equalization transistors require silicon area and their contribution to the total area can be non-trivial. CACTI's wire area calculation is based on the width and spacing of wires from the ITRS roadmap [7]. Sometimes employing very large repeaters or tristate buffers make it impossible to lay wires with minimum standard spacing. For such wires, CACTI calculates both repeater/buffer area and the area due to wire pitch and picks the maximum of both to model interconnect area.

14.4 Usage

CACTI takes a list of parameters such as cache size, block size, associativity, and technology from a user through a configuration file (cache.cfg) and outputs delay, area, and power values of the specified configuration. The configuration file also enables the user to describe the cache parameters in much greater detail. Some of the other non-standard parameters that can be specified in the cache.cfg file include the following:

- Number of read ports, write ports, and read–write ports in a cache
- Technology type (HP—high performance; LSTP—low standby power; LOP—low operating power; LP-DRAM—embedded DRAM; COMM-DRAM—commodity DRAM)
- Input/output bus width
- Operating temperature (which is used for calculating the cache leakage value)
- Custom tag size (that can be used to model special structures such as branch target buffer, cache directory)
- Access mode (fast—low access time but power hungry; sequential—high access time but low power; normal—less aggressive in terms of both power and delay)

- Cache type (DRAM, SRAM, or a simple scratch RAM such as a register file which does not need the tag array)
- NUCA bank count (by default CACTI calculates the optimal bank count value. However, the user can force the tool to use a particular NUCA bank count value)
- Number of cores
- Cache level—L2 or L3 (core count and cache level are used to calculate the contention values for a NUCA model)
- Design objective (weight and deviate parameters for NUCA and UCA)

CACTI is available for download from

`http://www.hpl.hp.com/research/cacti/`

14.5 Future Directions

Memory hierarchies are continuously evolving and it is highly likely that solid state non-volatile memories will soon dominate the entire hierarchy in future systems. Industry has already started embracing FLASH memories as an alternative to storage disks [5]. Moreover, upcoming memory technologies such as *Phase Change Memories* [4, 16] and *Memristors* [17] are becoming increasingly popular and there has been a lot of work from both industry and academia that try to leverage these technologies to improve system performance [3, 10, 15, 20]. To accelerate research on upcoming technologies, future versions of CACTI will have functionalities to support FLASH, PCRAM, and Memristor-based memories.

14.6 Summary

CACTI supports modeling the entire range of memory hierarchy components from register files, buffers, and various types and levels of caches to main memory. By modeling all these memories accurately we can model a significant portion of a proposed machine and quantitatively evaluate new architectural ideas impacting the memory hierarchy. CACTI's analytical models have been validated to be within 12% of high fidelity SPICE models. Combined with cycle-accurate instruction-level simulation, this can enable accurate overall system optimization, including the effect of machine organization and microarchitecture trade-offs on cycle time, power, and area.

Acknowledgments The authors would like to thank all the researchers that have contributed to CACTI over the past 15 years, including Steve Wilton, Glenn Reinman, Premkishore Shivakumar, David Tarjan, and Shyamkumar Thoziyoor.

References

1. Chen, X., Peh, L.: "Leakage power modeling and optimization in interconnection networks." In: *Proceedings of ISLPED*, pp. 90–95, Seoul, Korea, August 25–27, (2003).
2. De, V., Borkar, S.: Technology and design challenges for low power and high performance. In: *Proceedings of ISLPED*, pp. 163–168, San Diego, CA, USA, August 16–17, (1999).
3. Dong, X., Muralimanohar, N., Jouppi, N., Kaufmann, R., Xie, Y.: Leveraging 3D PCRAM Technologies to Reduce Checkpoint Overhead for Future Exascale Systems. In: *Proceedings of SC* (2009). Article No.:57, Portland, OR, USA, November 14–20, (2009).
4. Freitas, R.F., Wilcke, W.W.: Storage-class memory: The next storage system technology. *IBM J Res Devel* **52**(4/5):439–447, July (2008).
5. Fusion-io: ioDrive. 2010 http://www.fusionio.com/
6. Horowitz, M.: Timing models for MOS circuits (1999). Stanford University, Stanford, CA, USA (1983).
7. ITRS: "International Technology Roadmap for Semiconductors, 2007 Edition." http://www.itrs.net/Links/2007ITRS/Home2007.htm (2007).
8. JEDEC: JESD79: Double Data Rate (DDR) SDRAM Specification. JEDEC Solid State Technology Association, Virginia, USA (2003).
9. Kim, C., Burger, D., Keckler, S.: An adaptive, non-uniform cache structure for wire-dominated on-chip caches. In: *Proceedings of ASPLOS*, pp. 211–222, San Jose, CA, USA, October 5–9, (2002).
10. Lee, B.C., Ipek, E., Mutlu, O., Burger, D.: Architecting phase change memory as a scalable DRAM alternative. In: *Proceedings of ISCA-36*, pp. 2–3, Austin, TX, USA, June 20–24, (2009).
11. Muralimanohar, N., Balasubramonian, R.: Interconnect design considerations for large NUCA caches. In: *Proceedings of ISCA*, pp. 369–380, San Diego, CA, USA, June 9–13, (2007).
12. Muralimanohar, N., Balasubramonian, R., Jouppi, N.: Optimizing NUCA organizations and wiring alternatives for large caches with CACTI 6.0. In: *Proceedings of MICRO*, pp. 3–14, Chicago, IL, USA, December 1–5, (2007).
13. Narendra, S., Borkar, S., De, V., Antoniadis, D., Chandrakasan, A.: Scaling of stack effect and its application for leakage reduction. In: *Proceedings of ISLPED*, pp. 195–200, Huntington Beach, CA, USA, August 6–7, (2001).
14. Peh, L.S., Dally, W.: A delay model and speculative architecture for pipelined routers. In: *Proceedings of HPCA*, pp. 255–266, Nuevo Leone, Mexico, January 20–24, (2001).
15. Qureshi, M.K., Srinivasan, V., Rivers, J.A.: Scalable high performance main memory system using phase-change memory technology. In: *Proceedings of ISCA-36*, pp. 24–33, Austin, TX, USA, June 20–24, (2009).
16. Raoux, S., Burr, G.W., Breitwisch, M.J., et al.: Phase-change random access memory: A scalable technology. *IBM J Res Devel* **52**(4/5) pp. 465–479, July (2008).
17. Strukov, D.B., Snider, G.S., Stewart, D.R., Stanley, R.: The missing memristor found. Nature 453, 80–83, (2008).
18. Sutherland, I.E., Sproull, R.F., Harris, D.: Logical effort: Designing fast CMOS circuits. In: Morgan Kaufmann (1999).
19. Yoshida, H., De, K., Boppana, V.: Accurate pre-layout estimation of standard cell characteristics. In: *Proceedings of DAC*, pp. 208–211, San Diego, CA, USA, June (2004).
20. Zhou, P., Zhao, B., Yang, J., Zhang, Y.: A Durable and energy efficient main memory using phase change memory technology. In: *Proceedings of ISCA-36*, pp. 14–23, Austin, TX, USA, June 20–24, (2009).

Chapter 15
Thermal Modeling for Processors and Systems-on-Chip

Kevin Skadron, Mircea Stan, and Wei Huang

Abstract Chip power density and consequently on-chip hot spot temperature have been increasing steadily as a result of non-ideal technology scaling, leading to severely thermally constrained designs. In this chapter, we review a chip- and package-level thermal modeling and simulation approach, HotSpot, that is unique because it is compact, correct by construction, flexible, and parameterized. HotSpot is important for temperature-aware design, especially during early pre-RTL stages of SoC and processor designs. Several case studies further illustrate the necessity of thermal simulations and the usefulness of HotSpot.

15.1 Temperature-Aware Design

An unfortunate side effect of miniaturization and the continued scaling of CMOS technology is a steady increase in power densities. The resulting difficulties in managing temperature, especially local hot spots, have become one of the major challenges for designers. High temperatures have several detrimental effects on VLSI systems such as microprocessors and systems-on-chip (SoC). First, the device carrier mobility is degraded at higher temperatures, resulting in slower devices. Second, leakage power is escalated due to the exponential increase of sub-threshold current with temperature. Third, the interconnect resistivity increases with temperature, leading to more severe power grid IR drops and longer interconnect RC delays, causing performance loss and complicating timing and noise analysis. Finally, elevated temperatures accelerate interconnect and device aging, while package reliability can be severely affected by local hot spots and higher temperature gradients. For all these reasons, in order to fully account for the thermal effects, it is important to model chip temperature in an accurate but also efficient way at all stages of the design. In particular, it is crucial to take thermal effects into account as early as possible in the design cycle, since early high-level thermal-aware design decisions can significantly improve design efficiency and reduce design cost.

K. Skadron (✉)
Department of Computer Science, University of Virginia, Charlottesville, VA 22904, USA
e-mail: kadron@cs.virginia.edu

R. Leupers, O. Temam (eds.), *Processor and System-on-Chip Simulation*,
DOI 10.1007/978-1-4419-6175-4_15,

Figure 15.1 shows a basic ASIC/SoC design flow adapted to become *temperature-aware*. Temperature profiles are needed at both functional-block level and standard-cell level during the design flow. Similar arguments also apply to microprocessor design flows. From the flow it is clear that it is very important to be able to estimate temperature at different granularities and at different design stages, especially early in the design flow. The estimated temperatures can then be used to perform power, performance, and reliability analyses, together with placement, packaging design, etc. As a result, all the design decisions use temperature as a guideline and the design is intrinsically thermally optimized and free from thermal limitations. We call this type of design methodology *temperature-aware* design. The idea of temperature-aware design is essential because the operating temperatures are properly considered during the *entire* design flow instead of being determined only after the fact and at the end of the design flow.

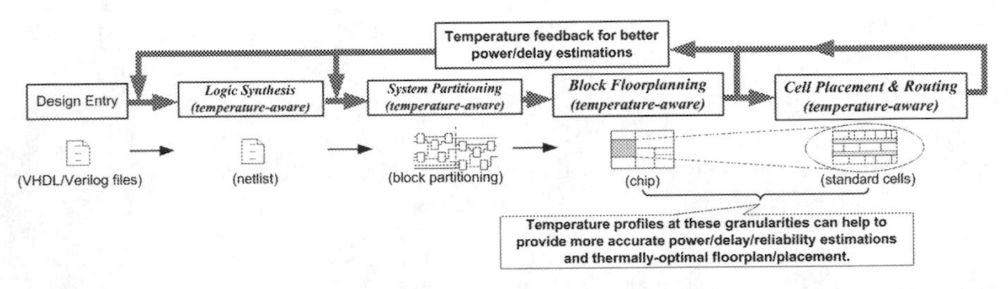

Fig. 15.1 An example of temperature-aware ASIC/SoC design flow [1]

It is important to recognize that power-aware design is insufficient for temperature management, because temperature also depends on the pattern of power dissipation in space and time, packaging and cooling choices, and the behavior of other system components. Furthermore, sometimes lower energy or low-power solutions can actually lead to higher temperatures (e.g., solutions that have higher local power densities), while a temperature-aware design can allow cheaper packages, quieter fans, and higher reliability. Also, low-power techniques tend to conserve power when activity is low, while thermal effects are a problem when activity is high. Thermal management techniques must therefore be designed differently from energy management techniques, even if they use some of the same underlying design "knobs" to control power. In recent years, research on temperature-aware design in the computer architecture and circuits communities has rapidly expanded, and more often we see chip architects and circuit designers branching out to collaborate with the design automation and thermal engineering communities.

15.2 Compact Thermal Modeling

The thermal model used to estimate temperatures is the key element for a temperature-aware design methodology. Figure 15.2 shows how the thermal model helps to close the loop for accurate power, performance, and reliability estimations.

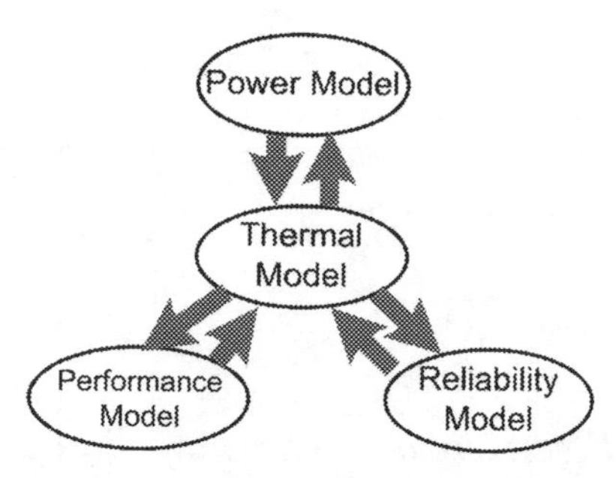

Fig. 15.2 Interactions among thermal model and power, performance, and reliability models [1]

For example, the power model first provides estimated power to the thermal model. The thermal model in turn provides estimated temperatures to the power model, and so on. After a few iterations, both power and temperature estimations converge; at that point, temperature-aware power estimation is achieved. Similarly, temperature-aware performance and reliability estimations can also be achieved.

Due to the huge computational requirements, it is almost impossible to model temperature and analyze the thermal effects of a system together with the environment in their full details. Using detailed numerical analysis methods, such as finite-element models (FEM), is time-consuming and costly, hence appropriate to model temperatures only in special cases. The best trade-offs are offered by compact thermal models (CTMs) with reasonably accurate temperature predictions at different levels, e.g., circuit level, die level, package level [2].

A top-down hierarchy of compact thermal models can help designers at different levels of abstraction. As listed below, there are several desirable features that increase the usefulness of such a compact thermal model at a particular level.

15.2.1 Correct by Construction

A fully by-construction and parameterized compact thermal model allows chip designers and computer architects to explore new design alternatives and evaluate different thermally related design trade-offs at their corresponding design levels before the actual physical design becomes available. More importantly, with the aid of the parameterized compact thermal models, designers at different design levels can have more productive interactions and collaborations during early design stages. This leads to early discovery and considerations of potential thermal hazards of the system. True *physical* parameterization requires that the models be constructed solely on design geometries and material properties.

15.2.2 Boundary-Condition Independence (BCI)

A crucial feature of compact thermal models is boundary-condition independence (BCI). By achieving BCI, assumptions about the environment (e.g., ambient temperature) do not affect the construction of the actual model. Prior *package-level* compact thermal models in [3, 4] achieve BCI by finding a thermal resistance network with minimum overall error when used with different boundary conditions [4–6]. At the chip and block level, we also need to find a way to construct a thermal model that is BCI.

15.2.3 Analogy of Electrical and Thermal Phenomena

The compact model must be easy to construct. One possibility is to utilize a well-known analogy of heat transfer and electrical phenomena, as shown in Table 15.1. In this analogy, heat flow that passes through a thermal resistance is analogous to electrical current; temperature difference is analogous to voltage. Similar to an electrical capacitor that accumulates electrical charges, thermal capacitance defines the capability of a structure to absorb heat. The rationale behind this analogy is that electrical current and heat flow can be described by a similar set of differential equations (there is no thermal equivalent of electrical inductance though). A compact thermal model is essentially a thermal RC circuit. Each node in the circuit corresponds to a block at the desired level of granularity. The heat dissipation of each block is modeled as a current source connected to the corresponding node. Solving this thermal RC circuit gives the temperatures of each node.

Table 15.1 Analogy of thermal and electrical quantities

Thermal quantity	Unit	Electrical quantity	Unit
P, Heat flow, power	W	I, Current	A
T, Temperature difference	K	V, Voltage	V
R_{th}, Thermal resistance	K/W	R, Electrical resistance	Ω
C_{th}, Thermal capacitance	J/K	C, Electrical capacitance	F

15.2.4 Detailed Temperature Information

A compact thermal model should also provide thermal information at the desired level of abstraction. For example, for package-level compact thermal models, previous studies [7, 8] have shown that the information of temperature distribution across the package is required. Using only a single junction-to-case thermal resistance leads to an inferior package design; instead, multiple nodes are needed on the package surfaces. Similarly, a compact thermal model at the silicon level should consist of enough nodes for detailed temperature distribution information across the die. In addition, both static and transient temperatures need to be modeled.

15.2.5 Granularity

A compact thermal model needs to match the spatial and temporal granularities to the level of abstraction (lower granularities at higher levels, and higher granularities at lower levels) in order to hide the details of the lower levels and make sure that the compact thermal model itself is no more complex than necessary. Modeling at finer granularity introduces unnecessary details and makes the computation slower. For example, system-level package compact thermal models, such as the DELPHI models [3, 4], hide the lower level details of the package structures, including the die, the thermal attach, and the solder balls, mainly because these details are intellectual properties of the vendors, but also because they would increase the complexity of the model without significantly improving the simulation accuracy. Similarly, a compact thermal model at the die level should hide the lower level details of the die, such as the actual circuit structures and physical layout.

15.2.6 Pitfalls

In order to construct a reasonably accurate yet compact thermal model for processor and SoC designs, there are several points worth considering:

- *Aspect ratio*. We found that a block with high aspect ratio cannot be modeled as one node. This is because the heat spreading in one direction dominates the other directions, resulting in a significantly non-uniform temperature distribution within the block, and hence a single temperature is no longer representative to the entire block. The solution is to divide the block into several sub-blocks with aspect ratios close to unity [9]. This also applies in the depth dimension, i.e., if a block occupies a tiny chip area (e.g., $100 \times 100\,\mu m$), whereas the chip is $700\,\mu m$ thick, it is better to further divide the chip into multiple sub-layers or to combine the tiny block with adjacent blocks to make the aspect ratio in the depth dimension closer to unity.
- *Package components*. It is tempting to simplify the thermal package (including thermal interface material (TIM), heat spreader, heat sink, and the secondary heat transfer path) into only one or a few nodes and try to have a simple thermal boundary condition for the silicon die. The result, however, may be inaccurate and yield misleading silicon temperature estimations. The right thing to do is to analyze the geometries and thermal properties of each package components and understand their impact on the die temperature before simplifying them [10].
- *Granularity*. In most cases, it is not necessary to model temperature at granularities that are so fine as to be computationally infeasible, such as nanometers and nanoseconds. This is because, in time, the thermal time constants (milliseconds or more) are much longer than electrical time constants; and the lateral heat spreading makes a small heat source (size less than a few hundred microns on each side) relatively cool [11]. We show some analysis and examples related to this issue in later sections of the chapter.

15.3 The *HotSpot* Thermal Modeling Approach

In this section, we present a compact thermal modeling method, named *HotSpot*, which takes a structured assembly approach of constructing a physical compact thermal model by first modeling the silicon die and other packaging components as a collection of simple 3D shapes, and then assembling them into more complex compact thermal models according to the overall structure. This modeling approach is fully parameterized and satisfies most of the above desirable features as shown later in the chapter.

15.3.1 Overview

When constructing a compact thermal model with HotSpot, one needs to first identify the different layers of the design that are made of different materials. This requires the designer have some prior detailed knowledge of the (intended) design structure. These layers are then stacked on top of each other as shown in Fig. 15.3. The layers can, for instance, represent heat sink, heat spreader, silicon substrate, on-chip interconnect layer, C4 pads, ceramic packaging substrate, solder balls, etc. Usually, the surface that generates heat is the surface of the silicon substrate layer.

Each layer is then divided into a number of blocks. For example, in Fig. 15.4c, the silicon substrate layer can be divided according to architecture-level blocks (only three blocks are shown for simplicity) or finer granularity, depending on what the die-level design requires. For other layers that may require less detailed thermal information, one can simply divide that layer as illustrated in Fig. 15.4a. The center shaded part in a layer shown by Fig. 15.4a is the area covered by another adjacent layer such as the one shown in Fig. 15.4c. This center part can have the same number of nodes as its smaller neighbor layer or collapse those nodes into a single node, depending on the accuracy and computation overhead requirements. The remaining part at the periphery in Fig. 15.4a is then divided into four trapezoidal blocks, and each is assigned one thermal node. Each block in each layer has a vertical (depth) thermal resistance and several lateral resistances, which model vertical heat transfer to its neighbor layers and lateral heat spreading/constriction within the layer itself, respectively. Figure 15.4b shows a side view of one block with both the lateral and the vertical resistances. Vertical (depth) resistances can

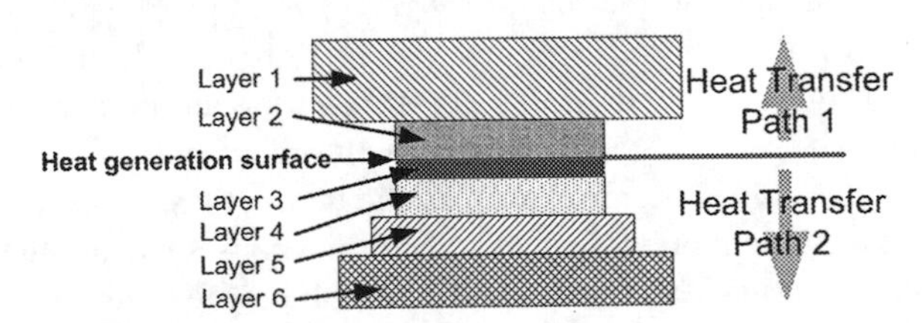

Fig. 15.3 The stacked layers of different materials in the HotSpot modeling approach. Heat generating surface and major heat transfer paths are also shown [12]

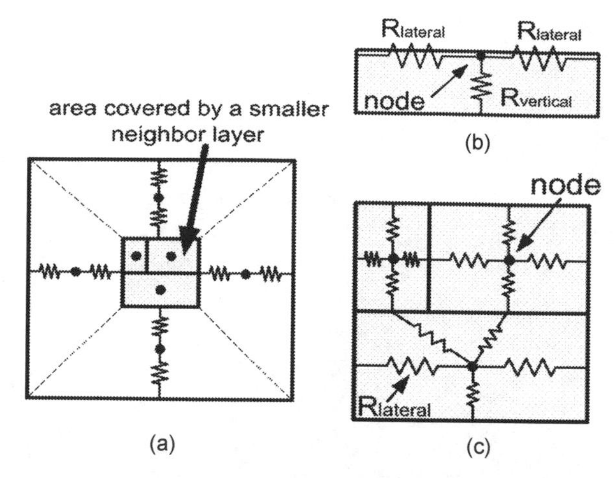

Fig. 15.4 (**a**) Area division of larger layers (*top view*). (**b**) Side view of one block with its lateral and vertical (depth) thermal resistances. (**c**) A layer, for example, the silicon die, can be divided into arbitrary number of blocks if detail thermal information is needed (*top view*) [13]

be calculated by $R_{\text{vertical}} = t/(k \cdot A)$, where t is the thickness of that layer, k is the thermal conductivity of the material of that layer, and A is the cross-sectional area of the block. Calculating lateral thermal resistances is not as straightforward as the depth resistances. This is because of the complex nature of modeling heat spreading and constriction. One can consider the lateral thermal resistance of one block as the spreading/constriction thermal resistance of the other parts within a layer to that specific block.

For layers that have surfaces interfacing with the ambient, i.e., the boundaries, we assume that each surface has a constant heat transfer coefficient h. The corresponding thermal resistance can then be calculated as $R_{\text{convection}} = 1/(h \cdot A)$, where A is the surface area. Strictly speaking, these convection thermal resistances are not part of the compact thermal model, because they include the information of the environment. If the environment changes, i.e., the boundary conditions change, the value of these convection resistances also change. On the other hand, for a particular design, the values of all the other thermal resistances shown in Fig. 15.4a–c should not change if the compact thermal model is BCI.

The HotSpot modeling approach first described in [14] has been successfully used in many academic as well as industry research activities.

15.3.2 Primary and Secondary Heat Transfer Paths

In HotSpot, we also consider lateral heat flow within a layer by adding lateral thermal resistances to achieve greater accuracy of temperature estimation. Figure 15.5 shows a modern single-chip CBGA package [15]. Heat generated from the active silicon device layer is conducted through the silicon die to the thermal interface material, heat spreader, and heat sink and then convectively removed to the ambient.

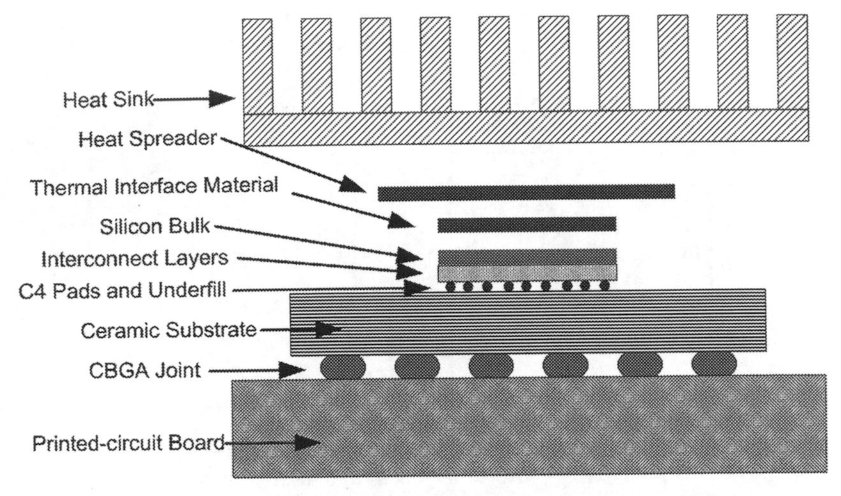

Fig. 15.5 Packaging components in a typical CBGA package, adapted from [15]

In addition to this primary heat transfer path, a secondary heat flow path exists due to conduction through the interconnect layer, I/O pads, ceramic substrate, leads/balls to the printed-circuit board. HotSpot models all these layers in both heat flow paths. Although the primary path removes most heat (more than 90%) in advanced cooling packages, the secondary path can also be essential in cooling solutions where forced convection or heat sinks are not used, as in most low-power SoCs.

15.3.3 Functional Units Versus Regular Grids

Layers such as the silicon die, the interface material, and the center part of the heat spreader can either be divided naturally according to functional unit shapes or be divided into regular grid cells, depending on the needs of the designer. For example, if a chip designer needs to estimate average temperatures for each functional block a thermal model at the functional unit granularity is best in that case. However, for a package designer, the temperature gradients across silicon die and other package layers are important metrics to evaluate the reliability of the package. In this case, a grid-like thermal model is more suitable, since it provides more detailed estimations of maximum and minimum temperatures, whereas the functional block-level thermal model does not. HotSpot offers a choice between a possibly irregular functional unit partitioning and a regular grid of variable granularity.

15.3.4 Model Validation

We performed a number of validations for the HotSpot thermal model—with detailed finite-element modeling software, thermal test chip [1], infrared thermal imaging [16], and actual on-chip temperature sensors [17]. We have also validated that the HotSpot thermal modeling approach is boundary-condition independent [13].

15.4 Case Studies

In this section, we provide several brief case studies illustrating how to use a compact thermal modeling approach such as HotSpot in various aspects of thermal analysis and temperature-aware design for processors and SoCs.

15.4.1 A SoC Thermal Analysis Example

To illustrate the importance of temperature-aware design early in the design process, we show the thermal analysis together with the temperature-leakage closed loop for a SoC design [9]. We use InCyte®, a commercial early design estimation tool[1] to reconstruct an SoC design based on the published 180 nm design data in [18]. We use HotSpot 4.0 for the thermally self-consistent leakage analysis of this SoC design.

We picked logic and memory modules similar to those in [18] from InCyte's incorporated IP libraries and came up with an early SoC design whose total power was almost identical to data reported in [18]. InCyte also outputs a preliminary floorplan for the design. Notice that InCyte estimates leakage power of each block at a constant temperature. The estimated temperature map of this 180 nm design is shown in Fig. 15.6.

Because InCyte does not yet include the temperature dependence of leakage, whereas sub-threshold leakage is exponentially dependent on temperature, we double-check to see whether the thermally self-consistent leakage power causes thermal problems to this 180 nm SoC design. Using HotSpot and the simple leakage model in [10] to iterate the temperature-leakage loop as shown in Fig. 15.7, after convergence we find that the final total leakage is only a negligible 546 μW for this design with the selected package. Therefore, the above temperature estimation is quite accurate without considering the temperature-leakage loop.

However, if we re-design this SoC design in a 90 nm technology, there are two design possibilities: (1) We can scale both the area and active power of each individual blocks and thus maintain the same function and complexity. This means that the total power of the entire design is also scaled accordingly, and the power density remains roughly the same due to area scaling. Therefore, we can use a cheaper thermal package for less overall power consumption and keep the chip below the 85°C thermal constraint. (2) Since the ITRS [19] projects that the die size and power remain the same, if not increase, across different technology nodes, we can

[1] http://www.chipestimate.com/

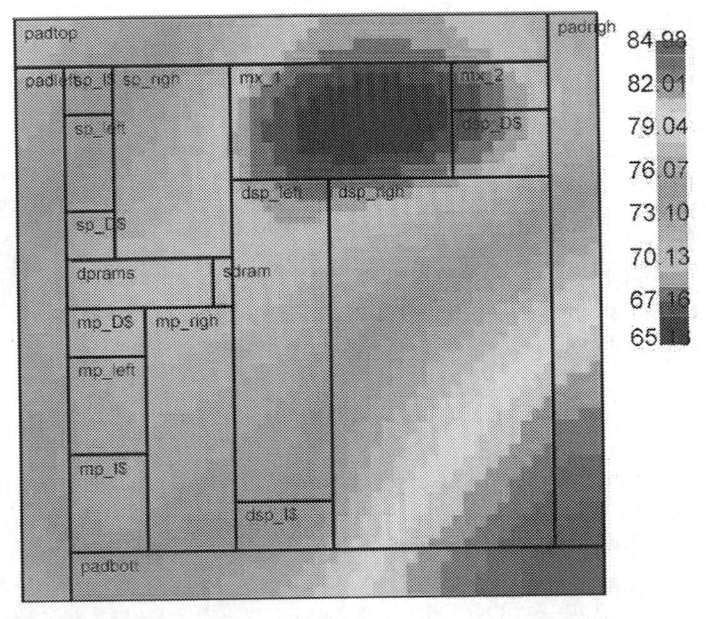

Fig. 15.6 Estimated temperature map of an SoC design at 180 nm technology, based on data in [18] and InCyte. Temperatures are in degree celsius [9]

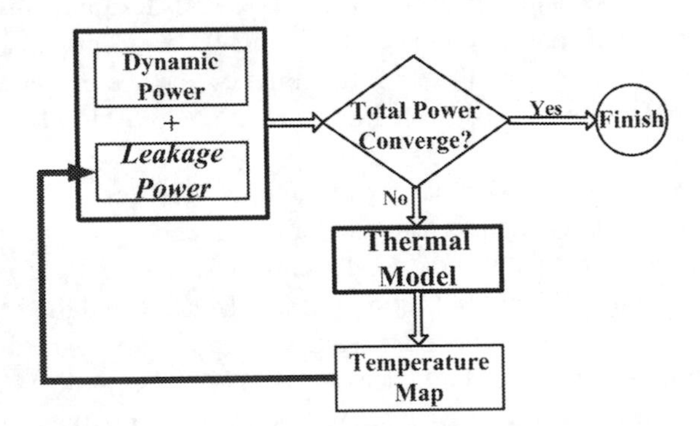

Fig. 15.7 A thermal model closes the loop for leakage power calculation [10]

alternatively assume that the total active power and chip area remain the same as those in the 180 nm design. Assuming a floorplan similar to that in 180 nm technology, this is equivalent to adding more parallelism (such as more processing cores and higher memory bandwidth) to the die and designing the chip for higher throughput by burning more power. In this case, with the same thermal package, after iterating with the leakage-temperature loop, the hottest on-chip temperature exceeds the thermal threshold and eventually causes thermal runaway! The reason is twofold: (1) at 90 nm, a greater fraction of total power consumption is caused by leakage [19], and (2) the sub-threshold leakage power's dependency on temperature is stronger at

90 nm than at 180 nm (using leakage model coefficients in [20, 21]). The results are listed in Table 15.2.

Table 15.2 As technology scales, temperature dependence of sub-threshold leakage power becomes more problematic. Without early-stage thermally optimized design flow, thermal runaway can happen even for low-power SoC designs

	Active power	Avg. temp rise	Hottest temp rise	Actual leakage	Leakage error at const temp
180 nm, orig. design	1.7 W	30.77°C	59.98°C	546 μW	116%
90 nm, scaled power area	316 mW	6.19°C	25.42°C	24 mW	66%
90 nm, same power area	1.7 W	>35.08°C	Runaway	>277 mW	>680%

The above SoC design example shows that it is crucial to incorporate thermal estimations (such as leakage-temperature dependence) early in the design process in order to locate potential thermal hazards that are too costly to fix in later design stages. At this early design stage, possible solutions to the SoC design at 90 nm can be as follows: (1) circuit designers can choose IPs that have high-Vt transistors and use reverse body-bias or sleep transistors for non-critical paths to reduce leakage; (2) architects can consider using dynamic voltage and frequency scaling (DVFS), migrating computation [20, 22], more parallelism, and temperature-aware floorplanning techniques [23], etc., to reduce hot spot temperatures. Alternatively, (3) package designers need to consider the possibility of adding a heatsink or a fan. Trade-offs among portability, cost, performance, and temperature have to be made.

15.4.2 Exploring Design Choices with Parameterization in HotSpot

As mentioned before, HotSpot is parameterized in terms of geometries and material properties of silicon, package components, and cooling solutions. Achieving full parameterization of compact thermal models is important in many ways. It allows designers at all design stages to freely explore all the possible thermally related design spaces. For example, the heat spreader and heat sink are two important package components for high-performance VLSI designs. While a large heat spreader and heat sink made from high thermal conductivity materials can reduce the temperature of silicon die, they also significantly increase the total price of the system. Therefore, exploring design trade-offs between hot spot temperatures and package cost is crucial. With parameterized compact thermal models, this exploration can be done easily and efficiently by simply sweeping the size of the heat spreader and heat sink with different material properties to achieve the desired package design point. On the other hand, building package prototypes or detailed thermal models greatly slows the design process and increases the design cost.

For illustration, we first present an example analysis to show the strength of using parameterized compact thermal models to efficiently investigate the impact of thermal interface material on across-silicon temperature differences. Figure 15.8 shows the relationship between the thickness of the thermal interface material (TIM), which glues the silicon die to the heat spreader. We plot the temperature readings from a compact thermal model with 40 × 40 grid cells on silicon. This analysis is based on an Alpha 21364-like floorplan shown in Fig. 15.9a. Average silicon die temperature is also plotted in Fig. 15.8 for reference. The total heat generated from the silicon surface is 40.2 W, the die size is 15.9 × 15.9 × 0.5mm, and the thermal conductivity of the thermal interface material is 1.33 W/(m-K).

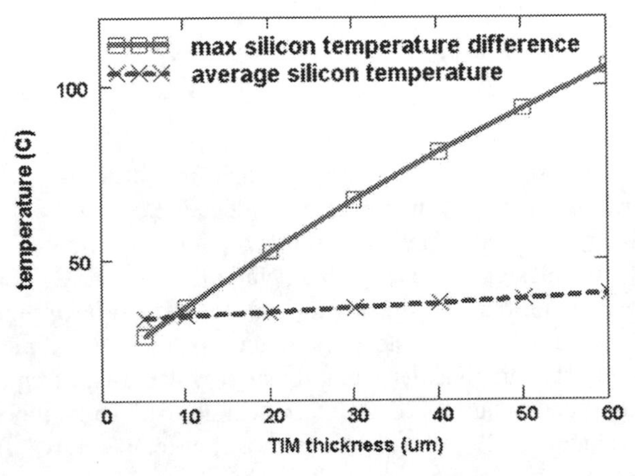

Fig. 15.8 The impact of thermal interface material (TIM) thickness to silicon die temperature difference. Average silicon die temperature is also plotted as reference [10]

As can be observed from Fig. 15.8, although the TIM thickness does not have an obvious impact on the average die temperature, thicker TIM results in poor heat spreading which leads to large temperature differences across the die. Such large temperature differences may be disastrous to circuit performance and die/package reliability. Using a better heat sink will only lower the average silicon temperature but will not help to reduce the temperature difference. From this analysis, which has been easily performed by our parameterized model, we can reach the conclusion that using as thin as possible a thermal interface material is one of the key opportunities for package designers to consider and more important than, for example, using a larger heatsink. In some recent work [4, 24], the importance of measuring and modeling thermal interface's impacts on the entire package has been also discussed, although not at the same silicon die level as the above example shows.

Another example of utilizing our parameterized compact thermal model is the investigation of a dynamic thermal management technique (DTM) at the micro-architecture level called migrating computation (MC) [22]. It is obvious that two silicon-level functional units that run hot by themselves will tend to run even hotter when adjacent. On the other hand, separating them will introduce additional communication latency that is incurred regardless of operating temperature. This suggests the use of spare units located in cold areas of the chip, to which computation can migrate only when the primary units overheat. In this study, the floorplan of the Alpha 21364 core is carefully changed to include an extra copy of integer register file (see Fig. 15.9), which is usually the hottest spot on the silicon die for this design. We also model the fact that accessing the secondary integer register file entails extra power and extra access time due to the longer distance. With this new floorplan, we can shift the workload back and forth between the two register files when the one in use overheats, with a little performance overhead (11.2% slower). The changes in silicon floorplan can be easily adapted into corresponding parameterized compact thermal models, and thus the temperatures of functional units can be analyzed efficiently to investigate the usefulness of the migrating computation DTM technique. By doing this, packaging complexity and cost can be significantly reduced and still yield almost the same operating temperature and performance as with much more expensive and complicated thermal package.

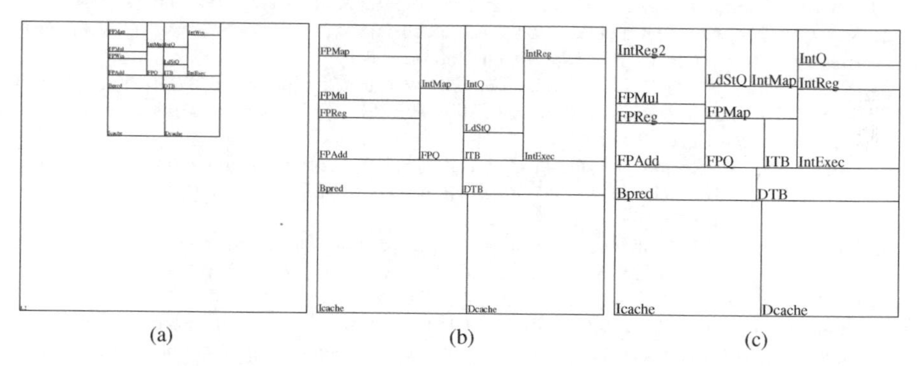

Fig. 15.9 (**a**) Approximated floorplan of Alpha 21364 microprocessor. (**b**) Close-up look of the 21364 core, with only one register file on the *top-right corner*. (**c**) 21364 core with an extra copy of register file at the *top-left corner* [14]

15.4.3 Other Examples of HotSpot Usage by the Temperature-Aware Research Community

There are many other applications of the HotSpot thermal model. For example, temperature-aware floorplanning for both 2D [23] and 3D chips [25]; temperature considerations for reliability-aware design [21]; exploration of the thermal constraint on the design space for chip multiprocessor architectures [26, 27]; dynamic thermal management at the architecture level [14]; on-chip temperature sensor placement [28, 29]; temperature-aware high-level synthesis [30]; temperature-aware supply voltage planning in SoC [31]; thermal-aware SoC test scheduling [32], etc. We encourage interested readers to explore the plethora of research topics related to temperature-aware design.

15.5 Conclusion

In this chapter, we reviewed the important concept of temperature-aware design, especially during early stages of SoC and processor designs. We then introduced an accurate yet fast thermal modeling approach—HotSpot. HotSpot is unique and useful because it is compact, correct by construction, flexible, and parameterized. Since its introduction, HotSpot has been adopted and applied in many research areas in processor, SoC, and package design.

References

1. Huang, W., Stan, M.R., Skadron, K., Sankaranarayanan, K., Ghosh, S., Velusamy, S.: Compact thermal modeling for temperature-aware design. In: *Proceedings of the ACM/IEEE Design Automation Conference (DAC)*, pp. 878–883, San Diego, CA, USA, June (2004).
2. Sabry, M.-N.: Compact thermal models for electronic systems. *IEEE Trans Components Packaging Technol* **26**(1), 179–185, March (2003).
3. Rosten, H.I., Lasance, C.J.M., Parry, J.D.: The world of thermal charaterization according to DELPHI—part I: Background to DELPHI. *IEEE Trans Components Packaging Technol* **20**(4), 384–391, December (1997).
4. Lasance, C.J.M., Rosten, H.I., Parry, J.D.: The world of thermal charaterization according to DELPHI—part II: Experimental and numerical methods. *IEEE Trans Components Packaging Technol* **20**(4), 392–398, December (1997).
5. Lasance, C.J.M.: The influence of various common assumptions on the boundary-condition-independence of compact thermal models. *IEEE Trans Components, Packaging, Manufacturing Technol–Part A* **27**(3), 523–529, September (2004).
6. Vinke, H., Lasance, C.J.M.: Compact models for accurate thermal charaterization of electronic parts. *IEEE Trans Components, Packaging, Manufacturing Technol–Part A* **20**(4), 411–419, December (1997).
7. Rosten, H., Lasance, C.: Delphi: The development of libraries of physical models of electronic components for an integrated design environment. In: *Proceedings on Conference of International Electronics Packaging Society (CIEPS)*, (1994).
8. Bar-Cohen, A., Elperin, T., Eliasi, R.: θ_{jc} charaterization of chip packages—justification, limitations and future. *IEEE Trans Components, Hybrids, Manufacturing Technol* 12, 724–731, December (1989).
9. Huang, W., Skadron, K., Sankaranarayanan, K., Ribando, R.J., Stan, M.R.: Accurate, pre-RTL temperature-aware processor design using a parameterized, Geometric thermal model. *IEEE Trans Comput* **57**(9), 1277–1288, September (2008).
10. Huang, W., Humenay, E., Skadron, K., Stan, M.: The need for a full-chip and package thermal model for thermally optimized IC designs. In: *Proceedings of the ACM/IEEE International Symposium on Low Power Electronic Design (ISLPED)*, pp. 245–250, San Diego, CA, USA, August (2005).
11. Huang, W., Stan, M., Sankaranarayanan, K., Ribando, R., Skadron, K.: Many-core design from a thermal perspective. In: *Proceedings of the ACM/IEEE Design Automation Conference (DAC)*, Anaheim, CA (2008).
12. Huang, W., Stan, M.R., Skadron, K., Ghosh, S., Velusamy, S., Sankaranarayanan, K.: Hotspot: A compact thermal modeling methodology for early-stage vlsi design. *IEEE Trans Very Large Scale Integration (VLSI) Syst* **14**(5):501–513, May (2006).
13. Huang, W., Stan, M.R., Skadron, K.: Parameterized physical compact thermal modeling. *IEEE Trans Components Packaging Technol* **28**(4), 615–622, December (2005).

14. Skadron, K., Stan, M.R., Huang, W., Velusamy, S., Sankaranarayanan, K., Tarjan, D.: Temperature-aware microarchitecture. In: *Proceedings of the ACM/IEEE International Symposium on Computer Architecture (ISCA)*, pp. 2–13, San Diego, CA, USA, June (2003).
15. Parry, J., Rosten, H., Kromann, G.B.: The development of component-level thermal compact models of a C4/CBGA interconnect technology: The motorola PowerPC 603 and PowerPC 604 RISC microproceesors. *IEEE Trans Components Packaging Manufacturing Technol–Part A* **21**(1), 104–112, March (1998).
16. Huang, W., Skadron, K., Gurumurthi, S., Ribando, R.J., Stan, M.R.: Differentiating the roles of IR measurement and simulation for power and temperature-aware design. In: *Proceedings of the IEEE International Symposium on Performance Analysis of Systems and Software (ISPASS)*, Boston, MA, USA, April (2009).
17. Velusamy, S., Huang, W., Lach, J., Stan, M.R., Skadron, K.: Monitoring temperature in fpga based socs. In: *Proceedings of the IEEE International Conference on Computer Design (ICCD)*, pp. 634–637, San Jose, CA, USA, October 2005.
18. Stolberg, H., Moch, S., Friebe, L., Dehnhardt, A., Kulaczewski, M., Berekovic, M., Pirsch, P.: An soc with two multimedia dsps and a risc core for video compression applications. In: *Digest of Papers, IEEE International Solid-State Circuits Conference (ISSCC)*, San Francisco, CA, USA, February (2004).
19. The International Technology Roadmap for Semiconductors (ITRS), (2007).
20. Heo, S., Barr, K., Asanovic, K.: Reducing power density through activity migration. In: *Proceedings of the ACM/IEEE International Symposium on Low Power Electronics and Design (ISLPED)*, pp. 217–222, Seoul, Korea, August (2003).
21. Srinivasan, J., Adve, S.V., Bose, P., Rivers, J.A.: The impact of technology scaling on lifetime reliability. In: *Proceedings of the International Conference on Dependable Systems and Networks (DSN)*, Florence, Italy, June (2004).
22. Skadron, K., Sankaranarayanan, K., Velusamy, S., Tarjan, D., Stan, M.R., Huang, W.: Temperature-aware microarchitecture: Modeling and implementation. *ACM Trans Arch Code Optim* **1**(1), 94–125, March (2004).
23. Sankaranarayanan, K., Velusamy, S., Stan, M.R., Skadron, K.: A case for thermal-aware floorplanning at the microarchitectural level. *J Instr-Level Parallelism* **7**, October (2005).
24. Lasance, C.J.M.: The urgent need for widely-accepted test methods for thermal interface materials. In: *Proceedings of the 19th IEEE SEMI-THERM Symposium*, pp. 123–128, San Jose, CA, USA, (2003).
25. Han, Y., Koren, I., Moritz, C.A.: Temperature-aware floorplanning. In: *Proceedings of Workshop on Temperature-Aware Computer Systems (TACS)*, Madison, WI, USA, (2005).
26. Li, Y., Lee, B., Brooks, D., Hu, Z., Skadron, K.: CMP design space exploration subject to physical constraints. In: *Proceedings of the IEEE International Symposium on High Performance Computer Architecture (HPCA)*, Austin, TX, USA, (2006).
27. Chaparro, P., Gonzalez, J., Magklis, G., Cai, Q., Gonzalez, A.: Understanding the thermal implications of multicore architectures. *IEEE Trans Parallel Distributed Syst* **18**(8), 1055–1065, (2007).
28. Memik, S.O., Mukherjee, R., Ni, M., Long, J.: Optimizing thermal sensor allocation for microprocessors. *IEEE Trans Comput Aided Design* **27**(3), 516–527, March (2008).
29. Sharifi, S., Simunic Rosing, T.: Accurate temperature sensing for efficient thermal management. In: *Proceedings of the IEEE International Symposium on Quality Electronic Design (ISQED)*, San Jose, CA, USA, (2008).
30. Memik, S.O., Mukherjee, R.: An integrated approach to thermal management in high-level synthesis. *IEEE Trans VLSI* **14**(11), 1165–1174, November (2006).
31. Yang, S., Wolf, W., Vijaykrishnan, N., Xie, Y.: Reliability-aware SOC voltage islands partition and floorplan. In: *Proceedings of IEEE Annual Symposium on VLSI (ISVLSI)*, Montpellier, France, (2006).
32. He, Z., Peng, Z., Eles, P., Rosinger, P., Al-Hashimi, B.M.: Thermal-aware SoC test scheduling with test set partitioning and interleaving. *J Electron Testing Theory Appl* **24**(1–3), 247–257, (2008).

Chapter 16
Rapid Technology-Aware Design Space Exploration for Embedded Heterogeneous Multiprocessors

Marc Duranton, Jan Hoogerbrugge, Ghiath Al-kadi, Surendra Guntur, and Andrei Terechko

Abstract Multicore architectures provide scalable performance with a hardware design effort lower than for a single core processor with similar performance. This chapter presents a design methodology and an embedded multicore architecture focusing on boosting performance density and reducing the software design complexity. The methodology is based on a predictive formula computing performance of heterogeneous multicores, which allows drastic pruning of the design space for few accurate simulations. Using this design space exploration methodology for high definition and quad high definition H.264 video decoding, the resulting areas for a multicore system in CMOS 45 nm are 2.5 and 8.6 mm^2, respectively. These results show that heterogeneous chip multiprocessors are cost-effective for embedded applications.

16.1 Introduction

Emerging video, graphics, and modem applications demand always higher performance from embedded systems but economics dictates a comparable or decreasing cost, translated into a similar or lower silicon area. Due to the increasing diversity of functions an embedded system should be able to perform, fixed hardware solutions are more and more replaced by programmable solutions, pushing the flexibility into software. To boost performance density, embedded chips often deploy domain-specific multicore architectures where each core runs a relatively independent application (e.g., audio decoding), requiring little synchronization and communication with other cores. However, if a demanding application with sufficient task-level parallelism (TLP), such as high definition (HD) H.264 video decoding,

M. Duranton (✉)
NXP Semiconductors, Corporate I&T / Research, High Tech Campus 32, 5656 AE Eindhoven, The Netherlands
e-mail: marc.duranton@gmail.com

R. Leupers, O. Temam (eds.), *Processor and System-on-Chip Simulation*,
DOI 10.1007/978-1-4419-6175-4_16, © Springer Science+Business Media, LLC 2010

exceeds the compute power of a standalone core, this application should be split into tasks and run on several cores in parallel.

Although general purpose and server processors have already adopted multicore architectures, embedded chip multiprocessors (eCMP) still compete with function-specific hardware accelerators in terms of performance density and energy efficiency. Therefore, each core of the eCMP has to be optimized for the set of applications the system will have to process, and the global system should be supplemented by "generic" coprocessors for tasks that are still inefficiently performed by the cores. Of course, the specialization should not lead to a too much specialized eCMP; it should be usable for one or several domains of application so that its development cost will be shared by all use cases. On top of that, the widening design productivity gap encourages core replication in favor of designing each time a different function-specific hardware.

The problem of architecting an efficient eCMP is highly complex and the number of design choices to choose becomes huge. All parts are linked together and optimizing, one element could lead to the loss of efficiency elsewhere: it is not enough to make only local optimizations; global optimization is also required. Tools that allow to navigate in the parameter space and to choose an optimum are key to the success and to the short time to market. One approach for this design space exploration (DSE) is to combine domain knowledge, analytical approximations, and execution-driven simulations. Simulations should be fast, reliable for systems where a lot of parameters are still variable. The rest of this chapter will describe a methodology of DSE that combines different approaches to architect heterogeneous multicore systems.

16.2 Characteristics of Embedded Multimedia Applications

The targeted multimedia applications for the considered eCMP include audio/video codecs (H.264, for example), video enhancement applications, motion-compensated frame rate conversion, and 3D graphics. Image resolution of video applications tends to grow from standard definition (0.5 million pixels) to high definition (2 million pixels) to quad high definition (8 million pixels) and even to super hi-vision (32 million pixels). *Data partitioning* of a task-level parallel application, also known as single-program multiple data [5], scales well with the growing video resolution from standard definition to high definition to super high definition.

Many of these applications contain task-level parallelism (TLP) [3]. Modern video codecs, as well as the frame rate conversion algorithms, feature tasks with variable execution time and inter-task dependencies. For example, Fig. 16.1 shows three video frames split in macroblocks of pixels (0,0), (0,1), (1,0), etc. Red arrows denote dependencies between macroblocks in the H.264 deblocking filter. Macroblock (1,2), for instance, depends on macroblocks (1,1) and (0,2). On the right in Fig. 16.1 is presented an execution trace of the macroblocks, showing variable execution time of macroblocks depending on the video content. The variation in execution time of the tasks may reach an order of magnitude, for example, in the

H.264 video decoder the PicturePrediction kernel's execution time varies between 250 and 3500 cycles on a TriMedia VLIW [16]. *Dynamic task scheduling* enables better load balancing in the context of variable execution time multimedia tasks [16] and also simplifies (in-field) software extensibility and upgradeability. Fast and fair synchronization between such tasks is key in achieving high performance.

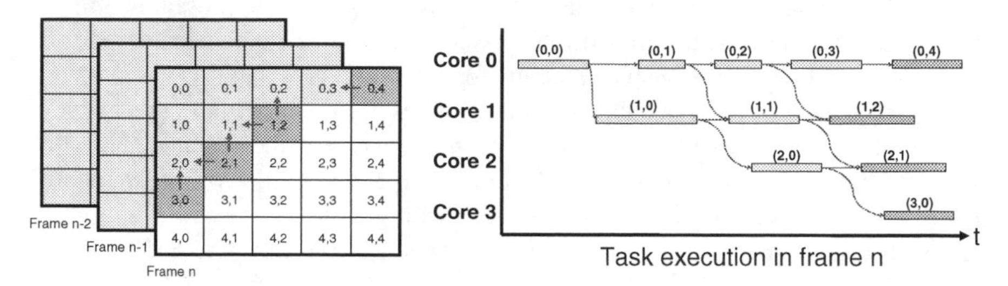

Fig. 16.1 Task-level parallelism in multimedia workloads

16.3 Architecture Constraints and a Proposed Architecture Template

The template of the architecture of the proposed eCMP, called Ne-XVP for Nexperia eXtreme Versatile Processor, is shown in Fig. 16.2. It is built around VLIW cores, which together with a highly optimizing ANSI C/C++ compiler [9] efficiently

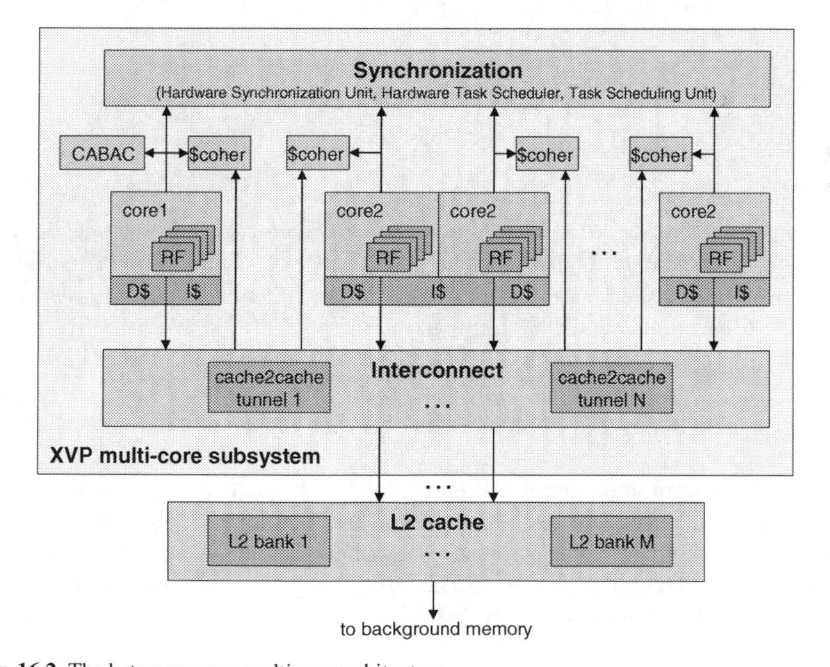

Fig. 16.2 The heterogeneous multicore architecture

exploit the instruction-level parallelism (ILP). The VLIW cores based on the TriMedia TM327x processors [21] feature an ISA tuned for multimedia processing including SIMD operations and super-operations with more than two source operands and/or more than one destination operand. Moreover, the cores have advanced instruction and data caches with prefetching and collapsed operations to efficiently address the memory bottleneck without compromising ease of programmability. A previous work [10] demonstrated that shared memory multicores enable performance scalability with reasonable software design effort.

Hill [8] advocates that Amdahl's law necessitates asymmetric multicores, where one larger core combats the sequential bottleneck, whereas the remainder of silicon real-estate is covered with small efficient cores. Ne-XVP heterogeneous multiprocessor extends the asymmetric architecture with generic coprocessors speeding up task and data management to improve utilization of the compute and memory resources.

Data partitioning, dynamic task scheduling, and shared memory with hardware synchronization primitives deliver good performance for complex multimedia applications while keeping the program design effort minimal. To support these programming model choices, Ne-XVP features a set of hardware synchronization primitives in the *hardware synchronization unit* and a cache-coherent shared memory using a *hardware coherence monitor*. Moreover, different programming models are supported, e.g., the ACOTES programming model [1].

Although there are many known methods to improve silicon efficiency, the Ne-XVP architecture template only uses those that do not compromise the advantages of the multicore architectures, including high scalability, low design effort, and ease of application software development. The evaluated methods are

1. (micro-)architectural downscaling;
2. heterogeneous multicore architectures;
3. multithreading;
4. hardware accelerators; and
5. conjoining.

A typical TriMedia TM3270 core with 128 KB data cache, which was dimensioned for a standalone instantiation in a SoC, is not adequate for a multicore architecture. Instead of keeping one large core, *(micro-)architectural downscaling* of the core can improve the performance density of the final eCMP. To downscale the microarchitecture of a VLIW processor, cache sizes, internal buffering, bypasses, etc., can be changed. On top of that, synthesizing for lower clock frequency can help improve performance density without changing the instruction set architecture.

Amdahl's law, on the other hand, indicates that the sequential part can quickly become the bottleneck for a multiprocessor. Furthermore, single-thread performance is of high importance, because some functions cannot be parallelized due to their sequential nature (e.g., context-adaptive binary arithmetic coding (CABAC) [6]). Amdhal's law and the single thread performance motivate for a *heterogeneous multicore architecture*, where at least one core is more powerful than the others in order to accelerate the sequential code, while the other more efficient cores tackle the parallel workload. This is known as asymmetric multicores which are superior to homogeneous multicore systems in terms of both performance density [8] and energy efficiency [14].

The explorations are restricted to two types of cores—core1 and core2. More cores type in the architecture would compromise ease of hardware design and verification. Moreover, Kumar [12] suggests that having beyond two types of cores does not significantly contribute to higher performance density.

Hardware multithreading [20] refers to the idea of duplicating processor's state enabling quick context switches between several software threads. Multithreading hides memory latency (by switching to another task when the current one is waiting due to a cache miss), decreases operation latencies, improves code/data sharing, and simplifies the pipeline bypasses and the register file [20]. Hoogerbrugge [10] demonstrated a good potential of hardware multithreading for embedded video applications, resulting in significant performance density boosts.

Hardware accelerators drastically improve silicon efficiency. There are three types of accelerators: generic (e.g., like the hardware synchronization unit—HSU), domain-specific, and function-specific. The hardware task scheduler (HTS) [2] and the CoreManager [13] are examples of hardware task schedulers for video and software-defined radio applications, respectively. In particular, it accelerates applications with a repetitive dependency pattern between n-dimensional tasks. Repetitive dependency patterns are found in H.264 video standard as well as in other coding standards, e.g., the transform coefficient prediction in VC-1. Similar patterns are also present in the motion estimation algorithms, which are widely used in video post-processing such as motion-compensated frame rate up-conversion. The HTS is part of the synchronization block of the Ne-XVP architecture in Fig. 16.2. The architecture also supports traditional fixed-function accelerators. For example, the CABAC speeds up highly sequential context-adaptive binary arithmetic coding of the H.264 video compression standard, attached to core1 in Fig. 16.2.

Multiprocessors duplicate many resources, such as the instruction decode, instruction cache, registers, and interrupt vectors. Although, it simplifies hardware design, over-duplication decreases performance density. *Conjoining* [11] or resource sharing has been investigated in several multiprocessor architectures. On top of conjoining effects in a single multithreaded core, sharing of an instruction cache among several cores [15] was also evaluated. The motivation for such sharing stems from the fact that the applications are data-partitioned and cores often run the same code. Figure 16.2 shows an instruction cache shared between two core2 cores. The data sets of parallel tasks are often disjoint and, therefore, L1 data caches were not shared to avoid cache thrashing effects. Furthermore, floating point units of the cores can be shared at the expense of a performance drop on, for example, 3D graphics applications. More details on the architecture can be found in [19].

As it can be guessed from the previous sections, the heterogeneous architecture template has many parameters (e.g., the number of cores, core configurations, kind of multithreading, and technology) and covers a huge design space. Finding the adequate instantiation of the template for a given workload is obviously a challenge.

This challenge is addressed by an advanced design space exploration methodology based on performance calculation and simulation.

16.4 Dimensions and Parameter Space

In order to evaluate the performance density improvements from the presented architectural techniques for the targeted multimedia applications at the HD and QuadHD resolutions, we defined the following design space:

1. number of "small" cores: [1..15];
2. how many cores share one instruction cache: [1,2,3,4];
3. number of VLIW issue slots: [1,2,3,4,5];
4. number of load/stores: [1, 2];
5. number of registers: [128, 96];
6. data cache size: [1K to 128K in powers of 2];
7. instruction cache size: [1K to 128K in powers of 2]; and
8. number of foreground threads and total number of threads: [(1,1), (1,2), (1,4), (2,2), (2,4)].

Parameters from 3 to 8 relate to a single core. The minimum number of cores in the DSE was two—one "big" core and one "small" core. The sharing factor for the instruction caches equals to the number of small cores sharing one instruction cache. The multithreading option enumerates pairs, composed of the number of foreground threads and total number of threads. Details on multithreading types used are explained in [10].

The variation of parameters for each core creates a population of 6,400 different cores. Hence, there exist 40,953,600 core pairs (one "big" core and one "small" core) to choose from. From these pairs, there are 1,105,747,200 heterogeneous multicores in the design space, assuming that the "small" cores can be instantiated multiple times. Exhaustive simulations of this large design space require 2,303,640 years, if each simulation takes 1 h and is executed on 20 machines in parallel (therefore the importance for a fast simulator that can be executed in parallel). The main focus of the DSE methodology is to quickly explore the large design space.

16.5 Design Space Exploration Methodology

16.5.1 Technology-Aware Design

During technology-aware design [17] a technology model feeds the architect with the clock frequency, area, and power of processor architectures, enabling optimization of execution time, performance density, and energy efficiency instead of traditional cycle counts. Due to the large number of design points the Ne-XVP DSE flow requires a very quick lookup of area and clock frequencies of all hardware components, including cores, accelerators, and networks. Furthermore, the model should be able to estimate how the architecture performs in future IC technology nodes. To satisfy these requirements, the MoodOrgan [18] tool supporting an advanced methodology for technology modeling was developed.

MoodOrgan provides two modes of operation:

1. characterization and
2. selection.

During the first phase, see Fig. 16.3, the architect creates a database of hardware components for a given process technology. Once the database is created, the architect selects and configures components for the processor. The MoodOrgan tool then can compute the delay, area, and power of the processor composed of the selected and configured components. The selection phase, therefore, is fast and can be used for design space explorations, web-portals to the model, etc.

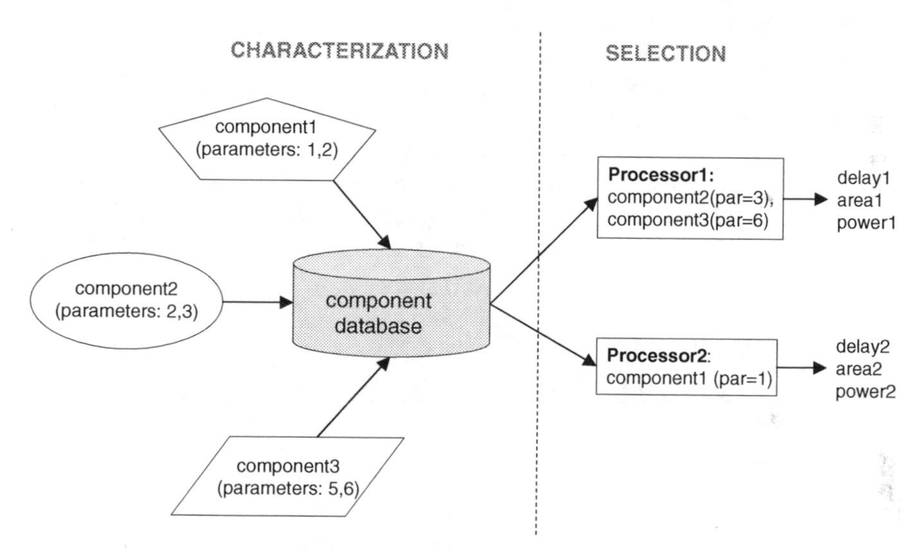

Fig. 16.3 The MoodOrgan methodology's modes of operation

The MoodOrgan tool facilitates the creation of the component database, possibly accompanied by RTL synthesis of components. Furthermore, the tool provides an API to select components and assign values to their parameters. Once a processor is composed of pre-characterized components (e.g., Processor1 composed of component 2 and component 3 with parameters set to 3 and 6 in Fig. 16.3), the API functions allow to compute delay, area, and power of the processor (e.g., delay1, area1, and power1 in Fig. 16.3).

Hardware components are basic modeling elements out of which the architect constructs a complete processor model. A hardware component can be any hardware block, for example, a register file, an ALU, an SRAM memory, a cache controller, a network, and a chip. Naturally, hardware components may have parameters, e.g., bit-width of the ALU.

Each hardware component has three attributes—delay, area, and power in the given process technology. The attributes can be omitted. For example, if a block does not influence the clock frequency of the processor, its delay component can be set to 0. The architect may choose from *three views* to describe a component:

1. analytical;
2. empirical; and
3. RTL.

The analytical view is convenient for describing new components, for which only a rough estimate of physical properties can be captured in a simple formula. In the case of the analytical view, the architect should provide functions for the delay, area, and power of the component. For example, the delay, area, and power of an MMIO unit can be captured in three formulas as a function of the number of registers:

$$\text{Delay(registers)} = 0. \quad (1)$$

$$\text{Area(registers)} = 0.23 * \text{registers}/32 \quad (2)$$

$$\text{Power(registers)} = 0.012 * \text{registers}/32 \quad (3)$$

Delay is expressed in ns, area—in mm^2, and power—in mW/MHz. The delay of the MMIO component is 0, indicating that the architect does not expect it to be limiting the clock frequency of the whole processor.

The view based on empirical data points is a good way of specifying laid out components, fabricated IPs, and for which delay, area, and power are already available. An example of an empirical component is shown below for our hardware task scheduler (Table 16.1).

Table 16.1 The hardware task scheduler—an empirical component

NCORES (parameter)	Delay (ns)	Area (mm^2)	Power (mW/MHz)
1	0	0.0318687	0.0342
2	0	0.0366746	0.0402
4	0	0.0477864	0.0522
8	0	0.0685097	0.0762

As soon as a component is developed, the existing analytical view can be replaced with a more accurate RTL description. The specification of the component will then include paths to RTL files and timing constraints. By this way the architect can gradually improve accuracy of the processor model, keeping the model fully compatible with the rest of the flow (simulator, compiler, web-based portal, etc.). Note that for modeling physical properties of a component, the corresponding RTL does not have to be fully verified. For the RTL component MoodOrgan will synthesize the RTL and extract delay, area, and power during the characterization phase for all possible parameter combinations.

As previously mentioned, the attributes of hardware components correspond to a particular process technology node (e.g., TSMC 90 nm LP), in which the hardware component was characterized. MoodOrgan incorporates delay, area, and power scaling factors enabling the evaluation of hardware components in a different technology node (e.g., 32 nm). Matlab scripts from the PSYCHIC design flow description language [4] were used to generate the scaling factors.

From our production experience, the scaling factors are different for gate-limited components (e.g., ALUs), interconnect-limited components (e.g., register files), and embedded SRAMs (e.g., caches). The methodology for generating the scaling factors involves layouting the functional units (for gate-limited scale factors) and the register file (for interconnect-limited scale factors) of the TriMedia 3,270 processor at 65 nm process technology. The resulting statistics such as the total number of cells, number of rows, effective utilization factor, and the list of cells on the critical path are used in conjunction with the PSYCHIC scripts to determine the delay, area, and power for another process node. The ratios between the other node's results and the 65 nm results give the scale factors for the three component types (gate-limited, interconnect-limited, and SRAMs).

16.5.2 Performance Computation

Essentially, Ne-XVP DSE methodology is built around the computation of heterogeneous eCMP system performances. Based on the computation, only a selected subset of the eCMPs will be simulated to verify their computed performances. The formula used for computation of the eCMP performances is shown below:

$$\begin{aligned} E &= E^{\text{seq}} + E^{\text{par}} \\ &= D^{\text{big}} N^{\text{big}}_{\text{seq.cycles}} + \frac{N^{\text{par}}_{\text{instr}}}{\dfrac{N^{\text{par}}_{\text{instr}}}{D^{\text{big}} N^{\text{big}}_{\text{par.cycles}}} + \dfrac{N^{\text{par}}_{\text{instr}}}{D^{\text{small}} N^{\text{small}}_{\text{par.cycles}}} \text{speedup}(C_{\text{small}}, icsf)}, \end{aligned} \tag{16.1}$$

$$E = D^{\text{big}} N^{\text{big}}_{\text{seq.cycles}} + \frac{1}{\dfrac{1}{D^{\text{big}} N^{\text{big}}_{\text{par.cycles}}} + \dfrac{1}{D^{\text{small}} N^{\text{small}}_{\text{par.cycles}}} \text{speedup}(C_{\text{small}}, icsf)}. \tag{16.2}$$

Execution time E of the application is the sum of the execution time in the sequential E_{seq} and in the parallel E_{par} parts of the application. The sequential execution time equals to the product of the "big" core's clock period D_{big} and the number of cycles of the "big" core used for running the sequential part.

The second term of the formula defines the execution time of the parallel code running on both "big" core and several "small" cores. *icsf* stands for the instruction cache sharing factor and equals to the number of "small" cores sharing one instruction cache. The `speedup()` function captures TLP limits of the application, task creation overhead, and other multicore effects. One of the assumptions of this formula is, though, that the speed of the "big" core is independent of the speed of the "small" cores, which slightly reduces the accuracy of our performance computation. Terechko [19] gives further explanations on the formulas.

16.5.3 Design Space Exploration Flow

The cycle count spent in the sequential and parallel parts on the "big" core and the "small" cores, as well as clock periods of the two core types, is required to compute performance of the eCMP. Figure 16.4 presents the simplified DSE flow, with the formula in the center of the figure. The phases "cycle counts of the cores," "multicore speedup() simulations," and "delay, area, power of the cores" provide to the formula step with the necessary ingredients to compute the performances of all eCMPs in the design space. After that step, only few eCMPs having about the real-time performance are selected for simulation to verify the predictions of the formula. The smallest eCMP in silicon area reaching the real-time performance according to the simulation is the final result of the DSE.

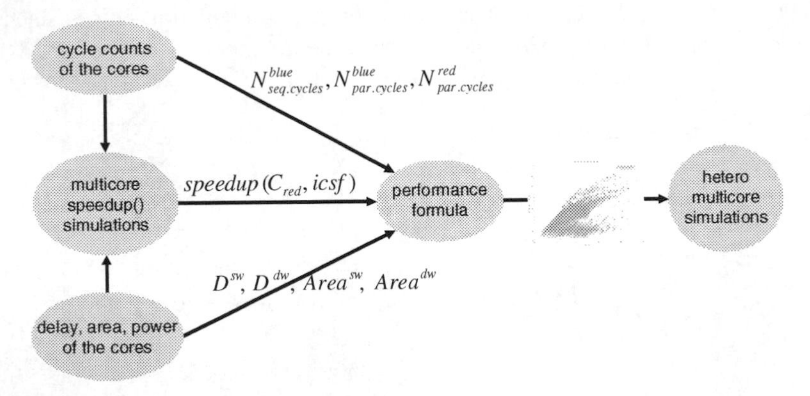

Fig. 16.4 Design space exploration flow (simplified)

The "cycle counts of the cores" step simulates each core out of 6,400 with a breakdown of the cycle counts in the sequential and parallel parts. Using the plot from Fig. 16.5, this step selects cores from the Pareto curve (the thick black line) for forming couples, because the cores below the Pareto curve perform worse at the same area cost. The "small" cores do not run the sequential code. For them, a similar graph to the one in Fig. 16.5 but excluding the time spent on the sequential code is built. The cores from the two Pareto curves are used to create the final eCMP. Note that the performance is measured in 1/s, which is computed using the cycle counts from the xvpsim simulator and cycle periods from the MoodOrgan technology model. MoodOrgan also provides area estimates of the cores.

The speedup curves for cores differ due to the limited TLP and ILP in the application, as well as various requirements for data and instruction memories. Consider,

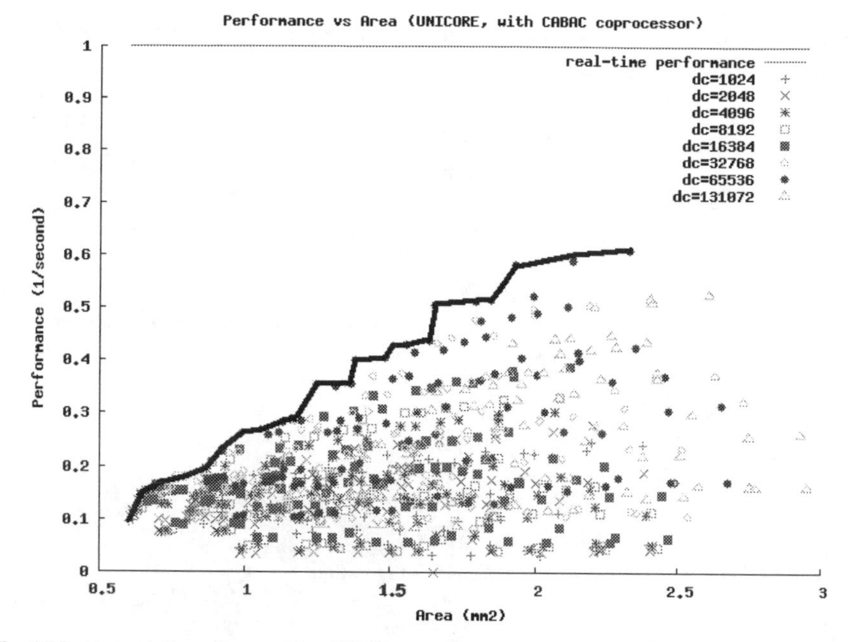

Fig. 16.5 A population of cores (simplified)

for example, Fig. 16.6. The baseline in this figure is a 5 issue slot single-threaded machine with 64 KB instruction and 64 KB data cache. 1slot core refers to the core with baseline parameters, except for the single issue slot architecture. 1kdata$ is a baseline core with a 1 KB data cache instead of 64 KB. Finally, several cores with varying multithreading configurations multithreaded(1,2), multithreaded(2,2), multithreaded(2,4) are shown, which differ from the baseline in terms of foreground and background threads counts. Figure 16.6 shows that speedups for multithreaded cores quickly saturate due to limited TLP for the standard definition (SD) image. More details on the technology modeling can be found in [18].

Finally, the performances of the heterogeneous eCMPs are computed using formula (1). Figure 16.7 shows the calculated performances of various eCMPs in green crosses. The blue crosses indicate the eCMPs selected for simulation. Note that only a small fraction of the eCMPs (typically, much less than 1%) is selected for simulation. The dimensions and the position of the selected region can be easily controlled, allowing for probing and/or calibration of the formula. This way, the methodology explores a wide range of possible eCMPs without exploding the exploration time.

The described DSE flow allowed us to explore 1,105,747,200 design points by simulating only about 0.0007% of them. The total simulation count of approximately 8,000 includes 6,400 for the "cycle counts of the cores," about 1,200 simulations for the "multicore speedup() simulations" phase, and, finally, 400 simulations of the heterogeneous multicores in the neighborhood of the cross-over between the calculated Pareto curve and the real-time performance.

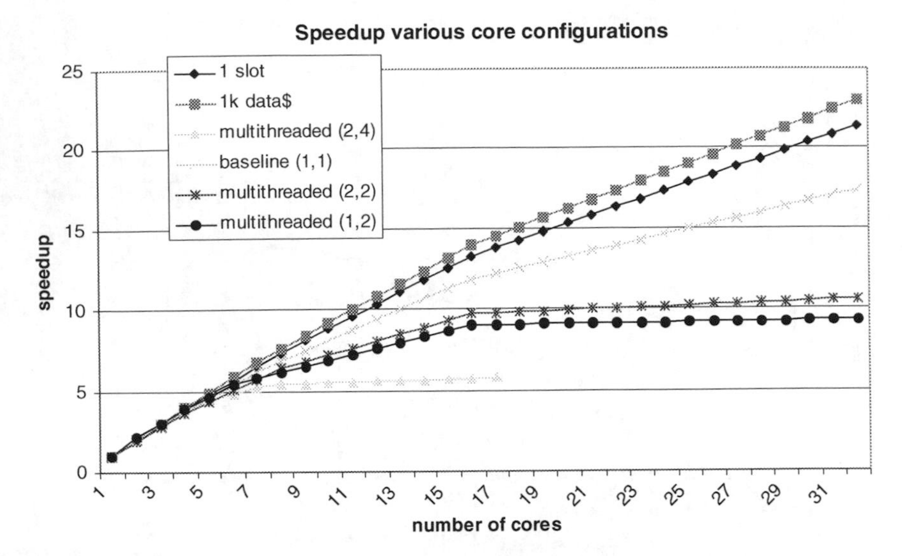

Fig. 16.6 Speedup() simulations for various core configurations

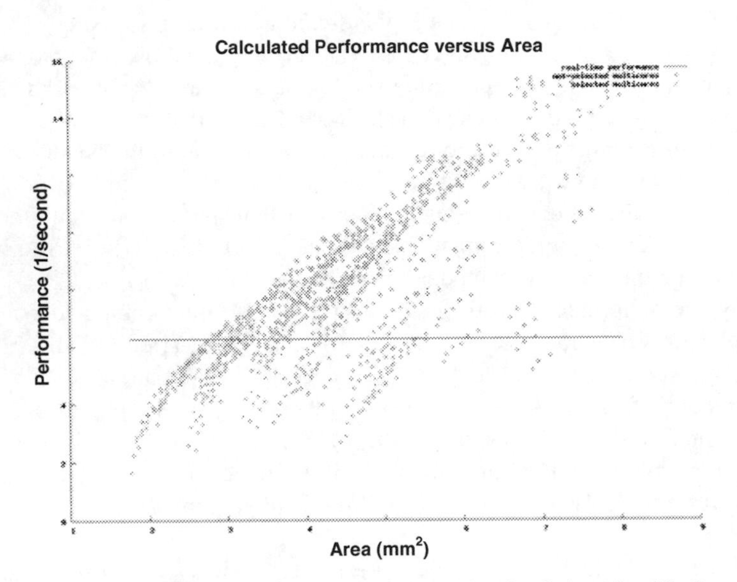

Fig. 16.7 Calculated performance and multicores selected for simulation (simplified)

16.6 Use Case: DSE for the Ne-XVP Architecture

The hardware architecture modeled in the Ne-XVP simulator consists of an heterogeneous multicore, the hardware task scheduler, the task scheduling unit, the hardware synchronization unit, and interconnect infrastructure. Note, that a particular instance of the architectural template may exclude certain hardware components. For example, if the relevant set of applications requires no task scheduling acceleration for irregular task graphs, then the task scheduling unit can be omitted.

Our *multicore simulator* structure shown in Fig. 16.8 is based on the production-quality execution-driven simulator of the TriMedia cores. The simulator is extended with instruction and data caches supporting various coherence protocols. Further additions are a model of conflicts of the L2 cache, and an hardware task scheduler and hardware synchronization units coprocessors. The multiple clock generators enable different clock frequencies (obtained from the technology model) for each core and memory.

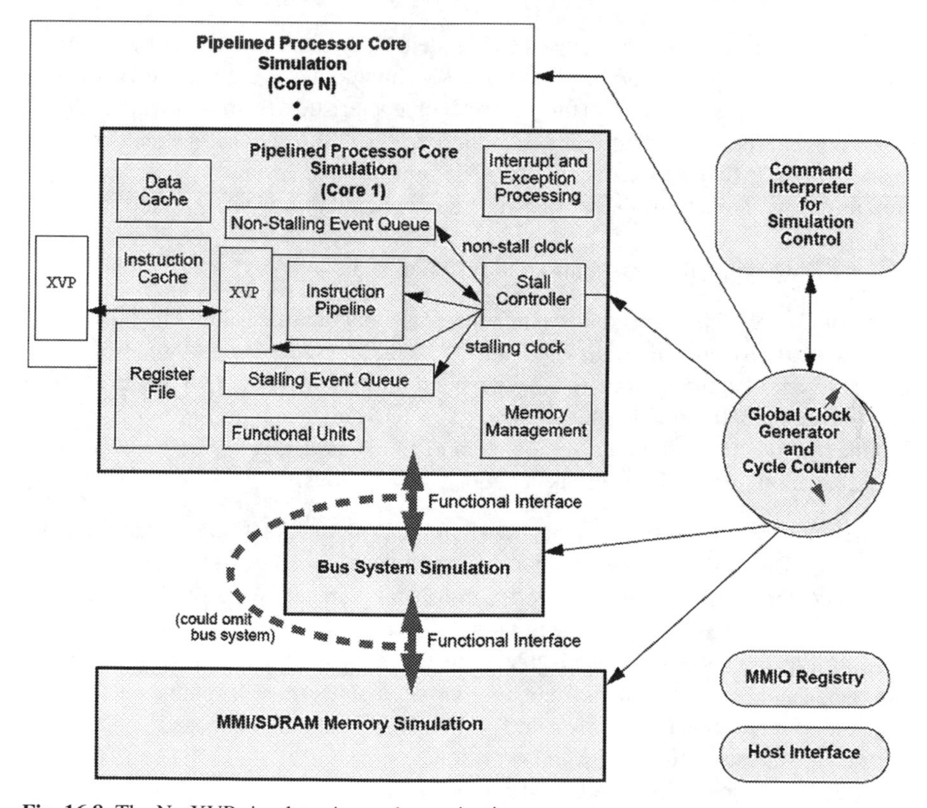

Fig. 16.8 The Ne-XVP simulator internal organization

The limitations of the simulator environment are

1. missing instruction and data prefetch engines in the cache controllers, making the results pessimistic;
2. the memory controller and an off-chip SDRAM are modeled by a fixed delay of 66 ns, which is representative for memory hierarchies of embedded ICs; and
3. the CABAC coprocessor and the related synchronization overhead are not modeled.

The *H.264 video decoder* [10] is used as the main driver application for the DSE. The H.264 software underwent rigorous manual optimizations for TriMedia, including code restructuring and insertion of custom TriMedia operations (SIMD, collapsed loads). The state-of-the-art TriMedia production compiler TCS version 5.1, featuring major ILP optimizations [9], is used to compile applications.

The following video sequences are used for H.264 video decoding:

- crowdrun_420_50.264 (Quad HD 3840 × 2160p, 48 frames, 25 fps) and
- tractor_1_50.264 (HD 1920 × 1080p, 48 frames, 25 fps).

The MoodOrgan technology tool [18] modeled physical properties of the cores, hardware task scheduler, CABAC coprocessor, interconnect, etc. The scaling factors for SRAMs are derived using SRAM characteristics from fabricated chips in various low-power technologies. For the current DSE, the scaling factors only between the CMOS 90 nm and CMOS 65 nm were measured. They are used to extrapolate characteristics of hardware components into CMOS 45 nm process technology.

For HD decoding the best configuration found had two cores:

1. A "big" core with 32 KB data cache and 32 KB instruction cache, 96 registers, 4 issue slots, 1 load/store unit, and a subset static interleaved multithreading with two foreground threads and two background threads. In CMOS 45 nm, the clock frequency reaches 662 MHz and the area = 1.15 mm^2.
2. One "small" core which has the same characteristics as the "big" core, so in this case, the optimal architecture is actually homogeneous.

The total area of this multicore subsystem is 2.45 mm^2 in CMOS 45 nm, including the CABAC coprocessor, memory concentrator, cache to cache tunnels, and coherence coprocessors. This solution is comparable in size with an NXP solution based on a dedicated hardware accelerator.

Multithreading is especially advantageous for multicores with a few cores, where the background threads compensate for the smaller instruction and data caches. Note that the subset static interleaved form of multithreading proposed in [10] was selected as the most efficient for this workload.

Figure 16.9 shows simulated eCMP performances and the real-time deadline for H.264 video decoding at HD resolution. The eCMP in Fig. 16.9 varies in terms of cache sizes, multithreading options, etc., which is not plotted.

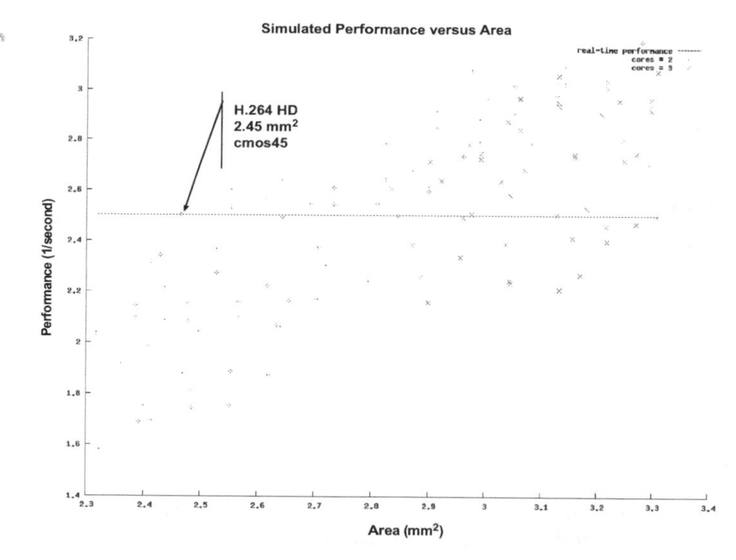

Fig. 16.9 Simulated multicores for H.264 HD decode

For Quad HD decoding the best configuration had eight cores:

1. The "big" core with a 64 KB data and 64 KB instruction cache, 96 registers, 5 issue slots, 1 load/stores, and the interleaved multithreading with two foreground threads. In CMOS 45 nm the clock frequency reaches 616 MHz and the area is 1.4 mm^2.
2. There are 7 "small" cores, each with a data cache of 32 KB, an instruction cache of 64 KB shared among every 4 cores, 96 registers, 4 issue slots, 1 load/store unit, and interleaved multithreading with two foreground threads. In CMOS 45 nm, the clock frequency reaches 662 MHz and the area is 0.99 mm^2.

The total area of this multicore subsystem is 8.6 mm^2, including the CABAC coprocessor, memory concentrator, cache to cache tunnels, and coherence coprocessors. Embedded SoC typically has a silicon real-estate in the order of 1 cm^2, which would be big enough to accommodate a multimedia eCMP subsystem of 8.6 mm^2.

16.7 Conclusions

In order to reach the silicon efficiency of fixed-function hardware, embedded chip multiprocessors need to be tuned to the application domain, and complemented with the right set of accelerators. Design space exploration is required to define which parameters of the architecture will lead to the best results. However, the space to explore is gigantic, and fast simulations approaches together with a methodology based on performance calculation are required. This chapter shows that using only

0.0007% simulations of the total design point count, the described flow enables exploration of huge multicore design spaces in a few weeks of simulation time. Moreover, our integrated technology modeling enabled design decisions based on relevant metrics such as silicon area and execution time.

Acknowledgments The authors would like to thank Philip Christie and his team for their contribution to this work and the ACOTES project (European project IST-034869) for his valuable contributions to the programming model of the Ne-XVP architecture.

References

1. ACOTES programming model, submitted to the *International Journal on Parallel Programming*, Springer, New York, (2009).
2. Al-Kadi, G., Terechko, A.S.: A hardware task scheduler for embedded video processing. In: *Proceedings of the 4th International Conference on High Performance and Embedded Architectures and Compilers, Paphos*, Cyprus, January 25–28, (2009).
3. Azevedo, A., Meenderinck, C., Juurlink, B.H.H., Terechko, A., Hoogerbrugge, J., Alvarez, M., Ramírez, A.: Parallel H.264 decoding on an embedded multicore processor. In: *Proceedings of the 4th International conference on High Performance and Embedded Architectures and Compilers HiPEAC* 404–418, (2009).
4. Christie, P., Nackaerts, A., Kumar, A., Terechko, A.S., Doornbos, G.: Rapid design flows for advanced technology pathfinding, invited paper. In: *International Electron Devices Meeting*, San Francisco, CA, (2008).
5. Darema, F.: SPMD model: Past, present and future. In: *Recent Advances in Parallel Virtual Machine and Message Passing Interface: 8th European PVM/MPI Users' Group Meeting*, Santorini/Thera, Greece, September 23–26, (2001). Lecture Notes in Computer Science 2131, p. 1, (2001).
6. H.264/MPEG-4 Part 10 White Paper, two page summary of MPEG CABAC, available at `http://www.rgu.ac.uk/files/h264_cabac.pdf`. Accessed October 2009.
7. Hill, M.D.: What is scalability?. *ACM SIGARCH Comput Arch News Arch* **18**(4), 18–21, December (1990).
8. Hill, M.D., Marty, M.R.: Amdahl's law in the multicore era. *IEEE Comput* 7, 33–38, July (2008).
9. Hoogerbrugge, J., Augusteijn, L.: Instruction scheduling for trimedia. *J Instruct Level Parallelism* **1**, (1999).
10. Hoogerbrugge, J., Terechko, A.: A multithreaded multicore system for embedded media processing. *Trans High-Performance Embedded Arch Compilers* **4**(2), (2008).
11. Kumar, R., Jouppi, N., Tullsen, D.: Conjoined-core chip multiprocessing. In: *37th International Symposium on Microarchitecture, Micro-37*, Portland, Oregon, December (2004).
12. Kumar, R., Tullsen, D.M., Jouppi, N.P.: Core architecture optimization for heterogeneous chip multiprocessors. In: *Proceedings of the Parallel Architectures and Compiler Techniques*, Seattle, USA, September 16–20, (2006) pp. 23–32, (2006).
13. Limberg, T., Winter, M., Bimberg, M., et al.: A heterogeneous MpSoc with hardware supported dynamic task scheduling for software defined radio. In: *Design Automation Conference*, San Francisco, CA, July, (2009).
14. Mogul, J.C., Mudigonda, J., Binkert, N., Rangana-than, P., Talwar, V.: Using asymmetric single-ISA CMPs to save energy on operating systems. *IEEE Micro* **28**(3): 26–41 (2008).
15. Rock: A SPARC CMT Processor, In: *HotChips*, Stanford, (2008).
16. Själander, M., Terechko, A., Duranton, M.: A look-ahead task management unit for embedded multi-core architectures. In: *Proceedings of the 2008 11th EUROMICRO Conference on Dig-*

ital System Design Architectures, Methods and Tools. IEEE Computer Society Washington, DC, USA, pp. 149–157, (2008), ISBN:978–0-7695–3277–6.

17. Technology-Aware Design, IMEC, Leuven, `http://www.imec.be/tad/`. September, (2009).
18. Terechko, A., Hoogerbrugge, J., Al-Kadi, G., Lahiri, A., Guntur, S., Duranton, M., Christie, P., Nackaerts, A., Kumar, A.: Performance density exploration of heterogeneous multicore architectures. In: *Invited Presentation at Rapid Simulation and Performance Evaluation: Methods and Tools (RAPIDO'09), January 25 2009, in conjunction with the 4th International Conference on High-Performance and Embedded Architectures and Compilers (HiPEAC)*, Paphos, Cyprus, January 25–28, (2009).
19. Terechko, A., Hoogerbrugge, J., Al-Kadi, G., Guntur, S., Lahiri, A., Duranton, M., Wüst, C., Christie, P., Nackaerts, A., Kumar, A.: Balancing programmability and silicon efficiency of heterogeneous multicore architectures, to be published In: *ACM Transactions on Embedded Computing Systems*, special issue on Embedded Systems for Real-time Multimedia, (2010).
20. Ungerer, T., Robic, B., Silc, J.: A survey of processors with explicit multithreading. *ACM Comput Surv (CSUR) Arch* **35**(1), pp. 29–63, March (2003).
21. van de Waerdt, J.-W., Vassiliadis, S., Sanjeev, D., Mirolo, S., Yen, C., Zhong, B., Basto, C., van Itegem, J.-P., Dinesh, A., Kulbhushan, K., Rodriguez, P., van Antwerpen, H.: The TM3270 media-processor. Philips Semiconductors, San Jose, CA, Microarchitecture, (2005). MICRO-38. In: *Proceedings of the 38th Annual IEEE/ACM International Symposium*, 16–16 November (2005), pp. 331–342; Barcelona, ISBN: 0–7695–2440–0.

Part IV
Embedded Systems Simulation

Chapter 17
IP Modeling and Verification

Emre Özer, Nathan Chong, and Krisztián Flautner

Abstract This chapter presents the infrastructure for modeling intellectual property (IP). After a brief introduction to ARM IP and architecture, the modeling and simulation strategies are presented for the RISC CPUs and system-on-chips (SoCs) based on ARM IP. We will explain how the initial investigation of the processor architecture and microarchitecture for performance and power is accomplished before the real design phase. This will be followed by the discussion of different modeling techniques used in the processor design stage such as RTL simulation, emulation, and FPGA prototyping. Then, we will look into the system modeling/simulation frameworks from programmer's and system designer's perspectives and discuss their trade-offs. Finally, we will discuss the verification strategies for validating the processor design from ad hoc test techniques, random coverage strategies, assertion-based verification, and formal verification methods.

17.1 ARM IP and Architecture

ARM Ltd. is a UK-based semiconductor IP company offering processor, peripheral, and physical IP. The ARM 32-bit RISC CPU family is the most widely used architecture in embedded systems. It is the architecture of choice for more than 80% of the high-performance embedded products in design today. When ARM pioneered the concept of openly licensable IP for the development of 32-bit RISC microprocessor-based SoCs in the early 1990s, it changed the dynamics of the semiconductor industry. By licensing, rather than manufacturing and selling its chip technology, the company established a new business model that has redefined the way microprocessors are designed, produced, and sold. *ARM powered* microprocessors are pervasive in the electronic products, driving key functions in a variety of applications in diverse markets, including automotive, consumer entertainment, imaging, microcontrollers, networking, storage, medical, security, wireless, smartphones, and most recently netbooks/smartbooks.

E. Özer (✉)
ARM Ltd, 110 Fulbourn Rd., Cambridge, CB1 9ND, UK
e-mail: emre.ozer@arm.com

R. Leupers, O. Temam (eds.), *Processor and System-on-Chip Simulation*,
DOI 10.1007/978-1-4419-6175-4_17,

ARM IP consists of three main categories: (1) processor IP, (2) fabric and peripheral IP, and (3) physical IP. Processor IP includes general-purpose RISC processors, secure cores, graphics, and video processors. The fabric and peripheral IP includes the debug and trace units, AMBA interconnect, L2 cache controllers, memory controllers (DDR, LPDDR2), DMA engines, LCD and interrupt controllers, general-purpose I/Os (GPIOs), and UARTs. The physical IP provides silicon-proven IP for CMOS and SOI (silicon-on-insulator) processes such as memory and register file compilers, logic, and interface libraries for different foundries and processes including SOI.

The ARM architecture consists of different ISAs and architecture extensions such as 32-bit ARM, 16-bit Thumb, 32-bit Thumb2 ISAs, Java acceleration (Jazelle), security (TrustZone), SIMD (Neon), and floating point (VFP). The evolution of the ARM architecture is shown in Fig. 17.1. The ARM *Cortex* processor core family are the implementations of the most recent version of the ARM architecture, namely ARM v7 [1]. ARM's *Cortex* processors are classified into three distinct categories. The first one is microcontroller class or M-class processor family that includes *Cortex M0* [2], *M1*, and *M3* [3]. The M-class processor cores consume very small amount of power in the range of microwatts. While *Cortex M0* and *M3* are aimed at SoCs, *Cortex M1* is specifically designed for implementations in commercial FPGAs such as Actel, Altera, and Xilinx. The second category is the embedded real-time processor family or R-class which includes *Cortex R4*. The R-class also consumes relatively low power in the range of milliwatts with higher performance delivery. The final category of the ARM processor family is the A-class processor cores which include *Cortex A5* [4], *Cortex A8*, multi-core *Cortex A9*, and *Osprey* cores. *Cortex A5*, *A8*, and *A9* cores are delivered as soft core IP while *Osprey* core comes as a hard macro for 40nm technology capable of achieving 2 GHz clock frequency [5].

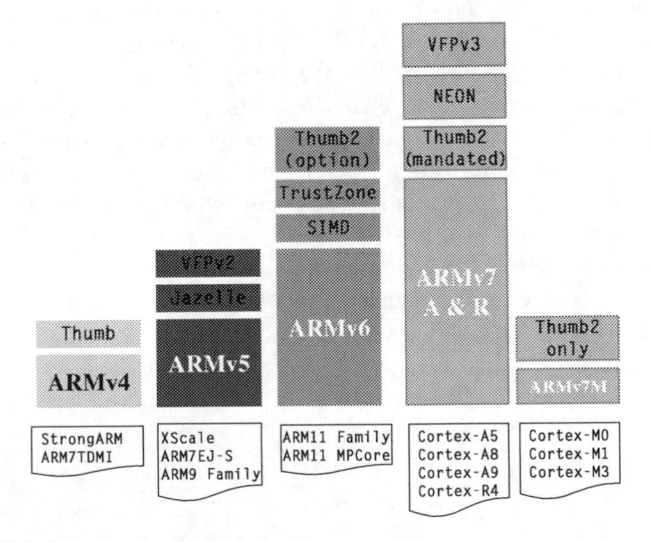

Fig. 17.1 ARM architecture evolution

17.2 Processor IP Modeling

ARM processor simulation/modeling strategy consists of two stages: (1) modeling for initial investigation and (2) modeling for design. Modeling for initial investigation involves modeling the underlying architecture and analyzing the performance of the processor microarchitecture implementing the architecture. Modeling for design requires simulating the microarchitecture at the logic or RTL level in order to do an accurate performance and power analysis.

17.2.1 Initial Investigation Modeling

The main purpose for initial investigation is to implement instruction set simulator (ISS) to model the architectural features and a performance modeling framework to model the microarchitectural blocks. The ISS is necessary to model all the features of the instruction sets, memory, and device behaviors as specified in the architecture for which the microarchitecture of the processor core under design is implementing. When designing the microarchitecture for a given architecture, there are certain parameters that are unknown to the designers such as cache associativity, cache line size, branch predictor type, size, pipeline depth, issue width. At the beginning of the initial investigation, designers know only the performance, power, and area (PPA) targets, which are, in general, driven by the market and customer requirements at which the processor is aiming. Performance, power, and area are defined by clock frequency and IPC (instructions per cycle), mW/MHz, and millimeter square, respectively. Given the PPA numbers, designers outline the processor microarchitecture with expected pipeline depth and number of execution units. Once the outline is shaped, designers are ready to model the microarchitecture to find answers to several unknowns. The initial investigation of a processor core consists of two distinct stages: instruction set simulator (ISS) and trace-based model (TBM).

17.2.1.1 Instruction Set Simulator

Instruction set simulation (ISS) is the first product-level simulator implemented for each processor core product. It simulates the instruction set architecture (ISA) of the processor without knowledge of timing. With every new architecture generation, the ISS may need to be rewritten because of new instructions added to the architecture specification and major changes in the architecture specification. Even within the same architecture generation, the instruction sets of different processor cores may differ (e.g. one supports Thumb ISA (instruction set architecture) and another supports Thumb2 ISA).

The ISS models are models in which only functional behavior of the processor is modeled. They are not cycle accurate and therefore allow fast simulation speeds. The ISS models at ARM use binary translation techniques in which ARM instructions are translated into the host machine instructions on the fly, and the translations are cached to enable fast simulation of ARM code. The simulation speed of 100–500 MIPS can be achieved due to this fast binary translation.

17.2.1.2 Trace-Based Modeling

We use a trace-based modeling (TBM) framework to model the performance of the processor microarchitecture. The TBM framework is written using a general-purpose programming language (normally C/C++). It is cycle-accurate, meaning that it models each cycle in detail. Each pipeline stage, caches, FIFOs, prefetch units, TLBs, branch prediction, and execution units (functional units) are modeled in detail. It is one level of abstraction higher than the RTL model hiding implementation details but is sufficiently detailed to make the meaningful performance analysis. In general, the TBM simulator can approach to 90–95% of the hardware implementation or RTL simulation in terms of simulation accuracy.

Although power at the pipelines and execution units is not modeled explicitly in the initial investigation stage, it is a first-order constraint from the initial design to the final product. When tuning the microarchitectural parameters in the trace-based modeling, the performance and power aspects are considered together. In most cases, power is favored to performance. For instance, simple replacement algorithms in caches (e.g., round-robin or random) are preferred over hardware-intensive cache replacement schemes (e.g., LRU(least recently used)) due to power concerns.

Power is modeled for the memory elements such as caches, scratchpads, and register files in the initial investigation. ARM provides in-house memory compilers including SRAM, register file, and ROM. Memory compilers provide configurable environment for high-density and high-speed single and dual-ported SRAM and register files. ARM memory compilers provide the user with a graphical user interface or command line to configure the number of words, number of bits per word, adding ECC (error correcting code), and enabling power gating and retention modes, clock frequency, foundry (e.g., TSMC, Samsung, IBM), and technology node (e.g., 65 nm, 40 nm). The output can be in the form of RTL, netlists, and data sheets with schematics. The data sheets provide data on the generated memory such as physical dimensions of the memory structure, read and write access time, and active and standby power. The active and standby power numbers can be plugged into the TBM model to estimate the power consumption of each memory structure. Thus, early microarchitectural exploration on the configuration of scratchpads, caches, TLBs, and the register file can be made for low power processor design.

The TBM does not model the OS (i.e., no OS booting) and therefore it is purely an application level simulator. The frontend of the trace-driven simulation is the ISS, which reads ELF binaries of applications, and spits out instruction traces to a pipe, which are then consumed by the trace-based simulator on the fly. The flow diagram of the TBM is shown in Fig. 17.2. The TBM also reads the power numbers (read and write access and leakage power) obtained from the ARM memory compilers regarding the memory elements in the microarchitecture. It can accurately estimate the power consumption of the caches, TBM, scratchpads, and the register file as the data on the number of accesses to these structures and the number of idle cycles can be derived from the traces of the applications. Then, the TBM generates the performance statistics of the overall processor microarchitecture and also the power statistics on the memory elements.

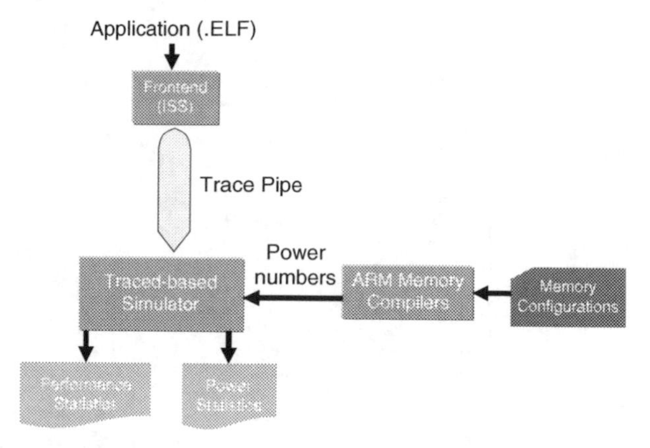

Fig. 17.2 Trace-based simulation framework

17.2.2 Processor Design Modeling

The TBM model in the initial investigation of the processor finalizes certain microarchitectural parameters with constant depth and size. Once this stage is over, the actual implementation and simulation can be performed. The design team uses synthesizable RTL (Verilog) to simulate the behavior of the processor core. It is important to note that there is no microarchitectural exploration done at the RTL level.

17.2.2.1 RTL Simulation

Simulation is required at the RTL level design for maintaining the functional correctness of the design. The processor design consists of modules or blocks where a module or block can be a pipeline stage or a group of pipeline stages. In general, each block is designed by a small design and verification team. In parallel to designing the block, the design team also creates testbench and test stimulus to verify the functional correctness of the block. Once the individual design and verification of the blocks are complete, the top-level simulation can commence.

The top-level design includes full integration of all modules and blocks, and the design is simulated for functional correctness by using the architecture and microarchitecture validation tests and simple benchmarks. The simulation of the top-level design requires a reference model to check whether the simulated RTL design gives the expected output as shown in Fig. 17.3. Herein, the ISS model helps the designers validate the processor core design in RTL. Both ISS and RTL models are fed with the same input stimulus. Each model independently executes the given input stimulus, the output of the models are then compared. In the case of a mismatch, both models are revised to find out which model has the incorrect implementation of the instruction under observation.

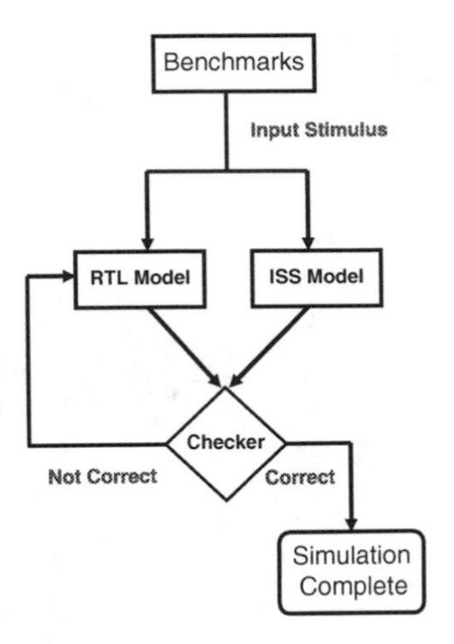

Fig. 17.3 Processor design simulation

Once the RTL design is simulated and verified against the ISS reference model using the architectural validation test benchmarks and other simple small applications, the design should be simulated with wider spectrum of applications (e.g., EEMBC [6] and SPEC2K [7]) running on the OS. Booting an OS and running real applications in the RTL model are extremely slow, therefore alternative simulation platforms are needed to simulate the processor design.

We use two different platforms to perform the fast RTL simulation of the processor: hardware emulation and FPGA simulation. The hardware emulator gives a middle ground between slow RTL simulation and fast FPGA simulation. It allows for early prototyping of very large processor designs, which would otherwise require partitioning the design to several FPGA platforms. It also provides better visibility and observability of the testchip being developed. The RTL simulation offers the full visibility of the processor design in terms of probing every signal at the cost of slow simulation. On the other hand, the FPGA platform provides the least visibility with the highest simulation speed. This is summarized in Fig. 17.4

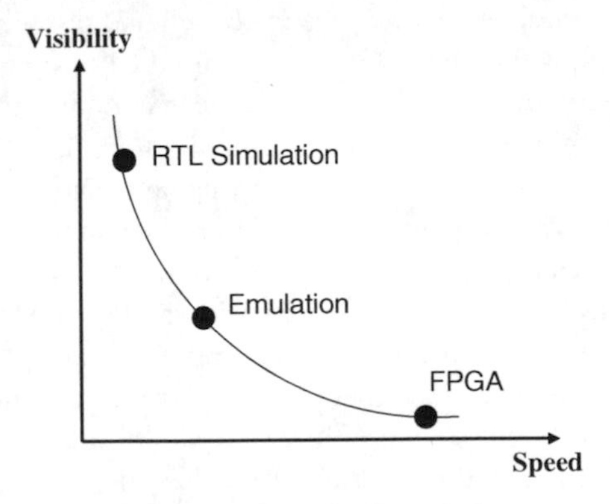

Fig. 17.4 Modeling methods: visibility versus simulation speed

17.2.2.2 Hardware Emulation

Hardware emulation mimics the design written in RTL using programmable logic in hardware specially designed for circuit emulation. Hardware emulation can be several orders of magnitude faster than the RTL simulation as the emulator hardware can perform simultaneous tasks. The emulator is useful for booting the OS and large applications and verifies functionality, detects bugs early in the design, and also allows concurrent development of hardware, software, and systems.

We use hardware emulators to emulate the processor core IPs. A hardware emulator is a modular, custom processor-based, compiled code emulation system. Design in RTL is compiled and allocated to an array of highly interconnected custom processors. Hardware emulators can emulate systems with hundreds of millions of gates and the emulation speed is in the range of MHz. Emulation can be done at full system level meaning that the processor IP and peripherals (i.e., memory and I/O) can be emulated. Alternatively, the peripherals in a separate logic board can be connected to the emulator that emulates the processor core only.

17.2.2.3 FPGA Simulation

One widely used technique for early design simulation and validation is to prototype the design on an FPGA. The growing density of FPGAs allows designers to map large IP blocks to FPGAs. FPGA platforms enable designers to perform much faster simulation than the RTL simulators and hardware emulators. We use FPGA boards to verify the design of our processor IP cores. The contemporary FPGA boards have the necessary support for DRAM and I/O to boot the OS (e.g., Linux) on the processor core under development on the FPGAs. However, one of the main challenges using FPGAs is to map a large processor IP design into a single FPGA. Even the largest FPGA (e.g., Xilinx Virtex 5 or 6 series [8]) may not accommodate a large processor design. In this case, the design can be partitioned into multiple FPGAs boards. However, the challenge is that I/O interfaces required to glue the partitioned designs are usually wider than the available I/O between the FPGA boards. Thus, the designers are faced with other ad hoc solutions such as time-multiplexing the I/O, which will increase the simulation time.

An early prototyping of the processor IP is done on FPGAs using in-house logic tile boards (e.g., RealView Versatile LT-XC5VLX330 [9]). A logic tile contains a Xilinx FPGA and provides an interface to stack more logic files to allow large processor designs to be partitioned into multiple FPGAs. The logic tile has also capability of being stacked on an emulation baseboard [10], which is another in-house board that contains memory and I/O functionality. The emulation baseboard provides a large amount of memory and peripherals and is also based on a large FPGA that implements the bus, memory system, and peripherals controllers such as flash memory, DDR SDRAM, AXI buses, serial ports, Ethernet, PCI interface, LCD controller, and general-purpose I/Os. A system platform can be realized by stacking the logic FPGA tile(s) to the emulation baseboard, which provides minimal functionality to boot OS (e.g., ARM Embedded Linux, Windows CE) and run applications on the processor core under development.

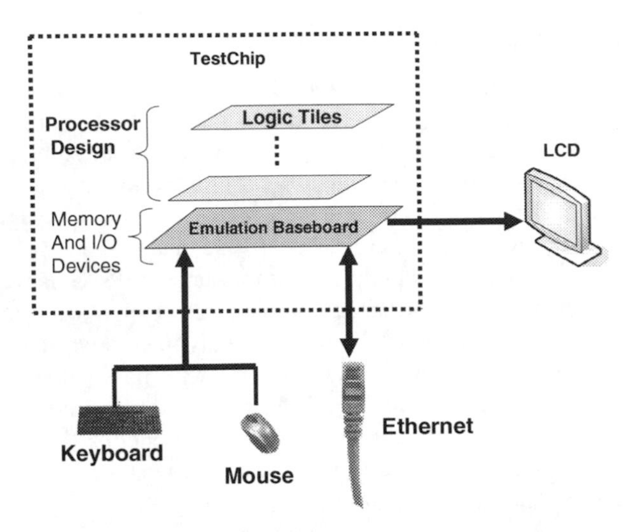

Fig. 17.5 Early FPGA prototyping with ARM versatile boards

An example of an early FPGA prototyping is shown in Fig. 17.5. The processor design is partitioned across several FPGAs each of which sits in a logic tile. Logic tiles are stacked on top of each other to have a coherent view of the processor under design. The stacked logic tiles can be stacked on the emulation baseboard seamlessly to appear as a complete testchip where the emulation baseboard provides memory and I/O functionality. LCD, keyboard, and mouse can be connected to the emulation baseboard to have the complete view of a standalone computer. The emulation baseboard can also be connected to a LAN through Ethernet interface where multiple users can run tests remotely.

17.2.2.4 Power Modeling

The accurate and reliable power modeling of the processor microarchitecture such as pipeline stages, execution units, and memory elements are performed at the netlist level after parasitics and load capacitance values are provided by layout extraction. These capacitance values depend on the foundry and technology node in which the standard cells are designed. Parasitics and load capacitance values along with the transistor switching activity allow designers to analyze the dynamic and static power consumption of their processor designs. The industry-proven power model tools are employed to perform the accurate power simulation.

17.3 System IP Modeling

System IP modeling at ARM involves in modeling a system-on-chip (SoC) consisting of an ARM CPU and other peripheral IP. Different IP blocks are put together to create a SoC for mobile phones, smartbooks, or set-top boxes. A typical SoC based on ARM CPU consists of the following IP blocks: a memory controller, scratchpad,

AXI interconnect, DMAs, UARTs, and a debugger. Modeling at the system level are essential to understand the bottlenecks in the system and identify IP blocks that can be optimized to achieve the expected system performance/power/bandwidth requirements. System simulation also enables the programmers to develop software for systems before the hardware platform becomes available.

System IP modeling has two perspectives: system designer's and programmer's views. The system should be able to provide detailed timing to analyze the bottlenecks in the communication fabric, memory, and I/O units from SoC designer's point of view. In the programmer's view, the system should be able to provide the sufficient functionality to boot an OS, develop applications, and run them on the system under development. In the programmer's world, the accurate timing model is not of primary concern. The programmer's view model is essentially similar to the virtual platform (Chapters 2 and 3).

17.3.1 System Designer's View Models

System designer's view (SDV) model provide a SoC modeling environment that can model timing and interactions between the system components accurately. System designer's view system model is performed in the RTL world in which the ARM components can be glued together in plug-and-play fashion to construct a model of an SoC. The basic components are processors, L2 controllers and caches, the graphics cores, dynamic and static memory controllers, bus fabrics, and I/O controllers. Each component is written in RTL and contains all the implementations details. The complete system can be simulated using RTL simulators, hardware emulators or FPGAs and is capable of booting OSes (e.g., Linux/WinCE/Android/Ubuntu) and run applications.

The SDV model differs from the programmer's view model in several ways. It is a cycle-accurate model that mimics memory and bus transactions very close to the real hardware implementation. Because the SDV model is cycle accurate, it is several orders of magnitude slower than the programmer's view model. Thus, it is to be used by the SoC designers rather than software developers. The SoC designers use the SDV model to understand the bottlenecks in the memory and I/O controllers such as memory bandwidth, memory and I/O latency, buffer sizes and delays, collisions and arbitration issues in the bus fabric between the processor cores and memory and I/O subsystems. It also allows system designers to analyze the system power consumption. Very accurate activity traces are collected by running OS and applications for each IP component and fed into a system-level power modeling tool, which has a power model for each IP block for a given technology and operating conditions. Thus, the SDV model enables designers to carry out a system-level performance and power analysis before taping out the testchip.

17.3.2 Programmer's View Models

Programmer's view (PV) models (also known as fast models) [11] provide virtual platform models with simulation speeds of ranges between 100 and 500 MIPS. The

PV models use dynamic binary translation techniques, similar to Chapter 9, to simulate systems in greater speeds. The PV model automatically generates the required interfaces for both standalone and integrated platforms. It is accurate to the system programmer's view in the sense that software written for the underlying system will generate the same results when run on hardware and on the PV model. However, the PV model does not accurately model bus transactions in terms of timing since simulation speed rather than timing accuracy is the primary goal. The PV model also allows early software development, hardware/software co-design, and system design exploration.

It comes with a graphical user interface (GUI) called *System Canvas*, which enables the programmer to realize systems by using graphical representations of components, ports, and connections between ports. *System Canvas* has a block diagram editor for creating the graphical representation of a system, which allows very fast modeling of systems consisting of several components. Examples of components are cores, buses, I/Os, memories. Components can be added to a system from a component database, and new components are defined by LISA+ code. The LISA+ language [12] is an enhanced version of LISA language that is used to describe the instruction set architectures. With LISA+, components and systems can also be described. The output of the PV model from *System Canvas* can run as a standalone system or can be hooked to a larger system as a component. The diagram of how the PV model works is shown in Fig. 17.6. The generated standalone system is capable of running the application binaries as if it is the real hardware.

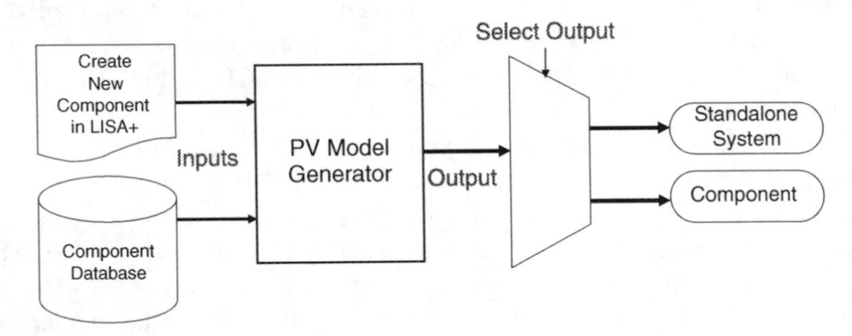

Fig. 17.6 ARM programmer's view modeling tool

17.4 Modeling for Verification

Modeling for processor core verification requires creating self-checking testbenches that are self-contained and can be run on different models such as ISS, RTL simulation, hardware emulators, and FPGAs. We classify the self-checking testbenches into two categories: *architectural validation suite (AVS)* and *device validation suite (DVS)*. AVS checks compliance to the ARM architecture reference manual [13] such as instructions set architecture (ISA), memory models, exceptions/interrupts, and different execution modes. DVS checks the features of the processor core outside of the architecture specification such as scratchpad memories, caches, bus interfaces. The validation strategy is to develop a large library of AVS and DVS that can be reused across processor cores so that new processor core designs can build a validation package quickly by selecting the appropriate AVS and DVS. Most of the effort for a new processor core goes to developing a DVS and sometimes altering AVS to deal with any new architectural changes such the ISA type, architecture version, and exception model.

17.4.1 Random Coverage Techniques

AVS and DVS are examples of directed test suites: each test is designed to exercise a particular function of the design. We measure this metric in two main ways: coverage of the (RTL) code and functional coverage of the design. RTL code coverage can be derived from our simulation tools; however, functional coverage is a manual process of defining the coverage points within the design and requires logging within a testbench to determine which points have been hit by a given simulation. Since directed tests are handcoded the rate that a design can be covered is a function of how quickly tests can be generated [14].

The idea behind constrained-random testing is to randomly autogenerate tests within certain constraints in order to increase the rate of design coverage. This is generally done at the block or top level of our designs. At the block level as shown in Fig. 17.7, this consists of placing the design in a testbench that can drive random stimulus to the design. The difficulty with this approach is that the random stimulus must be in some way "well behaved" so constraints are necessary to make sure this is the case (for example, if a signal is one-hot a constraint must ensure that the value is always one-hot although the exact value may be random). The testbench keeps track of the input stimulus driven to the design in a scoreboard and correlates this with output from the design. Errors can be flagged by monitors in the testbench.

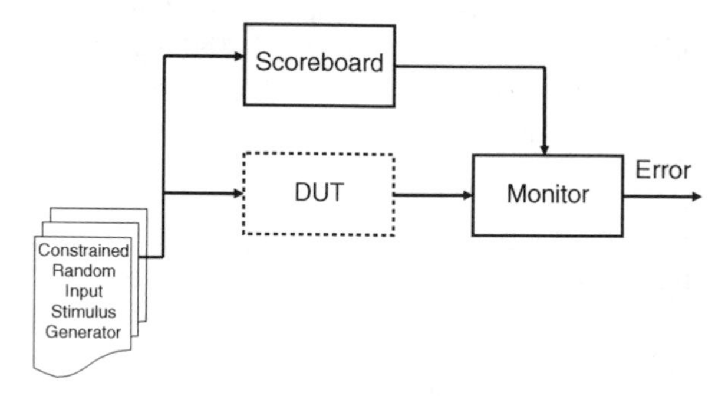

Fig. 17.7 Constrained-random testing for block-level designs

Random instruction set testing is an example of constrained-random testing at the top level. This is a tool that generates small, legal code sequences and checks programmable-visible state. We either check on a cycle-by-cycle basis by comparing the design against a golden model (such as an ISS) or at the end of the simulation.

17.4.2 Assertion-Based Verification

Assertion-based verification leverages designers to add assertions in the RTL in order to reduce the design verification time. Essentially, assertions are statements inserted into the RTL code to ensure that the code conducts the intended behavior. An assertion fails if illegal behavior is attempted. Assertions have been used in the high-level programming languages like C and Java to allow programmers to debug their code more efficiently. Similar to assertions in the high-level programming languages, assertions in the hardware description languages allow the hardware designers to verify their design more efficiently. Assertions increase observability in the design and controllability in terms of test coverage. The advantages of assertion-based verification are five fold: (1) faster verification, (2) reduction in the number of verification engineers, (3) high bug discovery rates, (4) reusability from module to system-level verification in the same project and across different projects, and (5) reducing the complexity of exhaustive formal verification. We use *OVL (Open Verification Library)* [15] and *SystemVerilog* [16] for assertion-based verification.

17.4.3 Formal Specification Models

The techniques above describe simulation-based verification: a scenario and stimulus are provided to the design and expected results are checked at the end of the simulation. We can vary the speed at which simulation is run using different targets (RTL simulation, hardware emulation, FPGA, and silicon) but we always run into the problem of state space explosion—an exhaustive walk over the state space of our models is not a practical solution (consider a 32-bit multiplier: if we wish to simulate every possible input then we face simulating 2^{64} possible scenarios).

Model checking is a complementary approach to verification that allows us to automatically and exhaustively check properties over a design. It works by building a symbolic representation of the design (a transition system) and checking properties using graph reachability. If a property passes then it holds exhaustively (over all scenarios). If a property fails then a counter-example (a scenario violating the property) is provided. Finally, a property may be inconclusive because the tool has not managed to conclusively pass or fail the property. In this case we may be satisfied with a bounded pass (the property holds for some number of cycles out of reset). *Merz* give a good introduction to model checking [17].

We use many different commercial tools, and our designers write properties (such as interface specifications and invariants) at the block and system level. We also use model checking as a tool for reasoning about specification models. The aim of a specification model is to unambiguously capture a specification such as the instruction set, the memory ordering model, or the exception model. Specification models differ from RTL designs (implementations) by virtue of the fact that they attempt to describe all possible legal implementation behaviors. So properties that hold over a specification model hold over all implementations (that correctly implement the specification).

Specification models have a number of unique advantages over traditional models: (i) they are amenable to model checking, allowing valuable high-level properties to be exhaustively proven; (ii) they exercise the specification in unusual ways which can result in a better specification; (iii) they can aid understanding of a specification since the model can be used as a what-if tool to explore different scenarios.

Our general strategy for modeling specification models is to use an atomic-guarded-command-language style: the state of the system is modeled by a number of data structures, and guarded commands are predicated state-to-state functions. A guard is a condition for the command firing and the effect of the command is to atomically update the state of some or all of the data structures. For example, a command for a cache-coherence model may model the issuing of a new read transaction and the guard could be that there is buffer space within the interconnect to accept the transaction. The state update as a result of this command would be the addition of the new transaction to some data structure that keeps track of live transactions.

Nondeterminism is an important distinction between specification models and implementation models. A specification describes desired behavior and does not necessarily prescribe an implementation choice (although it may indicate a preferred choice): it allows nondeterministic choice. In this sense, we see a specification as defining a set of possible implementation choices and a particular implementation is just one point in this set. Nondeterminism can be modeled in *SystemVerilog* using a free variable (a variable that is never assigned to). In this case, the model checker is free to assign to this variable as it wishes, and so we can model all possible scenarios.

Recent work has concentrated on building refinement checkers [18]: embedding the specification model as a dynamic checker within the RTL implementation environment in order to provide a link between the specification model and an implementation.

17.5 Conclusion

In this chapter, we have presented the simulation, modeling, and verification strategies and tools during the course of a processor IP design, system IP modeling, and modeling for verification. We have discussed initial investigation for microarchitecture exploration and different processor IP design modeling methods such as RTL simulation, hardware emulation, and FPGA prototyping. This was followed by the system IP modeling views both from programmer's and system designer's perspectives. Finally, we have presented the processor IP verification methodologies for validating the processor IP from ad hoc test techniques, random coverage strategies, assertions, and formal methods.

References

1. ARM Architecture Reference Manual ARMv7-A and ARMv7-R Edition, http://infocenter.arm.com/.
2. Halfhill, T.: ARM's Smallest Thumb, Microprocessor Report, (2009).
3. Yiu, J.: *The Definitive Guide to the ARM Cortex-M3*. Newnes, Burlington, MA, USA, (2009).
4. Halfhill, T.: ARM's Midsize Multiprocessor, Microprocessor Report, (2009).
5. ARM announces 'Osprey' A9 core as Atom-beater, EETimes, http://www.eetimes.eu/220000579.
6. The Embedded Microprocessor Benchmark Consortium, http://www.eembc.org.
7. SPEC CPU2000, Standard Performance Evaluation Corporation, http://www.spec.org/cpu2000/.
8. Xilinx, http://www.xilinx.com.
9. Logic Tile HBI-0172, RealView LT-XC5VLX330 UserGuide, http://infocenter.arm.com.
10. RealView Emulation Baseboard User Guide, http://infocenter.arm.com.
11. User Guide, Fast Model Tools Version 5.1, http://infocenter.arm.com.
12. LISA+ Language Reference Manual for Fast Models Version 5.1, http://infocenter.arm.com.
13. Seal, D.: *ARM Architecture Reference Manual*, 2nd ed. Addison-Wesley, Boston, MA, (2000).
14. SystemVerilog Reference Verification Methodology, http://www.eetimes.com/news/design/showArticle.jhtml?articleID=188703275.
15. Open Verification Library, Accellera, http://www.eda-stds.org/ovl/.
16. SystemVerilog, http://www.systemverilog.org/.
17. Merz, S.: *Modeling and Verification of Parallel Processes*. Springer, LNCS 2067, New York, NY pp. 3–38, (2001).
18. Bingham, J., Erickson, J., Singh, G., Andersen, F.: Industrial strength refinement checking. In: *International Conference on Formal Methods in Computer-aided Design (FMCAD)*, Austin, TX, November (2009).

Chapter 18
Configurable, Extensible Processor System Simulation

Grant Martin, Nenad Nedeljkovic, and David Heine

Abstract This chapter discusses the challenges of creating an instruction set simulator (ISS) and system modeling environment for configurable, extensible processors. Processor generation technology is highly automated and is used by designers to create application-specific instruction set processors (ASIPs) targeted to specific application domains. The main challenges for creating an ISS lie in the configurable and extensible nature of the processor, which is tailored by designers at design time, not fixed in advance. Thus simulation models need to be created on the fly when designers have configured a new version of the processor. They need to be created automatically and offer good performance with a variety of use modes. In addition, multiple speed-accuracy trade-offs are necessary to support these various use modes, such as instruction-accurate, cycle-accurate, and pin-level. Furthermore, these ISS models must themselves fit into system modeling and simulation environments, both generated along with the ASIP and commercial third party ESL/EDA tools. Again, multiple use models and trade-offs in simulation scope, speed, and accuracy must be supported. Standards to ease integration and model interoperability in this area have arrived only recently and are just beginning to make life easier for model providers.

18.1 Overview of Tensilica Processor and System Simulation Modeling Technologies

The fundamental simulation technology used for design with Tensilica processors is the instruction set simulator (ISS). Some of the history of developing the ISS is described in [1]. Since the Tensilica Xtensa processor is a configurable, extensible application-specific instruction set processor (ASIP) [5, 6], the ISS is not a static simulation model as one might have for a fixed instruction set architecture (ISA) processor. Nor is it completely created on the fly when a new configuration of the processor is generated. Rather, as discussed in [1], we employ a hybrid strategy,

G. Martin (✉)
Tensilica Inc, Santa Clara, CA, USA
e-mail: gmartin@tensilica.com

R. Leupers, O. Temam (eds.), *Processor and System-on-Chip Simulation*,
DOI 10.1007/978-1-4419-6175-4_18,

where the ISS consists of a configuration-independent portion and a configuration-dependent portion. These portions are made available in libraries—the independent portion in libraries shipped as part of the tools; the dependent portions in libraries created during processor generation and compilation of locally written instruction extensions in the Tensilica Instruction Extension (TIE) language [7]. The Tensilica ISS is available in two trade-offs of speed vs. accuracy: a cycle-accurate mode and an instruction-accurate, fast-functional mode called TurboXim. The ISS can run in either mode or switch dynamically between them depending on user choice, which allows a hybrid statistical sampling mode to be supported. In addition, the ISS can be run stand-alone from the command line, from the Xtensa Xplorer integrated development environment (IDE), and with or without debugger attachment.

The ISS supports optional clients for profiling and performance analysis, processor tracing, energy consumption analysis, memory leakage, and other analysis functions.

The ISS is also provided within two system level modeling environments. The first, XTMP, offers a C-based modeling library for single and multicore systems and was created several years ago. The second, XTSC, is SystemC based and came out 3 years ago. Both allow all modes of the ISS to be used in simulation and support its use in a system context. XTSC also offers a pin level access to the ISS. Finally, these environments have been used to interface Tensilica processor simulators to third party ESL system modeling and virtual prototype tools. The rest of this chapter will give further details about all these capabilities.

Many of the techniques used in the Tensilica ISS and system modeling environments are described generically in Chapter 8. In particular, the retargetable simulator generation and just-in-time cached compiled simulation methods are comparable.

18.2 The Instruction Set Simulator (ISS)

18.2.1 History and Design Choices in Developing ISS Models

When Tensilica started as a company, the ISS for a particular processor configuration including new instructions defined in TIE was generated completely from scratch and compiled each time as a new tool. This was later changed, so that the ISS has a configuration-independent part and configuration-dependent ISS libraries that are created during processor generation and within the TIE-compilation development loop. There is a strong correlation between the ISS configuration-independent and ISS configuration-dependent portions and the processor hardware structure. That is, much of the processor hardware is described in TIE and the corresponding Verilog processor description is generated by the TIE compiler. The rest of the processor hardware is described in hand-coded Verilog.

The hand-coded-RTL part of the processor is used primarily for the processor's infrastructure including the instruction-fetch engine, local-memory interfaces, the load-store unit(s), cache-memory interfaces, write buffers, store buffers, and the PIF (main bus) interface logic. Software models of these hardware components are the configuration-independent core of the ISS. Although these are configuration-independent, the processor hardware and the ISS are parameterized. For example, parameters include the presence or absence of certain local-memory interfaces (for instruction RAM, instruction ROM, data RAM, and data ROM) and their bit widths. The configuration-dependent part of the ISS matches the TIE description that is used to generate the rest of the processor hardware including the instruction semantics (for the processor's base ISA and all TIE instruction extensions), registers and register files, other TIE states, and exception semantics.

The Xtensa ISS has a cycle-accurate mode that directly models the processor pipeline. ISS instruction processing consists of three steps: stall computation, instruction issue, and instruction semantics for the activities occurring within each pipeline stage. The stall computation models pipeline interlocks. The instruction-issue step models the use of registers within each stage and sets up the computation for the stall functions of subsequent instructions. The TIE compiler generates the configuration-dependent parts of the ISS during the server-based configuration build or within the local (client-side) TDK (TIE development kit) loop. The generated *xtensa params* files provide additional configuration-specific parameters and the core and configuration-specific parts are connected to create the full ISS. The ISS uses function pointers and dynamically loaded libraries for efficiency.

Interfaces and signals provide communication between the core and the configuration-dependent parts of the ISS. For example, when an instruction wishes to load from memory, the ISS has an interface called *mem data in* that is called via the semantic functions generated by the TIE compiler. This function, which resides in the configuration-independent part of the ISS, then determines whether the memory reference refers to local memory, to a cache, or if it needs to be sent out via the processor interface to a system memory, based on the address mapping defined in the processor's configuration.

The ISS uses core signals to communicate between its configuration-dependent TIE part and its hard-coded, configuration-independent part, which represents the HDL-coded part of the processor. These core signals are used for special states, such as the *DBREAK* state. *DBREAK* allows the ISS user to set a watch on various memory locations so that a breakpoint occurs when the watched locations are accessed.

Instruction extensions written in TIE are modeled in the configuration-dependent part of the ISS; much of the fixed ISA is also written in TIE and is therefore modeled in this portion of the ISS. Some instructions are partially modeled in TIE for instruction decoding but the semantics are left blank and are modeled by the hard-coded ISS core. Special signals are used to signal execution of such instructions to the ISS. For example, synchronization instructions (ISYNC, DSYNC, and ESYNC) or cache-access operations such as dirty-line replacement tend to be hardwired in the ISS using this communication mechanism because they are very tightly bound to the processor infrastructure. As a consequence, users cannot provide their own cache model.

The Xtensa processor's configurable nature combined with the dynamic nature of the ISS results in some operational inefficiencies. For example, when an interrupt or memory access occurs, the ISS must dynamically check the settings of various configuration parameters (in this example, the interrupt settings and the cache parameters) to determine which part of the ISS logic to execute. Given a wide range of configuration parameters, the ISS code must check a number of Boolean and integer variables, resulting in a fair amount of branching code.

An optimized ISS for a fully fixed-ISA processor or a completely regenerated, configuration-specific ISS would not have this overhead. However, Tensilica has made a reasonable trade-off to create a flexible software system that matches and complements the flexibility in the Xtensa processor architecture. Of course, many fixed-ISA processors also have some variability that their ISSs must handle—for example, checking cache parameters. In theory we could move more of the overhead into the configuration-specific part of the ISS but this choice would make the ISS harder to maintain and would increase the amount of time taken to build new processor configurations. Although it is hard to estimate the cost of this overhead, one reasonable estimate is that the cycle-accurate mode of the Xtensa ISS runs at about half the speed that it would if it was fully regenerated for each new processor configuration. We believe this represents a good engineering compromise between speed and flexibility.

In contrast to the approach described in Chapter 17, Tensilica has emphasized the automatic generation of ISS and system models very early in the investigation phase for new micro-architectural features. Leaving the model generation until late in the development process for a new release of the processor generator would be inefficient and might cause unanticipated delay in the release process. Having this as an integral part of the development process leads to higher quality more quickly.

18.2.2 The Instruction Set Simulator Cycle-Accurate and Fast-Functional Modes

The Xtensa ISS has two major simulation modes. In the default cycle-accurate mode, rather than executing one instruction at a time, the ISS directly simulates multiple instructions in the pipeline, cycle by cycle. As discussed above, the simulator models most micro-architectural features of the Xtensa processor and maintains a very high degree of cycle accuracy. In this mode, the ISS can be used as a reference model for a detailed hardware simulation, thus serving as a verification tool.

Software developers always want more speed, so Tensilica expanded its ISS capabilities by adding a fast, functional, instruction-accurate simulation capability called TurboXim. This uses just-in-time compiled-code techniques to significantly improve simulation speed. Of course, the speed improvement varies widely depending on target code and processor configuration but, as a rule of thumb, 40–80X speed improvements or more have been seen on real applications. Because TurboXim performs just-in-time host-based code generation and compilation for specific target code on a specific processor configuration, it knows exactly what code to generate and avoids a lot of conditional processing. The instruction-accurate (as opposed to cycle-accurate) TurboXim allows software and firmware developers to trade off speed for accuracy. Although developers would prefer to have the speed while retaining 100% cycle accuracy, this is not practical.

In the TurboXim mode, the ISS does not model micro-architectural details of the Xtensa processor, but it does perform architecturally correct simulation of all Xtensa instructions and exceptions. Thus cycles lost due to pipeline stalls and replays, memory delays, and branch penalties are not taken into account, and each instruction is assumed to take a single cycle to complete. In addition, data and instruction cache behavior is not modeled—in effect, TurboXim assumes an infinitely large cache which always hits.

TurboXim reports the number of instructions executed but makes no attempt to predict the actual executed cycles that the target code might use on the target processor. It is possible to add the ability to predict cycle counts using table-lookup functions in a kind of "cycle-approximate" mode and this feature remains a future possibility. Meanwhile, it is possible to switch between cycle- and instruction-accurate simulations dynamically during a simulation run. This ability enables a kind of hybrid simulation where a sufficient amount of execution in cycle-accurate mode can be used to predict the overall cycle count of the complete simulation. Used in this manner, a hybrid simulation might run 99% in instruction-accurate mode and only 1% in cycle-accurate mode. Of course, some combination of both these methods would be possible. This hybrid simulation mode is similar to that described in Chapter 9, and in Chapter 11.

TurboXim works for the ISS running in stand-alone and debugging mode and also works for the ISS when executing within the system level modeling environments discussed later.

18.2.3 ISS Simulation Mode Switching

Because the ISS TurboXim mode and the cycle-accurate simulation mode are fully interoperable, the ISS provides several mechanisms for switching between the two modes. Statistical sampling is one of the most useful ones to allow the use of cycle-accurate simulation to generate predictions of the time required to run a complete application on a target processor configuration. Via a "sample" command that defines a sampling ratio (indicating the ratio between the number of cycles run in cycle-accurate mode and the number of cycles run in fast-functional mode) and a sampling period length (indicating the number of cycles in the cycle-accurate mode for each period spent), the user can control the sampling rate and precision.

In an experiment run using an MP3 decoder audio application on a Tensilica 330HiFi audio processor configuration [2], the use of statistical sampling provided estimates of execution speed that had well under 1% error and at simulation speeds that were 20× the speed of cycle-accurate simulation—within striking distance of a pure TurboXim simulation (about 25% slower).

There are a number of ways of controlling the switching between modes, including during interactive simulation with a mode command, from the xt-gdb debugger with a similar command, from within the Xtensa Xplorer integrated development environment (IDE) with a command, and from within a target application using special ISS simulation callbacks. This last method is especially useful if you want to run in fast mode during simulation of system or application initialization and bootup, and then switch to cycle-accurate mode for debugging, detailed profiling, or other studies of application behavior and performance.

18.2.4 ISS Use Models

Users may want to do many different things with the ISS. We can classify three major use models:

1. Software Development
2. Profiling
3. Debugging

For software development, the ISS can be invoked as a stand-alone program to simulate the execution of an application targeted for the Xtensa processor. This can be done either from the command line, using the xt-run command, or from the Xtensa Xplorer IDE, using the Run dialogue. When requested, the ISS provides a performance summary output, which includes cycle counts for various events that occurred during the simulated program execution, such as cache misses. The simulator can also generate execution traces, as well as target program profiling data, which can be analyzed with the xt-gprof profiler or Xtensa Xplorer. Comparisons of profiles from different Xtensa processor configurations can help users select an optimal set of configuration options. Finally, users can debug an Xtensa target program using either the GNU command-line debugger (xt-gdb) or the Xtensa Xplorer debugger. By default, the debugger automatically runs the simulator to execute the target program.

18.3 System Modeling Enviroments

As mentioned previously, the ISS is provided for use within two system simulation environments: XTMP, which uses a proprietary C-based modeling API, and XTSC, which is based on SystemC. These system simulation environments allow the easy use of ISS models for one or more configurations of the Xtensa processor and permit the creation of generic models that can be adapted to configuration characteristics with little or no model source-code modification. This is supported by two mechanisms: transaction level modeling (TLM) approaches that provide generic methods for classes of processor interfaces, and introspection via a number of model-query methods that allow all the relevant configuration characteristics of a particular configuration to be determined.

Examples of the TLM interfaces include TIE queue interfaces supported by standard methods that work for all queue sizes, local-memory interfaces with standard access methods that work for all memory widths (these local-memory interfaces are very similar to the system PIF memory interfaces), and TIE lookups, which support a common set of access methods for all lookup instances. Examples of the introspection routines include APIs that allow the simulator to determine configuration parameters, the presence of all configured interfaces, and access methods for the ports provided for these interfaces. When these configuration-specific interface methods and ports are combined with a number of generic devices (such as memory devices, connectors, routers, and arbiters), then system simulation models for multicore systems can be developed and also linked to third party models and system simulation tools.

18.3.1 System Simulation Use Models

The system simulation use models include single-processor system level simulation with models of various external devices, including memories and peripheral blocks; multi-processor system level simulation where multiple Tensilica cores may be mixed with third party processors and complex buses and hardware accelerators; mixed system level and HDL level simulation, where the ISS model can be interfaced using pin-level transactors and interfaces to Verilog simulation of hardware blocks; and virtual platform or virtual prototype simulation where a fixed system-level model of an architected system can be provided to users for software development and verification. In any of these cases, it may be useful at times to run the system model in cycle-accurate mode, which the ISS supports, or in fast-functional mode which the ISS supports via the TurboXim capability. See Chapter 2 for more about the virtual platform concept.

All of these use models at the system level parallel the three use models of the single ISS: software development and verification, profiling, and debugging. Of course, at the system level, especially when there are multiple heterogeneous processor cores, verification, profiling, and debugging are more complex than with a single ISS. In addition, the system level models of peripherals, buses, hardware accelerators and memories, and memory controllers can be very complex devices in their own right.

18.3.2 XTMP

Several years ago, Tensilica created and released the XTMP modeling environment to support users who wanted to model complex systems with one or more Tensilica processor cores within them. XTMP provided a proprietary simulation kernel, built on top of a platform thread execution capability, the capability to instantiate Tensilica core ISS models, and a basic library of components along with guidance and API calls to allow a user to develop their own component models. The simulation kernel in XTMP is offered in two forms on Linux platforms, using either quickthreads or SystemC threads (quickthreads is faster as it has less overhead than the SystemC simulation kernel). On Windows, XTMP uses Windows fibers. Components offered in the XTMP environment include connector models (which operate like routers, to direct transactions to appropriate memories and peripheral models), local and system memory models, queues and lookups, and a set of examples showing how to develop user memory-mapped models. XTMP uses a transaction-based modeling style that preceded SystemC TLM1 (and thus TLM2) by several years.

XTMP is implemented in C++, but in order to make it available to the widest possible user base, the API was provided using C interfaces. Users could choose to use either C or C++ in implementing their system models. Most chose to use straight C.

When the TurboXim fast-functional ISS capability became available, using it within the XTMP environment (and the XTSC environment described later) was important. Since in XTMP, ISS models need to interact with other system device models such as memories and queue models, as well as other user-defined memory-mapped devices, it was necessary to find mechanisms to match the inherent TurboXim ISS speed with fast access to data stored in memory models or memory-mapped registers in peripheral or other user-defined models.

Thus, in the TurboXim fast-functional mode, the core uses a fast-access protocol to request fast access to memory and then uses that access to perform data transfers to and from the memory. This allows it to read and write memory without yielding control of the simulation. Built-in components implement the fast-access protocol with very efficient data transfer mechanisms. Memory-mapped user devices by default use reasonably efficient peeks and pokes for instruction fetches and data transfers, but can implement more efficient mechanisms as well. Three mechanisms are available for fast access, ranging from debug-style peeks and pokes (slowest, but still faster than sending request-response transactions over a simulated interconnect network); callback fast access, where registered read and write routines for the device model can be invoked by the Turbo model for the ISS; and the fastest mechanism, raw access, where a pointer to the memory is given to the simulator for it to use directly for all reads and writes. The raw access is conceptually similar to OSCI's TLM2 DMI (direct memory interface). If no fast-access method is supported by a device model, the XTMP transaction posts can still be used, although they are the slowest to simulate.

A fast-functional core may optionally run in a relaxed simulation mode that allows multiple instructions to execute without yielding simulation control. This relaxed simulation mode has a different interleaving of events compared to a simulation of cores in lock-step, but results in a functionally valid simulation for properly synchronized systems. Some issues that arise in relaxed simulation mode but not in lock-step simulation can occur in hardware as well. The relaxed simulation mode is conceptually similar to OSCI TLM2's quantum-based out-of-order simulation.

When the cores are running in relaxed simulation mode, each core will perform as much work as it can until it has executed for relaxedCycleLimit (a control parameter) cycles or is otherwise forced to yield. It then schedules itself to execute again in the future based on the number of cycles that were simulated.

One problem that occurs in multicore systems is that incorrect synchronization between the cores can cause timing-dependent system behavior. Data races become a problem when different timings of events can cause incorrect system behavior. Simulating cores out of order can significantly change the relative timing of events. The different interleaving of events can expose latent system synchronization issues. Some of these latent synchronization issues can be bugs that do not show up in a lock-step simulation, but could show up in hardware. Others are issues that could never show up in hardware.

Cores that simulate out of order may have a delayed response to external inputs. The effect of setting an edge-triggered interrupt or the RunStall signal may be delayed by an arbitrary number of cycles up to the relaxation limit. Level interrupts should be held active for at least the number of cycles in the relaxation limit since a core only samples the interrupts when it is scheduled to execute.

Some poll-based synchronization can slow down a relaxed simulation. For example, in one form of poll-based synchronization the processor continually reads the value from a local-memory address until it is modified by an external device. When a core is allowed to execute out of order, it will continually read the same address with the same value because no other device can execute until the core yields control. XTMP includes target code simulation APIs to help address this busy-waiting performance issue.

Relaxed simulation should be used to verify the correct functionality of a system, not its timing-dependent behaviors. Care must be taken to interpret the results of relaxed simulations. Because one core can run for many instructions before yielding control to other system devices, if a core is pushing data into an external queue connected to a TIE output queue interface, it can easily fill the queue before it yields. Likewise, if a core is popping data from an external queue connected to a TIE input queue interface, it can empty the queue before it yields. Because of this behavior, relaxed simulations should not be used to determine the optimal size of such a queue.

Simulations may switch dynamically between the TurboXim fast-functional mode and the cycle-accurate mode. This allows the faster simulation mode to be used to fast-forward simulation to a desired point before switching to cycle-accurate mode to take profiling data. It also allows for sampled simulation where most simulation occurs in the faster simulation mode, but profiling data can be taken on a regular basis while the system is simulating in cycle-accurate mode.

18.3.3 XTSC

After several years of user experience with XTMP, the industry began to move to wider adoption of the OSCI SystemC language (standardized by the IEEE as 1666) as a standard system modeling language. User and customer interest in Tensilica support for SystemC grew to the point where the company developed a SystemC based modeling environment called XTSC.

XTSC went beyond XTMP's earlier support of SystemC using the SystemC simulation kernel to offer the needed threading capabilities. This earlier XTMP capability made it a little easier to link XTMP and SystemC simulation, but was quite limited and did not define equivalents to XTMP transactions in the SystemC world.

In XTSC, a series of fundamental SystemC-based classes and methods are defined to allow users to build native SystemC models incorporating the Tensilica ISS models. These include the following:

1. The xtsc_core class that wraps an Xtensa ISS in a SystemC module
2. TLM interface classes using XTSC transaction types used for intermodule communication. XTSC transactions include basic request and response transactions and specialized ones for queues, lookups, etc.
3. A set of additional core classes that play a supporting role in XTSC
4. A set of non-member functions used to set up and control the simulation run or perform various utility operations

XTSC offers a set of basic system/SoC modeling components, including memories, arbiters, routers, queue objects, lookup objects, general memory-mapped IO devices, and wires.

XTSC also offers a set of testbench models to make it easier to test system models built with various XTSC components. These include PIF (processor interface) master and slave components, memory masters, queue and lookup producer/consumer objects and drivers, wire sources, and sinks.

The xtsc_core class wraps an Xtensa Instruction Set Simulator in a SystemC module. It includes methods to allow TLM port binding to all the local memory ports, the Xtensa Local Memory Interface (XLMI) port, the processor interface (PIF) port, all user-defined TIE interfaces, and certain system-level input and output signals. The TIE interfaces include output queues, input queues, lookups, export states, and import wires. The xtsc_core class also includes methods for loading the core programs, loading simulation clients, stepping the core, probing the core state, querying the core construction parameters, and setting up the core for debugger control.

The XTSC TLM interface classes play the major role in defining the communication between Xtensa cores and various SOC component modules. An arbitrary TLM module can be connected to an xtsc_core module only if it is using the XTSC TLM interface classes.

The XTSC core library includes many non-member functions and macros, such as:

1. Functions to configure simulation timing
2. Functions to initialize and finalize simulation
3. Functions and macros to configure and perform text logging

4. Utility functions to dump a byte array, perform string-to-variable conversion, and safely copy and delete c-strings and arrays of c-strings

In late 2009, XTSC was extended to permit pin and signal-level integration of the Tensilica ISS within Verilog simulaton environments. This is supported at two levels: the first consists of adaptors or protocol converters that convert XTSC transactions at the nominal TLM level into SystemC signals produced and consumed by SystemC pin interfaces; the second level uses further converters to link SystemC pins and signals to appropriate Verilog pins and signals. This capability is supported with the Verilog simulators offered by Cadence, Mentor, and Synopsys, and its primary use model is to support ISS co-simulation with Verilog RTL and gate level models where SystemC models may not be available for a complete system simulation.

XTSC supports the same concepts of fast-functional simulation (TurboXim) as XTMP, discussed above. This included relaxed (out of order) simulation and fast access methods to model data using the same methods as in XTMP, but dealing with XTSC transactions directly rather than XTMP transactions.

The XTSC transactions were created with an eye to OSCI SystemC TLM1 transactions and were created before the availability of OSCI TLM2 proposals. Although conceptually similar in many ways to TLM2, XTSC works in a TLM2 environment by special PIF or memory to TLM2 transaction adaptors (conceptual zero-delay bridges).

18.4 Third Party Integration

Before the availability of the SystemC-based XTSC system modeling environment, it was difficult to integrate the Tensilica ISS into third party system level modeling (ESL) tools. Most of them used proprietary modeling approaches based on C/C++, or perhaps had begun the move to SystemC themselves. The concepts of transaction-level modeling, of which XTMP was a relatively early user, were not universally used in a consistent way in third party, commercial ESL tools—and are still not consistent, although OSCI TLM2 has helped to standardize thinking and concepts.

However, when XTSC became available, the use of a common underlying SystemC-based simulation infrastructure allowed easier integrations to occur—still of course requiring significant effort to translate Tensilica transactions to and from the commercial tool's notion of transactions. Growing customer interest in using system level modeling, especially for heterogeneous systems involving processors from different suppliers (such as an ARM for control processing and one or more Tensilica cores for dataplane processing), led to a growing demand for integrations with third party ESL system modeling tools. Working with the commercial vendors of these tools, and using either XTSC or XTMP as the integration environment, a number of ISS/system modeling integrations have now been carried out. Although most of them use the XTSC SystemC-based modeling API, some of them have used XTMP because they themselves are not SystemC based and greater performance can be achieved in cycle-accurate simulation using the simpler XTMP interface and simulation kernel (perhaps 2× faster).

Commercial vendors of ESL tools with announced integrations of the Tensilica ISS and models include CoWare, Carbon Design Systems (with the SoC Designer tool that started out in Rockwell Semiconductor/Conexant as Maxsim many years ago, then a spinoff Axys, then part of ARM, and finally now part of Carbon), VaST, Imperas, Virtutech, Synopsys (the old Virtio tools, now called Synopsys Innovator), and Mirabilis.

Some of these commercial ESL tools work in both cycle-accurate and fast-functional mode; some of them (such as Virtutech SIMICS) only in the fast-functional mode. Since the Tensilica ISS supports both such modes, it is easy for it to be used in either mode in a commercial ESL tool. The main issue in integration has been to adapt the particular Tensilica TLM modeling style and syntax to the commercial tool TLM styles, which differ from tool to tool. Since the semantics are usually fairly similar, this has not proved to be too difficult, although sometimes our approach and a commercial tool would have been built on quite different assumptions. For example, cycle-accurate simulation is an essential bedrock technology for configurable and extensible processors, since the question of whether a loop nest that an algorithm lives in for 95% of its cycles takes 30 cycles with particular TIE-based extensions vs. 25 cycles is an extremely important issue. Many of the commercial tools that support only a fast-functional mode are intended for software functional verification, not detailed performance analysis, and this has led to some speed-accuracy trade-offs in those tools that are a bit different than our models support. Nevertheless, the list of seven different commercial ESL tools above indicates that ways to build a suitable and useful integration have been found in all cases.

18.5 Modeling Standards

Our work in developing XTSC and in building or working with third party ESL tool providers on integration was greatly assisted by the existence of SystemC as a standard system modeling language and reference (proof of concept) simulator. However, OSCI has moved rather slowly in the past in evolving the standards or in adding new capabilities. In particular, with some bursts of more rapid activity, the TLM working group has moved disappointingly slowly in developing the TLM concepts and a reference standard. When we developed XTSC, TLM1 had come out from OSCI but although it was a useful marker against which we developed our own capabilities, it was inadequate to reflect the full complexity of our interfaces (or indeed, any real-world interfaces, especially those that had to operate in cycle-accurate and fast-functional modes).

By the time we released XTSC and began to work with third party ESL vendors on integrations, it would have been possible to take significant advantage of OSCI's TLM2 work, had it been available. However, it was not available, and even when released in the summer of 2008, remains incomplete and inadequate or under-specified in several key aspects. Thus our XTSC is not currently TLM2-compliant, although we are looking at adding such capabilities in the future. Even when we add them, deficiencies in TLM2 mean that we must guard-band our use of it and avoid or restrict its applicability to the full range of our configurations. For example, it does not standardize bus locking for memory-mapped bus-style interfaces, which means it is restricted when dealing with RCW (read-conditional-write) transactions which require bus locking. This is despite the fact that all buses of real interest on SoCs support some aspect of bus locking. It also has concepts as discussed above such as direct memory interface (DMI) and out-of-order time quantum-based simulation that are similar but not identical to concepts used in our ISS and XTMP/XTSC environments. Again, this means that some use of TLM2 is possible, but without a complete semantic mapping to our capabilities.

In general, useful industry standards in the EDA and ESL areas lag the most advanced uses and when they finally come out, retrofitting their use into working systems and integrations is costly and often of little benefit. In addition, OSCI, like many standards bodies, is a pay-to-play organization that is not set up to solicit opinion across the full range of the industry, including non-members, when developing new standards. This means that it comes out with proposed standards that are weaker than they would be if a wider range of input and influence was actively solicited.

18.6 Examples

Almost all of Tensilica's customers have used the ISS in both cycle-accurate and fast-functional mode as part of their configuration, extension, software development, profiling, and debugging processes. This is a natural for any processor IP vendor.

A good number of customers have used the XTMP and XTSC modeling environments or the third party ESL tool integrations. Since the release of XTSC, its usage has grown so that at this point well over half the customers who build system models are using the XTSC version, usually integrated with their own SystemC models or those provided by other suppliers. At least one quarter of SystemC-based system modelers are using commercial ESL tools into which our ISS has been integrated. This is especially true of the largest customers using the technology in many complex SoC and system developments.

One example where we have used the technology internally is an audio reference design that used two Tensilica processors: a 330 HiFi for audio decoding, and a 108mini for the control of the system, along with an AHB style bus, PIF2AHB bridge, memories, both local and system, DMA controller, UART, DAC, and a GPIO block. The system was fully designed in hardware and a reference XTSC model was created for it as an example of how to build and use system models. This XTSC model supported all the use cases discussed earlier for system models, albeit on a relatively straightforward design.

In building the XTSC model, the bus interconnect fabric was modeled using a combination of router and arbiter components. Queue models were used to communicate with the 330 HiFi output queue and output decoded samples that would normally go through an audio DAC to a speaker were instead collected to wavout files that could be played back. The rest of the system model closely followed the real system implementation. It can be run both in cycle-accurate and fast-functional mode and is a useful demonstration of building SystemC-based system models.

A second example of a system model was for a 2- or 3-core video system, 388VDO, described in [4]. For this system an XTMP system model was created, including the two processors (stream processor and pixel processor) configured with special video instructions, and an optional general control processor that by default was a Tensilica 330 HiFi. The model was developed to be run in either 2 core or 3 core mode in order to allow users to substitute a different control processor for the Tensilica control processor if desired. In addition, other system components, such as a hardware DMA engine, a bit transpose hardware block, a special interconnect network, and DDR-style system memory controller, were modeled. All of these models were written in XTMP and were developed to support both cycle-accurate simulation and fast-functional TurboXim simulation.

In general, running real codecs on real video streams, either decoding or encoding, the simulation model running in TurboXim mode was 20–30$\times$ faster than the model running in cycle-accurate mode. This allowed the simulation model to be used both for software codec development and verification on a functional level and detailed codec optimization using the cycle-accurate mode and various kinds of performance profiling.

We also supported codec porting to the 388VDO (VDO) system using a special synthetic single-processor model that incorporated the TIE instruction extensions for both the stream and pixel processor. The methodology followed was as follows, all using the single-processor XTMP model and the multi-core model:

1. Compile and profile the codec on the single-core ISS model with realistic input data. Assume a single access memory. Partition the profile by assigning functions to one of the video cores. Since in 130 nm technology, a reasonable performance target for video codecs is 200 MHz, aim to have the total amount of cycles per core to be less than 180 Mcycles/s, in order to keep a reasonable margin.
2. Identify hot spots: The execution profile reports how many processor cycles are spent in each function. Functions that are called frequently or require high processing bandwidth are considered as hot spots and are good candidates for acceleration by using the VDO-specific TIE instructions.
3. Instantiate VDO TIE intrinsics: VDO TIE instructions are available for use by C/C++ programs through intrinsic functions. Modify the application to call VDO TIE intrinsic functions in place of the software code that originally represented the hot spots of the application. The VDO TIE instructions from different cores should not be mixed in the same function because these will be executing on separate processors after multicore partitioning.
4. Verify TIE usage: After modifying the source code for TIE usage, compile and simulate the application. Verify the results of the TIE-enabled application against the results of the original source code. Profile the application again to check if optimization goals are reached for hot spots.

5. Implement DMA functionality with C functions. Verify TIE and DMA C code.
6. Compile and profile the codec on the single-core model with the DMA functionality now implemented with C functions. Assume a single access memory.
7. Partition functions for multicore. Completely separate the code destined for each core by placing them in different functions. Ensure that the functions destined for each core use only TIE instructions belonging to that core (i.e., Stream Processor TIE instructions or Pixel Processor TIE instructions in the 388VDO). Aim to have the total amount of cycles per core to be less than 180 Mcycles/s, to target a 200 MHz frequency for the system with reasonable margin. If it meets the performance, go to the following step; otherwise return to Step 2.
8. Data sharing for multicore: Each processor in the dual-core 388VDO Engine has separate local instruction and data memories. The 388VDO uses shared external system memory and a hardware DMA engine for data movement between cores. Because the single-core virtual model does not contain a DMA engine, data movement between the two cores can be simulated using memcopy functions. Those functions can then be replaced by DMA control functions when the code is ported to the VDO Engine.
9. Profile the multicore partition: Profile the application to make sure the cycle counts for the functions that will be executing on different cores is balanced. Note that the memcopy cycle counts may be removed since these functions will be performed by the DMA controller of the VDO Engine. These profile results will still be an approximation of the final performance since some overhead will be introduced for synchronization and communication between the two cores when the application is ported to the VDO Engine.

This proved to be an effective methodology for codec porting and was tested in practice by a team at the University of Thessaly in Greece as reported in [3]. They ported the AVS video decoder to the 388VDO using this methodology.

Many capabilities were added to the XTMP model of the 388VDO over time to support higher level analysis of video codecs running on the system. This included extensive gathering and analysis of statistics for memory accesses, cycle counts, DMA usage, and other features, on both frame by frame and time interval basis. Many additional experiments were run on different configurations of the DDR memory controller model to test architectural options. Early versions of the model were used in defining the instruction extensions used in the stream and pixel processors as well and to optimize their use.

Both the audio reference design and 388VDO demonstrated effective use of the ISS and system modeling technology.

18.7 Conclusion

Design of complex systems and SoCs with configurable and extensible processors demands a high level of support for both processor modeling and the use of those processors within a system context. This chapter has outlined Tensilica's approach to generating such models in an automated fashion and in building system modeling environments that allow users to make realistic design trade-offs as well as create, debug, and profile target code. Growth in system complexity will continue to increase the demands for such simulation capability.

References

1. Augustine, S., Gauthier, M., Leibson, S., Macliesh, P., Martin, G., Maydan, D., Nedeljkovic, N., Wilson, B.: Generation and use of an ASIP software tool chain. In: Ecker, W., Müller, W., Dömer, R. (eds.): *Hardware-Dependent Software: Principles and Practice*, pp. 173–202. Springer, Heidelberg (2009).
2. Bailey, B., Martin, G.: *ESL Models and their Application: Electronic System Level Design and Verification in Practice*. Springer, Heidelberg (2010).
3. Bellas, N., Katsavounidis, I., Koziri, M., Zacharis, D.: Mapping the AVS video decoder on a heterogeneous dual-core SIMD processor. In: *Design Automation Conference User Track*. San Francisco, CA (2009). http://www.dac.com/46th/proceedings/slides/07U_2.pdf. Cited 21 September (2009).
4. Ezer, G., Moolenaar, D.: MPSOC flow for multiformat video decoder based on configurable and extensible processors. In: *GSPx Conference*. Santa Clara, CA (2006).
5. Leibson, S.: *Designing SOCs with Configured Cores: Unleashing the Tensilica Diamond Cores Technology*. Elsevier-Morgan Kaufmann, San Francisco, CA (2006).
6. Rowen, C., Leibson, S.: *Engineering the Complex SOC: Fast, Flexible Design with Configurable Procesors*. Prentice-Hall PTR, Upper Saddle River, NJ (2004).
7. Sanghavi, H., Andrews, N.: TIE: An ADL for designing application-specific instruction set extensions. In: Mishra, P., Dutt, N. (eds.): *Processor Description Languages*. Elsevier-Morgan Kaufmann, San Francisco, CA, pp. 183–216, (2008).

Chapter 19
Simulation Acceleration in Wireless Baseband Processing

Timo Lehnigk-Emden, Matthias Alles, Torben Brack, and Norbert Wehn

Abstract The Monte Carlo method is state-of-the art to verify the performance of wireless communication systems. Statistical simulations for various signal-to-noise (SNR) operation points are mandatory. Bit error rates (BERs) of 10^{-9} or even lower require the simulation of tens to hundreds of thousands of blocks for a single SNR operating point. Therefore, system simulation in wireless baseband processing is a challenging task. For example, analyzing the error floor in DVB-S2 LDPC channel-decoding systems needs several weeks of simulation time for one specific set of code parameters. Design validation of hardware architectures is challenging as well for the same reasons. We will present different techniques for accelerating system simulation with emphasis on channel decoding techniques by distributed software simulation, accelerated simulation using the cell processor, and hardware-assisted acceleration using FPGAs. Moreover, we will show how design validation of such systems can be accelerated by rapid prototyping.

19.1 Motivation

19.1.1 Algorithms in Baseband Processing

Baseband receivers are an essential building block of all wireless communication systems (see Fig. 19.1). They can be decomposed into two parts: the inner and outer receiver [1]. The main task of the inner receiver is to acquire digital data from an analog-digital converter (ADC) and transform the sampled channel data back into symbols probably sent over the digital baseband. The outer receiver utilizes forward error correction (FEC) to exploit added redundancy in these symbols to increase

N. Wehn (✉)
Microelectronic Systems Design Research Group, University of Kaiserslautern,
Erwin-Schroedinger-Straße, 67663 Kaiserslautern, Germany
e-mail: norbert.wehn@eit.uni-kl.de

R. Leupers, O. Temam (eds.), *Processor and System-on-Chip Simulation*,
DOI 10.1007/978-1-4419-6175-4_19,

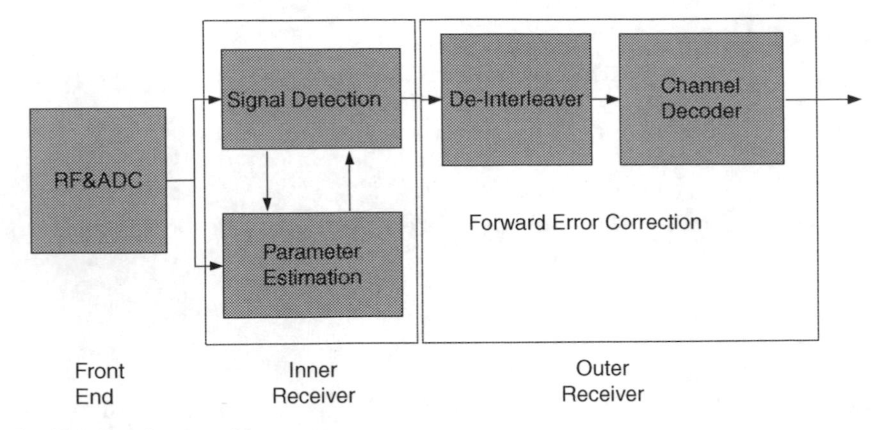

Fig. 19.1 Baseband receiver structure

overall communications performance. To carry out these tasks, both the inner and outer modems have to rely on numerous algorithms with differing complexities, like

- Eigen value decomposition
- DDS (direct digital synthesis)
- Matrix–matrix and matrix-vector multiplications and additions
- Complex multiplications and additions
- CORDIC (coordinate rotation digital computer) algorithms
- FFT (fast Fourier transformation)
- FIR (finite impulse response) filtering
- MAP (maximum-a-posteriori) or ML (maximum likelihood) estimators

All these algorithms have different properties and requirements regarding their implementation depending on their individual characteristics:

- *Computational versus memory-intensive algorithms*: A CORDIC algorithm needs many computational steps depending on the required precision while not utilizing much memory, while the DDS algorithm mostly relies on memory for the large LUTs (look-up tables).
- *DSP (digital signal processor) versus non-standard DSP algorithms*: While DSPs perform the FFT algorithm very efficiently, probabilistic algorithms are normally not supported by standard DSP processing.
- *Iterative approximation algorithms*: A classical algorithm of this kind is the Newton–Cotes algorithm for numerical integration, where a precision gain in the result is achieved by each iteration. More complex iterative approximation algorithms are used by modern channel decoders, e.g., LDPC decoders, for efficient MAP processing.
- *Complex interacting algorithms*: Very often data produced by one algorithm has to be used by another one and vice versa. In these cases, properties like memory bandwidth and complexity of the necessary control flow can become major obstacles.

With each new generation of wireless communication standards, the overall computational complexity of the outer receiver grows. Peak data rates have risen by a factor of 100 from 0.38 Mbit/s (3GPP/R99, WCDMA) in 1999 to 40 Mbit/s (3GPP/R7, HSPA evolved) in 2007 [2] and continue to grow fast in the foreseeable future. Especially the burden on the outer receivers has increased dramatically: up to 10 GOPS (giga-operations per second) are currently required to process 3GPP channel decoding alone [3]. For the fourth generation (long-term evolution—LTE) the complexity is rising by a factor of 10–100 GOPS. The whole baseband processing in LTE-A needs up to 2,000 GOPS at a data rate of 1 Gb/s [4]. The largest share of this computational complexity lies in the channel decoder. Finding the optimal hardware architecture is a key interest, but the design space for the receiver is quite large.

19.1.2 Challenges on Design Validation and Verification

To fulfill the requirements on implementation and communications performance for baseband receivers and to meet the evermore tightening time-to-market demands, engineers have to focus on novel approaches. The goal of the design-space exploration phase is to find the optimal trade-off of implementation cost versus communications performance. Employing system-level optimizations bears the largest potential. The fast validation of achievable system performance by extensive simulations and verification by rapid prototyping becomes mandatory. This need only increases the urge for fast simulation platforms, as an analytical evaluation is largely impossible in the face of non-bit-true transformations. Quantization, internal data representation, and low-level algorithmical optimizations which sacrifice communications performance for lower implementation costs are good examples, as the computational result of the decoder may be slightly different. Probabilistic methods like Monte Carlo-based simulation are the established state-of-the-art for this kind of *statistical validation*.

Here, not only the quality of the codes has to be evaluated but also the impact of all the high-level and low-level transformations. The goal is to *validate*, whether or not the system's communications performance is still within the specified and expected bounds. That necessitates not only the simulation of the bit error rate (BER) after the decoder for various SNR operating points. It is also often necessary to find out the error floor of a given implementation. Obviously, this requires simulations of hundreds of thousands of blocks.

Verification is another issue that adds to the complexity of the simulation. Once the receiver is prototyped, for example, in an FPGA platform, the verification of the system's functionality is the next step. The large set of service parameters of current communication standards (different block lengths, interleavers, code rates, etc.) makes this a challenging, high-dimensional problem. To verify the correctness of the implementation and for the co-simulation of the prototype a significant number of blocks has to be simulated for every possible parameter combination.

The simulation of communication systems like baseband receivers to evaluate their communications performance requires a complete model of a wireless transmission system including a random bit data source, a noisy channel, and a data sink for performance monitoring (see Fig. 19.2). In addition, the missing parts to a complete transceiver, which can be simplified in many cases, have to be present to form a complete transmission chain.

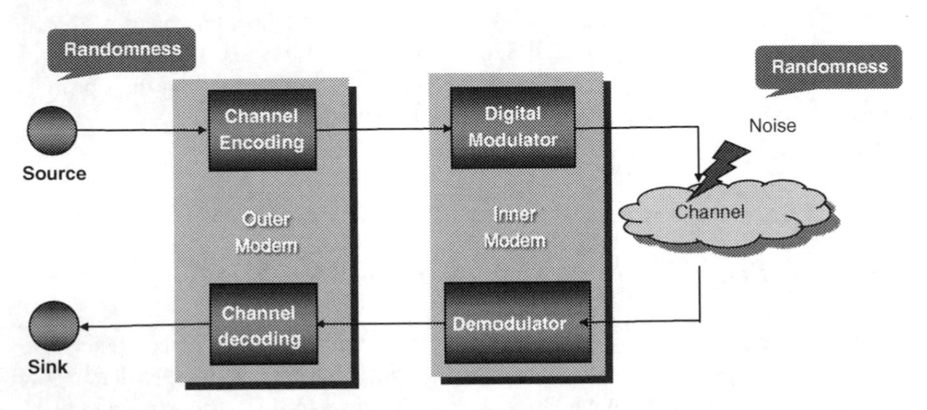

Fig. 19.2 Wireless transmission system for channel coding

The problem with simulation becomes obvious if the algorithms are examined more closely with respect to their software implementation: Many algorithms of a typical inner receiver (e.g., FFT, matrix–matrix operations) can efficiently exploit parallel processor architectures like SIMD (*single-instruction multiple-data*) and DSP-like co-processors. This is perfect for simulation acceleration on an architectural or component level. On the other hand, most of the outer receiver algorithms (e.g., MAP estimation) use non-standard DSP arithmetic. Also the large internal bandwidth requirements and very high complexity of these algorithms make software simulation difficult and time consuming on standard PC hardware.

As an example, let us consider a *complete DVB-S2 channel decoder* [5] and its analysis of its asymptotic error correction (error floor) capability for two different points in the design space. DVB-S2 is the next-generation satellite-based digital video standard employing low-density parity-check (LDPC) codes [6] with a codeword size of up to 64,800 bits.

The simulation down to the required bit error rates of 10^{-9} and below requires processing of tens to hundreds of thousands of blocks for a single SNR operating point. For one specific set of code and service parameters, the Monte Carlo simulation results shown in Fig. 19.3 took *several weeks to compute* on a standard PC. Two different decoder architectures were compared concerning their communications performance. In this case, acceleration on a system level by parallelizing the task of simulation itself yields a large improvement in simulation speed.

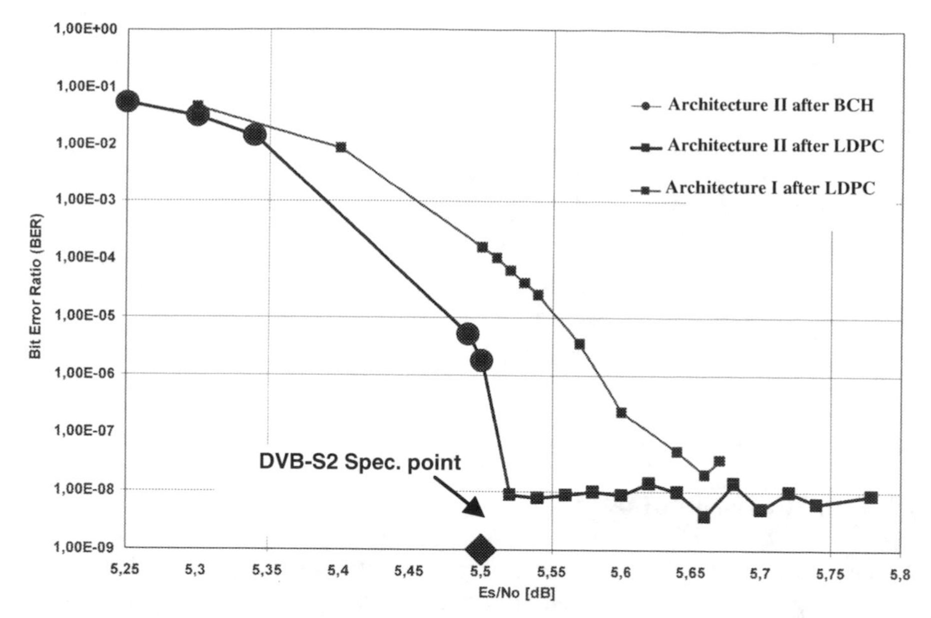

Fig. 19.3 DVB-S2 error-floor simulation results [5]

Another example is the simulation of LTE turbo-decoder services: *4 weeks of simulation time on four standard PCs working in parallel* on different SNR points were necessary to obtain the results required for communications performance evaluation. Considering only these two examples, the need for simulation acceleration for efficient communications systems simulation becomes obvious.

Verification of an actual target hardware implementation against the specifying software model is also a necessity. Without a fast software simulation environment, it is impossible to obtain a sufficient amount of verification data to co-simulate with the prototype implementation, making simulation acceleration indispensable for this application.

19.1.3 The Simulation Space

The simulation space encompasses two major groups: software- and hardware-accelerated simulation. *Software-based simulation* relies on software for standard PC hardware (including optimizations for SIMD, multi-core architectures, and multithread-capable operating systems), distributed computing clusters, or specialized accelerator platforms. *Hardware-based simulation* is mainly based on dedicated FPGA implementations of the complete simulation framework or the development of more flexible *rapid prototyping platforms* on FPGA (Table 19.1).

Table 19.1 The simulation space

Simulation space	
Software	Hardware
Standard PC simulation (including optimized multi-core, SIMD, and SMT provisions)	Dedicated FPGA-based implementations
Distributed computing clusters (e.g., BOINC)	Rapid prototyping platform
Specialized accelerator platforms (e.g., Cell, GPUs)	

The following two chapters examine the simulation space with more detail and with real-world examples.

19.2 Software Simulation

As mentioned before, there are various software platforms available for simulation acceleration. There also exist *different strategies* to exploit these platforms depending on the actual simulation problem. The key factor to finding the ideal simulation acceleration platform is matching the *level(s) of parallelism* it can efficiently support with the level(s) of parallelism the simulation problem can natively provide. This is what this chapter is about. It presents and analyzes different software platforms and shows their respective strengths and weaknesses in the context of *levels of parallelism*. As already introduced in Section 19.1.2 , simulation of *LDPC channel decoding for BER and error floor analysis* is used as an example throughout the chapter.

19.2.1 Level of Parallelism

We distinguish three levels of parallelism in this chapter: *data level*, *thread level*, and *program level*. These levels differ essentially in the provided *abstraction* of processing tasks, the *granularity* (measure of task size and execution time) of tasks that are processed in parallel, and the necessary *interaction* between tasks to solve the simulation problem. Parallelization on different levels is not exclusive and can be combined (Table 19.2).

Table 19.2 Levels of parallelism

	Abstraction	Granularity	Interaction
Data level	Low	Fine	High
Thread level	Medium	Medium	Medium
Program level	High	Coarse	Low

These properties lead to distinct requirements on the simulation platform and the simulation problem itself. For example, high interaction usually demands for high communication bandwidth, while parallelization by coarse-granular tasks benefits most from higher processing power. Parallelization of a simulation problem on a low abstraction level requires usually more knowledge and effort than any high-abstract problem division.

19.2.1.1 Data-Level Parallelism

Data-level parallelism is characterized by small tasks that can be executed in very limited time on a number of parallel processing elements. The high data dependencies naturally occurring on this level and the resulting demand for high communication bandwidth can only be satisfied by non-scalar processing architectures like

- *Superscalar* architectures (multiple instruction pipelines) as used in most commodity CPUs to exploit *ILP* (instruction-level parallelism)
- *Vector-oriented* architectures as used in the supercomputer domain
- *VLIW* (very long instruction words) architectures as used in most DSPs and some GPUs (operation-level parallelism)
- *SIMD* (single instruction, multiple data) architectures as used also in modern commodity CPUs (e.g., SSE) and the IBM cell processor

The simulation example we evaluate consists of a transmission chain including data generator, modulator, encoder, AWGN channel, DVB-S2 LDPC decoder, demodulator, and data sink for BER evaluation (see also Section 19.1.2). The simulation model is entirely written in ANSI-C.

As our primary simulation acceleration platform, we selected the *IBM Cell processor* [7] as a *native SIMD architecture*. For comparison, we evaluated the simulation model on a standard Intel Core 2 Duo CPU, with and without SSE2 SIMD optimizations by native libraries and appropriate compilation options.

The cell processor architecture (Fig. 19.4) consists of a 64 bit SMT-capable PowerPC processor element (*PPE*) for control purposes and up to eight so-called synergistic processors elements (*SPE*). A ring-based coherent *element interconnection bus* (EIB) connects all SPEs including their integrated SIMD-ALUs and the PPE. SPEs can directly access 256 KB of local SRAM; all transfers to the main memory and between the SPEs have to be commenced by the local DMA controllers.

Because the cell processor employs a multi-core architecture, mapping of the simulation chain to the cell processor actually involves not only the data level but also the thread level discussed later in this chapter in Section 19.2.1.2. The SMT-capability of the PPE is not utilized for this example. Our finally mapping for evaluation is shown in Fig. 19.5:

- One SPE (SPE #1) is solely used for data generation, mapping, additive white Gaussian noise (AWGN) channel, demapping, and BER evaluation (data sink). Because of the limited internal memory bandwidth, DMA transfers between data source, sink, and the LDPC decoder have to be minimized. It turns out, that the

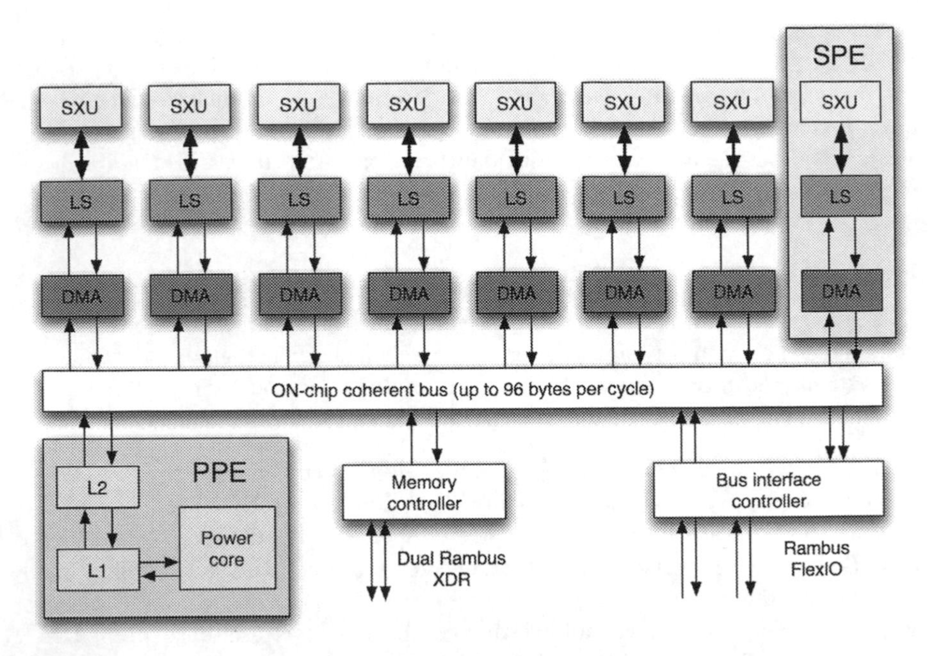

Fig. 19.4 The cell processor architecture

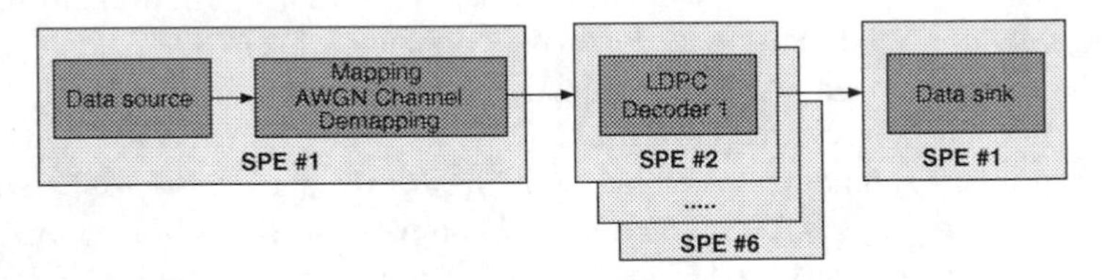

Fig. 19.5 Mapping of the simulation chain to the cell processor

workload of this SPE is dominated by the calculation-intensive AWGN channel, making this an ideal point of action for SIMD optimization.

- Five SPEs (SPE #2-#6) are used for the DVB-S2 LDPC decoding process itself. Input and output buffers connected to SPE #1 are interleaved to allow for overlapping computation and data transfer via DMA. On each SPE one LDPC decoding program is running. The LDPC decoding algorithm provides various degrees of freedom for effective utilization of the SIMD-ALU in these SPEs.

At first, we evaluate SIMD optimizations for the AWGN channel residing on SPE #1 because of its limiting effect on the overall data generation throughput. Calculation of the values for *AWGN channel* model is done by the *Box–Muller algorithm*, which transforms a uniform random number distribution to a Gaussian distribution with the help of trigonometric and square root functions. Uniform random number generation as well as processing of the Box–Muller algorithm is a very demanding task for any general purpose ALU.

Table 19.3 shows a throughput comparison of three different AWGN channel implementations with single precision: while even a standard PC implementation using a commodity CPU benefits significantly by SIMD optimizations, the cell processor as a native SIMD system outperforms the standard CPU by another order of magnitude.

Table 19.3 Simulation results for AWGN channel

Simulation results for AWGN channel		
Implementation	Architecture	Speed
Standard C code	Intel Core 2 Duo 2.2 GHz, 3GB RAM	0.5 Mbps
Optimized for SSE2 GNU scientific library		6 Mbps
Cell optimized using Monte Carlo library	IBM Cell 3.2 GHz, 256 MB RAM	72 Mbps

To optimize the processing of the LDPC decoder algorithm (SPE #2-#6) we now look onto different mapping strategies to exploit the SIMD-ALU of the SPEs effectively. LDPC decoding is an *iterative process* that relies on the so-called *message-passing* algorithm: a large number of messages representing probabilities or *beliefs* are transferred between two types of computational nodes: *variable nodes* and *check nodes*. The number of these nodes is determined by the codeword size and the code rate (e.g., for the DVB-S2 rate 1/2 LDPC code 64,800 variable nodes and 32,400 check nodes are present). These computational nodes can be implemented efficiently, but the message transfer of up to 226,800 messages per iteration is still a very challenging task.

A Cell processor's SPE instructions can operate only on the small local memory. Therefore, the DVB-S2 a priori codeword information has to be stored in the larger main memory with significantly lower bandwidth. Each SPE can process four single-precision float SIMD operations in parallel. These constraints given by the SPE architecture allow for two basic mapping strategies:

1. Decoding of *four* different codewords in parallel on each SPE. In this configuration, the LDPC decoding itself is done in a *fully serial* manner with *one* check node per codeword being processed at once.
2. Parallelized decoding of *one* codeword on each SPE. This configuration uses a *partially parallel* LDPC decoding approach where *four* check nodes per codeword are processed at once with SIMD instructions.

The *128-bit SPE architecture* relies strongly on correspondingly aligned data in local and main memory for efficient read, process, and write operations. Because of *data misalignment* emerging from the LDPC codeword generator, the first strategy requires costly re-ordering of codeword data before processing can commence. This strategy also suffers from a strong throughput limitation: decoding time in terms of required iterations can vary for different codewords and consequently the achievable throughput is limited by the maximum decoding time for four codeword.

The second strategy is not negatively affected by misalignment and varying decoding times, and data read and write operations from main memory can be streamed to the processing nodes even more efficiently by data pre-fetching.

These features give the partially parallel decoding approach a significant advantage over fully serial decoding. The low bandwidth to the main memory remains as the main bottleneck. The results achieved for the complete simulation chain on the cell processor including the SIMD-optimized AWGN channel and partially parallel LDPC decoders are given in Section 19.2.2.

19.2.1.2 Thread-Level Parallelism

Parallelism on the *thread level* (sometimes also called *process level*) is situated between data- and program-level parallelism in terms of abstraction, granularity, and interaction (see Table 19.2). Thread-level parallelism can also be efficiently combined with other levels of parallelism: Threads can contain data-level parallelism as in the cell processor example in Section 19.2.1.1, and programs can contain many threads or processes to exploit thread-level parallelism. Thread-level parallelism is supported by numerous architectures:

- Any *standard CPU* architecture capable of running a *multithreaded operating system* (e.g., WINDOWS or UNIX)
- *Multiple-processor* architectures (e.g., tightly coupled SMP systems with more than one CPU per main board)
- *Multi-core* architectures as used in many current commodity CPUs (e.g., Intel Core 2 Duo/Quad with more than one processor core per chip)
- *SMT* (simultaneous multithreading) or *hardware multithreading* architectures as used in many current commodity CPUs (e.g., "hyperthreading")

Multithreaded operating systems on a non-SMT single-core CPU cannot achieve any performance gain compared to purely sequential processing. Multithread programming can strongly decrease the programming effort by providing a higher abstraction level. Presented with the illusion that each thread has exclusive access to the CPU resources (*temporal multithreading)*, the programmer can split up complex tasks into simpler threads to reduce complexity.

Additionally, architectures capable of simultaneous or parallel multithreading do offer a potential speed-up. *SMT architectures* like the Intel Pentium 4 provide multiple register sets and caches to increase computation performance by utilizing their (superscalar) ALU more efficiently. This approach exploits *instruction-level parallelism* on the thread level and can lead to a proportional increase in processing speed, but some applications do not benefit from hardware multithreading at all. Multi-processor as well as multi-core architectures on the contrary have multiple fully featured processors and ALUs for parallel thread processing which increases calculation performance proportionally for most applications.

Using thread-level parallelism to increase performance for simulation acceleration implicates some challenges with respect to *synchronization* and *communication bandwidth*. Because of threads normally sharing resources, access to these resources has to be *synchronized to avoid data inconsistencies*. There are numerous mechanisms to achieve perfect synchronization, for example, by employing semaphores or barriers, but all these mechanisms will in turn limit the overall speed-up. For our cell processor implementation of the DVB-S2 simulation chain (see Fig. 19.5) thread synchronization across the decoding SPEs was mandatory for ordered access to the main memory. In addition, the data generation and BER monitoring threads on SPE #1 have to be synchronized to the decoding process.

The problem can be seen as a barrier synchronization problem, which limits the achievable decoding throughput for the first mapping strategy. Another issue is the *required communication bandwidth*. The medium of communication for threads may involve shared memory (cell processors main memory) or network structures (EIB between SPEs) or even both. In the presented simulation chain, the codeword data generation on SPE #1 and codeword decoding on SPEs #2–#6 always require storing and reading data from the main memory. The same holds for the BER monitoring. Even during decoding the main memory has to be accessed a huge number of times via the SPEs DMA engines because of the local memory being too small. This makes the communication bandwidth the limiting factor.

19.2.1.3 Program-Level Parallelism

Program-level processing involves large tasks with relatively low communication bandwidth requirements. As already stated in Section 19.2.2.2, tasks at the program level itself can utilize thread-level parallelism. Architectures providing parallelism at the program level include

- *Processor clusters* (e.g., loosely coupled multi-processor systems made from commodity hardware, like a Beowulf cluster)
- *Clusters of networked workstations* (often referred as "COW") using MPI (message passing interface) or PVM (parallel virtual machine) for communication
- *Grids of networked workstations* using a special grid-computing software like BOINC

As for the thread level, *standard CPU architectures in conjunction with an adequate multitasking operating system* can bring program-level parallelism to the programmer, but do not provide any speed-up. *Multi-processor architectures* do not only support thread-level parallelism as described in Section 19.2.2.2 but can also run complete programs in parallel with proportional performance gain. However, the focus of this section is on large-scale architectures that natively operate at the program level.

Table 19.4 presents highly scalable architectures that can achieve massive speed-ups: *processor clusters* were first introduced by *supercomputers* with dedicated pro-

Table 19.4 Architectures supporting program-level parallelism

Architectures supporting program-level parallelism			
Architecture	Max. # nodes	Scalability	Interaction
Processor cluster	Small–large	High	Low
COW	Large–huge	Very high	Very low
Grid	Huge	Nearly unlimited	Rare

cessing nodes setting up a *loosely coupled homogeneous multi-processing system.* Nowadays so-called *virtual supercomputers* consisting of hundreds of networked commodity processing nodes have taken over. *Clusters of networked workstations*, or COWs, differ mainly in the utilized processor nodes: While processor clusters offer one dedicated gateway to the whole system of identical nodes, a COW uses standard desktop PCs including terminal and display for each node. In addition, the network infrastructure of a COW consists of COTS (connection-oriented transport service) instead of dedicated hardware, limiting the available communication bandwidth. The largest scale systems for multiprogramming are *computing grids*, which normally consist of widely dispersed heterogeneous workstations communicating over inherently unreliable high-latency networks like the internet.

We selected a *grid-computing approach* for simulation acceleration and employed the open-source BOINC (*Berkeley open infrastructure for networked computers*) framework. It allows distributed computing on nearly any pre-existing heterogeneous computer infrastructure and supports all major hardware platforms and operating systems. BOINC relies on a *client-server-model*: the server divides the large primary task into smaller subtasks and distributes them to the available clients depending on current load and performance. The results from the clients are then collected by the server and combined to the overall solution to the primary task.

The simulation model of the DVB-S2 communications chain (see Section 19.2.1.1) was prepared to support program-level parallelism. Therefore, the server must be able to generate independent tasks that can be processed by the clients in adequate time. If the subtasks take very long to be processed especially on slow clients, this would hinder effective load balancing and result in large data loss if a client goes down. On the other hand, if the subtasks can be processed very fast, the communication overhead would become the bottleneck. Assuming small communications overhead and equally performing clients, the computational performance can increase proportionally.

To accelerate our simulation, the complete communications chain including data generation and BER monitoring has to run on every participating client. The subtasks exchanged between server and clients consist, among other things, of the SNR ranges, maximum number of iterations, and number of simulated blocks. The grid was actually very small and incorporated 10 standard desktop PCs with dual-core Intel CPUs. Section 19.2.2 presents the evaluation of our simulation chain performed on this grid.

19.2.2 Comparison and Results

This section presents the results for simulation acceleration of the DVB-S2 communications chain introduced in Section 19.2.1.1. To summarize, the simulation chain used for evaluation involves

- LDPC En-/decoder for DVB-S2 codewords, 64,800 codeword bits per block, 226,800 messages to exchange per iteration, code rate 1/2

- Monte Carlo simulation: data generator and monitor, AWGN channel, on average 100,000 blocks per SNR point, 10 SNR points in total

Derived from these figures, on each evaluated platform 64.8 gigabit of binary data were generated, encoded, decoded, and constantly monitored. Table 19.5 shows the comparison of the different platforms. The simulation on the Intel Core 2 Duo CPU with SSE2 optimization (see also Section 19.2.1.1) was most easy to set up, but took nearly one day to complete the simulation. This result is also used as benchmark for the other platforms. The cell processor simulation achieved a speed-up of only 2.4 times despite its high implementation effort. As discussed in Sections 19.2.1.1 and 19.2.1.2, the main bottleneck here was the unavoidable numerous DMA transfers to the main memory. If LDPC codes with smaller codewords as employed in the upcoming WiMAX or WIFI standards (<2,500 bit) will be used, this bottleneck will likely cease to exist. The codewords then could be completely stored in the local memories of the SPEs. The grid-computing approach using BOINC for simulation acceleration (see Section 19.2.1.3) yielded excellent results: with 10 computers containing 20 cores overall establishing the grid, the performance improved by a factor of eight. The increase in speed was approximately proportional to the core benchmark system with two cores. Providing nearly unlimited scalability with only medium migration effort and low cost, the use of grid computing proves to be a very effective way for large-scale simulation acceleration.

Table 19.5 Software simulation acceleration results

Software simulation acceleration results

	INTEL SSE2 optimized 2 cores, 2.2 GHz	Cell processor 6 SPE cores	BOINC 10 computers 20 cores
Parallelism levels	Data level Thread level	Data level Thread level	Data level Program level
Scalability	No	Minimal (up to 7)	Nearly unlimited
Effort	Low	High	Medium
Runtime	23 h 20 min	9 h 47 min	2 h 55 min
Throughput	0.772 Mbps	1.840 Mbps	6.172 Mbps
Improvement	1	× 2.4	× 8

19.3 Hardware-Accelerated Simulation

Hardware-accelerated simulation is normally based on dedicated implementations on FPGAs.[1] A more flexible approach is the use of a prototyping platform that can also enhance the reuse of already implemented simulation support systems. This becomes especially important if different communications systems have to be simulated in a short time period.

[1] Chapter 7 presents how to use FPGAs for architectural simulations.

19.3.1 Dedicated FPGA-based Simulation Accelerator

The application of a *dedicated FPGA-based simulation accelerator* is most convenient if a hardware description of the communication system under evaluation is already available. This is often the case if the communication system has later to be implemented on a hardware target, like an FPGA or ASIC, for the development of a product. In the case of the DVB-S2 LDPC decoder, the hardware description consisted of a *generic synthesizable VHDL model*. As an alternative, a *high-level synthesis tool like Catapult* [8] can be used to obtain a non-optimized hardware model very efficiently. In our case study, the decoder architecture was designed to process 90 check nodes and variable nodes in parallel. In comparison, the complete cell processor implementation from Section 19.2.1.1 was able to process only 24 check nodes at the same time.

The original DVB-S2 LDPC hardware model naturally featured no provisions for error rate simulation. Therefore, the random data generator, modulator, DVB-S2 LDPC encoder, AWGN channel, and BER monitor have to be implemented as a synthesizable VHDL model, too. Depending on the complexity of these simulation support systems, this can be a very demanding task (e.g., the implementation of a more advanced channel model like Rayleigh). Furthermore, a user interface to configure the DVB-S2 LDPC decoder simulation chain and to read out the simulation results must be integrated.

Table 19.6 shows the simulation acceleration results compared to benchmark: While only running at 135 MHz, the achieved speed-up was an impressive 117 times. Such performance gains are only obtainable by FPGA-based simulation because of the *very high bandwidth interfaces* between simulation support systems and the design under evaluation. This makes speed the key advantage of this acceleration approach. However, the very high implementation effort and the limited flexibility may put high constraints on its range of use.

Table 19.6 Hardware simulation acceleration results

Hardware simulation acceleration results		
	INTEL SSE2 optimized2 cores, 2.2GHz	FPGA Xilinx Virtex-4 LX-100 135 MHz, $P = 90$
Scalability	No	Yes (up to 360)
Effort	Low	Very high
Runtime	23 h 20 min	12.1 min
Throughput	0.772 Mbps	90 Mbps
Improvement	1	× 117

19.3.2 Design Validation and Verification by Rapid Prototyping

In contrast to dedicated FPGA-based accelerators, a *rapid prototyping platform* allows the developer to effectively validate and verify various designs without creating any specific provisions dedicated for simulation acceleration.

Figure 19.6 shows the structure and an actual realization of such a platform [9]. All elements required for simulation, such as data generation and BER monitoring, are implemented in software for maximum flexibility. To minimize the throughput bottleneck of the software–hardware interface, the software runs on a *hardwired PowerPC440* inside the FPGA and is connected to the application under evaluation by a processor local bus (PLB). *Dedicated co-processors* are implemented inside the FPGA to perform time critical tasks like AWGN generation to improve simulation speed further. User access can be implemented via any available standard interface like RS232, USB, ethernet, or PCI-express. This access offers also connectivity for evaluation tools like MATLAB and software-based simulation accelerators for verification purposes.

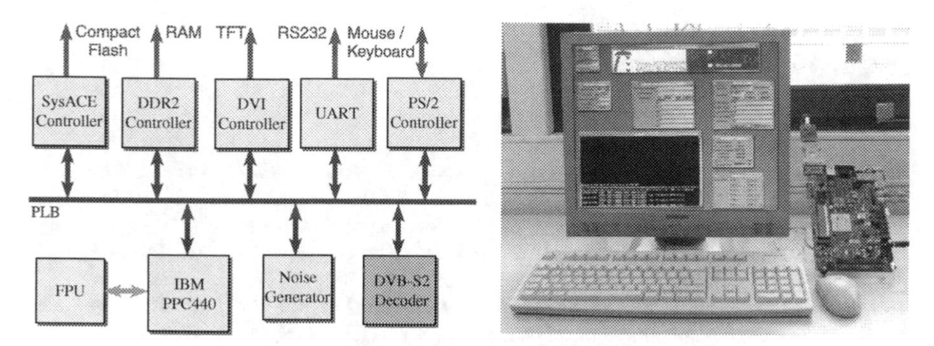

Fig. 19.6 FPGA-based rapid prototyping platform

Disadvantages of such a prototyping platform are the high initial building effort and the reduced speed-up compared to dedicated implementations.

A rapid prototyping platform also allows for *effective and efficient verification*. Combined with an ideally accelerated software simulator, reference data can be co-simulated using the FPGA implementation and an already validated software model. A bit-accurate comparison of input and output data to the application under test, e.g., the DVB-S2 LDPC decoder, sufficiently substantiates the specified functionality of the hardware realization.

19.4 Conclusion

Validation and verification of wireless communication systems become more and more challenging because of increasing complexity and performance requirements. Furthermore, the large design space of modern communication standards demands an effective evaluation of an enormous number of combinations in limited time. *Monte Carlo simulation*s are the only way of effectively coping with these challenges. While a huge amount of simulations is required to validate the communications performance, simulation acceleration on system and architectural level becomes mandatory.

Simulation acceleration can be achieved both software and hardware-based. *Software-based simulation acceleration* is provided by state-of-the-art PC architectures, specialized accelerator platforms like the SIMD-oriented multi-core cell processor and distributed software simulation systems like the BOINC grid-computing system. For hardware-based systems, dedicated FPGA-based accelerators as well as rapid prototyping platforms can be efficiently applied.

Our evaluation presents the advantages and problems accompanying the different platform and shows results using a DVB-S2 LDPC decoder simulation chain.

References

1. Meyr, H., Moeneclaey, M., Fechtel, S.: *Digital Communication Receivers*. Wiley, New York, NY, (1998), ISBN 0–471–50275–8.
2. Dahlgren, F.: In: *9th International Forum on Embedded MPSoC and Multicore, Technological Trends, Design Constraints and Architectural Challenges in Mobile Phone Platforms*, ST-Ericsson, http://www.mpsoc-forum.org/2009/slides/Dahlgren.pdf, October (2009).
3. Rowen, C.: In: *9th International Forum on Embedded MPSoC and Multicore, Technological Trends, Energy-Efficient LTE Baseband with Extensible Dataplane Processor Units*, Tensilica, USA, http://www.mpsoc-forum.org/2009/slides/Rowen.pdf, October (2009).
4. van Berkel, K.: In: *Proceedings of IEEE Conference Design, Automation and Test in Europe (DATE '09)*, Multi Core for Mobile Phone, Nice, France (2009).
5. Müller, S., Schreger, M., Kabutz, M., Alles, M., Kienle, F., When, N.: In: *Proceedings of IEEE Conference Design, Automation and Test in Europe (DATE '09), A Novel LDPC Decoder for DVB-S2 IP*, Nice, France, pp. 1308–1313, (2009).
6. (ETSI), E. T. S. I. Digital Video Broadcasting (DVB) Second generation framing structure, channel coding and modulation systems for Broadcasting, Interactive Services, News Gathering and other broadband satellite applications, EN 302 307 V1.1.2.
7. IBM Cell Processor, Cell Broadband Engine Programming Handbook, https://www-01.ibm.com/chips/techlib/techlib.nsf/techdocs/1741C509C5F64B3300257460006FD68D/$file/CellBE_PXCell_Handbook_v1.11_12May08_pub.pdf, May (2008).
8. High Level Synthesis Tool "Catapult C Synthesis", Mentor Graphics, http://www.mentor.com/products/esl/high_level_synthesis/catapult_synthesis/upload/Catapult_DS.pdf.
9. Alles, M., Lehnigk-Emden, T., Brehm, C., When, N.: In: *Proceedings of EDA Workshop 2009, A Rapid Prototyping Environment for ASIP Validation in Wireless Systems*, VDE, Dresden, Germany, pp. 43–48, (2009), ISBN 978–3-8007–3165–7.

Chapter 20
Trace-Driven Workload Simulation for MPSoC Software Performance Estimation

Tsuyoshi Isshiki

Abstract In this chapter, we introduce a method for accurately estimating the cycle counts of parameterized MPSoC architectures through workload simulation driven by program execution traces. Program traces are encoded in the form of *branch bitstreams* that capture the complete history of executed control flows, while the workload model called the *program trace graph* is automatically generated by analyzing the application programs at compile time. A collection of these trace-driven workload models that represent the behavior of processors in the MPSoC platform are coordinated by the workload simulator kernel for simulating the timing behaviors of the workloads affected by the interaction of MPSoC components (processors, busses, channels), thus achieving a cycle estimation accuracy of below 1% error while exhibiting a considerable simulation speedup.

20.1 Introduction

Software performance estimation plays a crucial role in system-level optimization of embedded systems not only for software performance tuning but also on the key architectural decisions especially at the early development phases. Current state-of-the-art embedded systems are rapidly adopting multiprocessor system-on-chips (MPSoCs) where processors and devices interact in a complex manner, creating a number of design challenges in software and architecture implementations for satisfying the requirements of the target system.

Recent ESL (electronic system level) design methodologies and tools [1] are aimed at enhancing the design productivity of these systems by providing a software development platform before the details of the MPSoC architecture are fixed, and facilitating the concurrent refinement of software and architecture implementations through system simulations and performance analysis. In the previous chapters, we have seen the latest techniques in improving the system simulation speed through

T. Isshiki (✉)
Department of Communications and Integrated Systems, Tokyo Institute of Technology, 2-12-1 S3-66 Ookayama, Meguro-ku, Tokyo 152-8552, Japan
e-mail: isshiki@vlsi.ss.titech.ac.jp

R. Leupers, O. Temam (eds.), *Processor and System-on-Chip Simulation*,
DOI 10.1007/978-1-4419-6175-4_20,

abstract hardware models such as SystemC TLM2.0 (Chapter 2) and state-of-the-art *instruction-set simulator* (ISS) technologies (Chapters 7–12), enabling a thorough design verification environment of the entire system. Adoption of such currently available design environment in the context of *design space exploration* during the early design phase is also feasible, although this requires the designers to commit sufficient resources into creating and maintaining the system simulation infrastructures (simulators, models, and tools) for this task. Simulation-driven design exploration also requires that the software must be written, to a certain extent, correctly and completely, including device drivers and OS (however, abstract they may be), in order to drive the system-level MPSoC simulation models. During the course of software and architectural refinements, a certain level of bug-insertion risk is unavoidable that may require additional painstaking debugging efforts and design iterations. Adding to the fact that the full-system simulation can still take a considerable amount of time in the complex MPSoC cases, the amount of design space in the exploration will be limited on the current ESL design tools.

In order to further speed up the embedded system simulation, a popular approach is to apply source-level timing annotation of the target processor for generating timing model during software execution [2–4]. Here, the timing information can include both static timing (obtained at compile time) and dynamic timing (caches and branch predictions) [3]. This approach is also applied to MPSoC modeling where the impacts of interconnect architecture and memory subsystems can be evaluated in detail [5, 6].

Statistical workload models is another effective approach for design space exploration which can cover a wide design space efficiently in a short amount of time, revealing key tradeoffs among a wide spectrum of design choices. However, deriving a proper workload is highly application dependent, and these models usually apply to data traffic but are difficult to generate accurate computation load models. Trace-path analysis method in [7] creates task-level timing models through application-specific profiling (MPEG-4 encoding) and manual instrumentation to achieve high cycle accuracy. At the MPSoC level, statistical workloads have been used to model MPSoC subsystems such as bus traffic [8] and NOCs [9] or for certain application domains such as networking [10].

In this chapter, we introduce a novel *trace-driven workload model* capable of precisely capturing data-dependent behavior of the application to achieve near-cycle-accurate performance estimation [11]. Program execution trace of an application on a given input data set is computed, through source-level instrumentation and native code execution, prior to the trace-driven workload simulation and efficiently encoded as *branch bitstream*. This branch bitstream is then used to steer the workload models in the form of *program trace graphs* generated for the target processor inside our parameterized MPSoC performance estimation framework. Our framework also includes automatic generation of accurate coarse-grain workloads directly from the application source code by using the precise execution profile information obtained from the same branch bitstream. These techniques are further combined with our MPSoC application development framework called *tightly coupled thread model* (TCT model) which provides a direct path from sequential programs to concurrent execution model by providing a simple programming model on the C language and a compiler for automatically generating communication instructions between threads. Thread synchronization protocols are modeled in the program trace graph representation that provides the interface between the workload models of individual threads and the workload simulator kernel for modeling the system-level interactions (buffered channel message passing) under the parameterized MPSoC architecture model.

20.2 Trace-Driven Workload Model

This section focuses on the key concepts of our trace-driven workload model. The first concept is the compact representation of the program execution trace using the *branch bitstream* which is simply a sequence of branch condition bits of the program execution. The second concept is the *program trace graph* that represents the accurate computational workload of the application software.

20.2.1 Branch Bitstream for Program Trace Representation

Consider a program whose interprocedural control flow graph (interprocedural-CFG or ICFG) is shown in Fig. 20.1. For clarity, branch operations are explicitly denoted as `bX` as well as calls to function `foo()` which are separated from the basic blocks (`BBX`). Branch edges are labeled as `T` (true) and `F` (false). A branch bitstream is

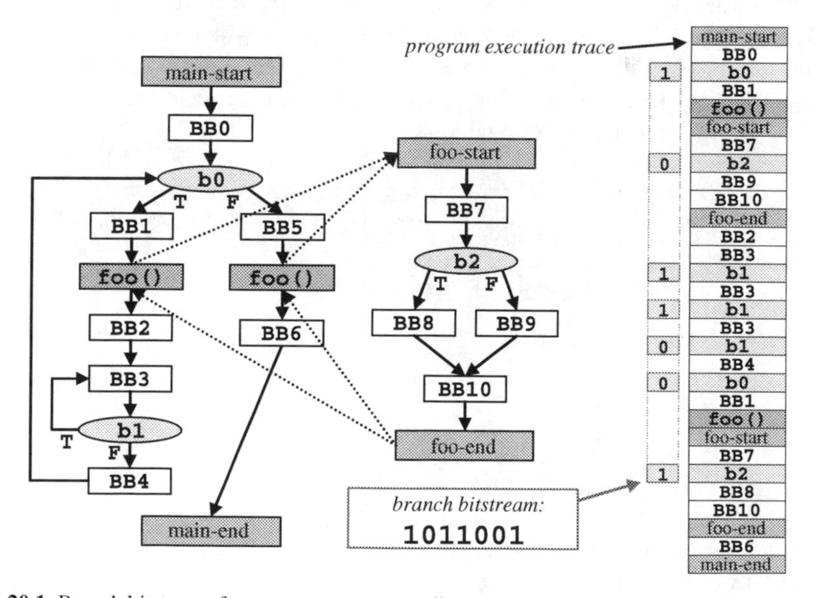

Fig. 20.1 Branch bitstream for program trace encoding

generated by simply recording the branch outcome (true or false) with a single bit in the order of program execution. On the program execution trace shown on the right side of Fig. 20.1, its corresponding branch bitstream is `1011001`. The first `1` indicates that b0 branched in the `T` path (which reaches the call `foo()`), the next `0` indicates that `b2` inside `foo` branched in the `F` path (then returns to `main`), the following `110` indicates that `b1` branched in the `T` path twice and then finally to the `F` path, the next `0` indicates that `b0` branched in the `F` path (which reaches the second call `foo()`), and the last `1` indicates that `b2` inside `foo` branched in the `T` path (then returns to `main`).

Although no information about the correspondence between the branch condition bits and branch operations is provided, this simple branch bitstream is sufficient for representing the *complete program execution trace*. Decoding the program trace from the branch bitstream is done by simply traversing the ICFG and reading 1 bit at a time upon reaching a branch operation to determine which branch path to traverse. Function calls can be handled by maintaining the call stack during the graph traversal. This capability of encoding the complete program trace further implies that even the *entire instruction trace can be completely reproduced from the branch bitstream* if the CFG accurately represents the control-flow structure of the actual target binary code.

Generation of such branch bitstream is done by a simple *source-level instrumentation* performed on the application source code using macro call at each branch operation to record the branch condition bits to a file (Fig. 20.2). The instrumented code is then compiled and executed on the host machine for fast native code execution. This instrumentation is immune from any code optimization during native compilation since the instrumented code section includes global side-effects whose execution order needs to be preserved, thus guaranteeing a correct order of branch condition bit sequence. On the other hand, the decision of which branch operation to instrument must be carried out in accordance to the actual binary code of the target processor, since the structure of the program graph must reflect the target binary code structure for precise performance estimation.

```
if(a > 0){…}                      if(_BC_(a > 0)){…}
else if(a < 0 || b){…}       =>   else if(_BC_(a < 0) || _BC_(b)){…}
c = (d != 0) ? c / d : 0;         c = _BC_(d != 0) ? c / d : 0;

#define _BC_(exp) (EMIT_BIT(exp), exp)
```

Fig. 20.2 Source-level instrumentation for branch bitstream generation

20.2.2 Program Trace Graph

For decoding the branch bitstreams more efficiently, we transform the ICFG in Fig. 20.1 to a *program trace graph* (PTG) as shown in Fig. 20.3. PTG is derived by degenerating the ICFG at the basic blocks while preserving the branch nodes and

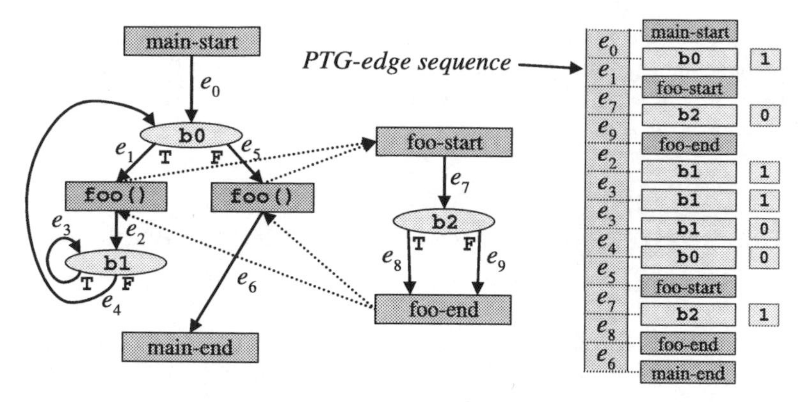

Fig. 20.3 Program trace graph (PTG)

call nodes. As illustrated in Fig. 20.3, PTG acts as a translator between the branch bitstream and the program trace in the form of *PTG-edge sequence.*

A *PTG-node* represents *function-start*, *function-end*, *branch,* or *call*, where each of these PTG-nodes corresponds to a location in the target binary code. A *PTG-edge* carries attributes about the program execution information, including *cycle count*, between the two PTG-nodes it connects to.

Accurate cycle count on the code segment represented by a PTG-edge is obtained by the code generator backend for the target processor in our compiler framework which is equipped with the full knowledge of its pipeline architecture to precisely predict pipeline hazards. This timing annotation can also be accomplished with a back-annotation tool framework as presented in [3]. Operations with data-dependent latencies are pre-characterized through additional source-level instrumentation for calculating the average latencies during the native code execution that are back-annotated in the PTG-edge cycle counts. An example of such data-dependent latency operation in our target processor is an integer division unit whose latency is determined by the difference in the number of significant bits between the two operands.

From the PTG-edge sequence ($e_0\ e_1\ e_7\ e_9\ e_2\ e_3\ e_3\ e_4\ e_5\ e_7\ e_8\ e_6$) in Fig. 20.3, the total program execution cycle count is calculated by accumulating the cycle count on this PTG-edge sequence as

$$\begin{aligned} T &= c(e_0) + c(e_1) + c(e_7) + c(e_9) + c(e_2) + c(e_3) \\ &\quad + c(e_3) + c(e_4) + c(e_5) + c(e_7) + c(e_8) + c(e_6)', \end{aligned} \tag{20.1}$$

where $c(e_i)$ is the cycle count on PTG-edge e_i. Another way of calculating the total cycle count is by counting the *occurrence* of each edge e_i in the PTG-edge sequence (denoted as $n(e_i)$) and accumulating the product of edge cycle count and its occurrence:

$$T = \sum c(e_i) \cdot n(e_i). \tag{20.2}$$

PTG-edge attribute (cycle count) is a very fine-grain workload model of the code segment that can be automatically and precisely calculated, and the process of translating the branch bitstream into PTG-edge sequence can be viewed as the *workload simulation* on a given application and input stimulus.

20.3 Coarse-Grain Workload Generation Using Graph Reduction Techniques

In this section, we give a simple graph reduction technique on the PTG for obtaining a *coarse-grain workload model* that is precise in terms of average cycle counts. The main purpose of this graph reduction technique is to reduce the size of the branch bitstream which will further speed up the cycle estimation calculation. Here, it is assumed that the branch bitstream obtained from the native execution is pre-scanned to obtain the total occurrences $n(e_i)$ on the entire PTG-edge sequence.

A *single-entry single-exit* (SESE) *region* is a distinct CFG substructure with entry- and exit-edge pair [12]. SESE regions on the PTG can be defined in the same manner as stated below:

- A *PTG-SESE region* $R(e_{\text{in}}, e_{\text{out}})$ is enclosed by its entry-edge e_{in} and exit-edge e_{out}, where e_{in} dominates e_{out}, e_{out} postdominates e_{in}, and e_{in} and e_{out} are cycle-equivalent [12].
- An *interior edge set* $E(e_{\text{in}}, e_{\text{out}})$ of PTG-SESE region $R(e_{\text{in}}, e_{\text{out}})$ is a set of PTG-edges that are each dominated by e_{in} and postdominated by e_{out}. By convention, an edge dominates and postdominates itself and thus $E(e_{\text{in}}, e_{\text{out}})$ contains both e_{in} and e_{out}.

One issue arises when dealing with SESE regions on the PTG; that is, SESE region's entry-edge and exit-edge on the CFG may be eliminated in the PTG since control-merge nodes are collapsed in the PTG, such as the case of if-else block's exit-edge inside function `foo` in Fig. 20.3. This issue is handled by simply restoring these CFG control-merge nodes on the PTG as shown in Fig. 20.4b. In general, the out-edge of *function-start* node and the in-edge of *function-end* node enclose the outermost SESE region of a function.

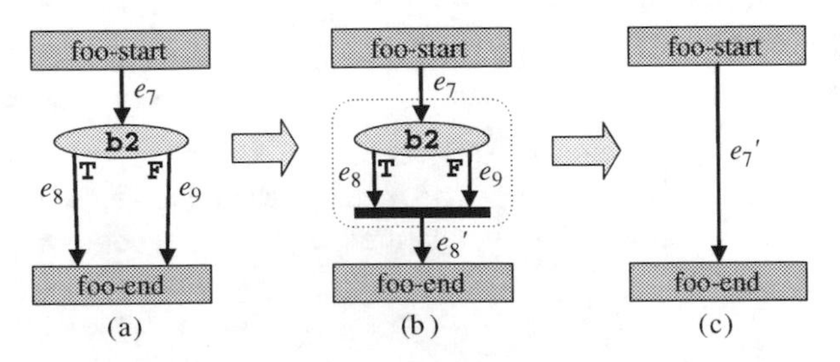

Fig. 20.4 PTG-SESE region reduction without call nodes

Here, we will first consider the PTG-SESE regions which do not contain any call nodes. What follows is that the precise average workload of such PTG-SESE region $R(e_{\text{in}}, e_{\text{out}})$ can be calculated by accumulating the total workload of its interior edge set $E(e_{\text{in}}, e_{\text{out}})$ and dividing it by the occurrence of its entry-edge e_{in}:

$$c(R(e_{\text{in}}, e_{\text{out}})) = \frac{1}{n(e_{\text{in}})} \sum_{e_i \in E(e_{\text{in}}, e_{\text{out}})} c(e_i) \cdot n(e_i). \tag{20.3}$$

In Fig. 20.4b, the PTG-SESE region $R(e_7, e_8')$ (covering the entire function `foo`) can be replaced with a single PTG-edge e_7', where $c(e_7') = (c(e_7) \cdot n(e_7) + c(e_8) \cdot n(e_8) + c(e_9) \cdot n(e_9))/n(e_7)$.

The average workload of non-recursive call nodes, on the other hand, can be calculated in bottom-up fashion on the call graph, starting from the leaf functions where equation 20.3 can be directly applied, and back-annotating the function's average workload at each of its call sites. Figure 20.5 illustrates this process on the main's PTG in Fig. 20.3, where the interprocedural edges to `foo`'s PTG are removed and the out-edges of these call nodes are back-annotated with the average workload of `foo` such that $c(e_2') = c(e_7') + c(e_2)$ and $c(e_6') = c(e_7') + c(e_6)$. Then, the outermost PTG-SESE region $R(e_0, e_6')$ can be reduced to a single PTG-edge e_0' whose cycle count $c(e_0')$ can be calculated by eq. (20.3).

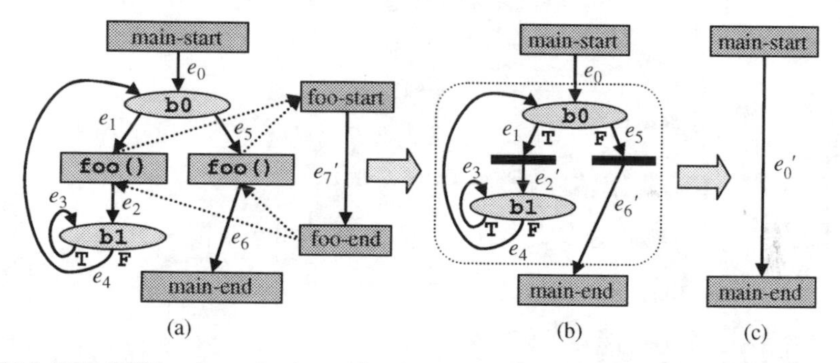

Fig. 20.5 PTG-SESE region reduction with call nodes

The significance of eq. (20.3) is that this simple formulation is applicable to any PTG-SESE region (without recursive calls) with arbitrary complex control flow structures. Since every function contains at least one PTG-SESE region, this also implies that the entire application program can always be reduced to a single PTG-edge, given that there are no recursive calls and that the PTG-edge cycle counts alone are sufficient for estimating the cycle count of the system. The latter assumption holds true for single-processor systems without any dynamic timing effects (such as caches) where the reduced PTG-edge contains the exact cycle count, whereas for MPSoC systems, we need to additionally consider the issues of system synchronization as discussed next.

20.4 MPSoC Software Behavior Modeling

Let us now move our focus to modeling the software behavior on the MPSoC environment, where various MPSoC components such as processors, busses, and channels affect the timing behavior of one another. While our PTG workload models driven by branch bitstream maintain the processor simulation clocks, these workloads require additional coordination by the system simulator kernel for controlling the workload execution which can be blocked during processor communications due to bus conflicts and buffer conditions or activated when such blocking conditions are resolved. In this section, we first address the MPSoC software execution model and then describe how the execution model is incorporated in our PTG workload model.

20.4.1 Tightly Coupled Thread (TCT) Model for MPSoC Application Development

The MPSoC application model employed in our simulation framework is based on the tightly coupled thread model (TCT model) [13, 14] that allows a simple coding style on the sequential C code for specifying "threads" without the need to specify any other parallelization directives, such as concurrency, synchronization, and communication, since they are all implicit in the TCT model (Fig. 20.6). Thread annotation on the C code is done by declaring thread scopes in the form:

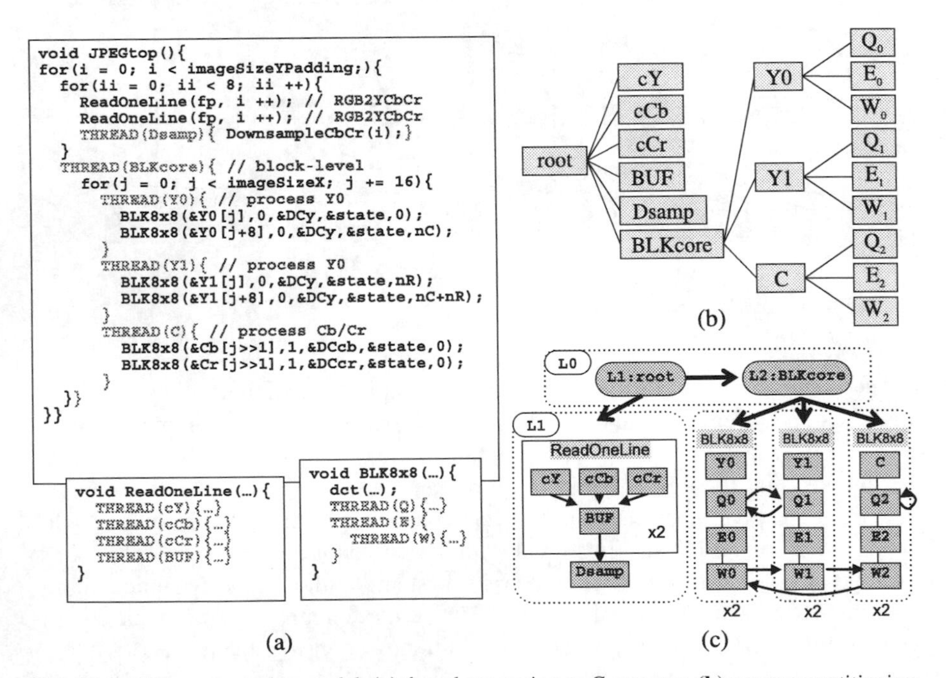

Fig. 20.6 TCT programming model: (**a**) thread annotation on C program, (**b**) program partitioning, and (**c**) hierarchical functional pipeline structure

```
THREAD(name){...}
```

where any C statements (including function calls and even nested thread scopes) can be included inside the thread-scope region as long as the thread scope forms a SESE region. Thread annotations can be inserted manually or through the MAPS framework [15] which assists the designer with rich program analysis capabilities to emit the thread annotated code semi-automatically. The TCT compiler performs a complete interprocedural dependence analysis to extract all data dependences between threads (including globals and pointer dereferences) and inserts message-passing instructions automatically (current TCT compiler assumes a fully distributed memory system without any shared-memory accesses). These tightly coupled threads operate in a functional pipeline manner, achieving high degree of parallelism. Also, the TCT compiler guarantees that the behavior of the parallelized code and the original sequential code is identical where race condition and deadlocks are automatically avoided. The TCT framework has been verified on a prototype MPSoC silicon (TCT-MPSoC) consisting of a six-processor array block where each processing element contains a dedicated communication module that executes the TCT communication protocol through a full-crossbar interconnect [14].

20.4.2 Thread Program Trace Graph (T-PTG)

A *thread program trace graph* (T-PTG) corresponds to the PTG that is enclosed by the SESE thread-scope region (Fig. 20.7). Each T-PTG is terminated by *thread-start* and *thread-end* nodes instead of *function-start* and *function-end* nodes in the normal PTG. In addition, three types of inter-processor communication instructions generated by the TCT compiler also appear in the T-PTG:

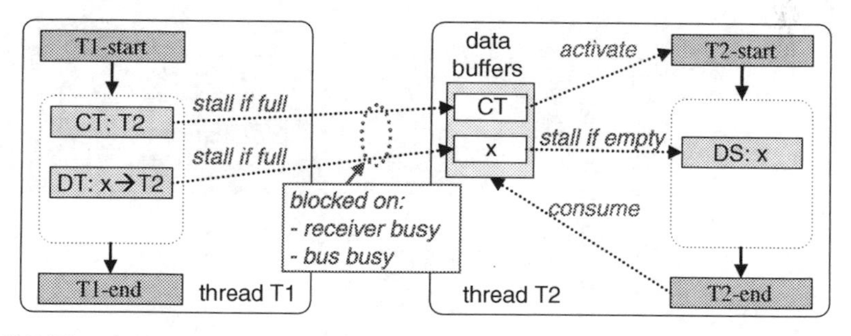

Fig. 20.7 Thread program trace graph (T-PTG)

- *Control token* (CT) sends an activation token to the target thread
- *Data transfer* (DT) sends data to the target thread
- *Data synchronization* (DS) checks if the data is received

Here, DT-node does not require a matching receives operation at the receiver end, and thus the data transfer occurs asynchronously with the receiver thread state. A finite-size buffer is allocated for individual data at the receiver, and the DT-node stalls while the receiver buffer for that particular data is full, where as DS-node stalls while the data buffer is empty. CT-node is modeled the same as DT-node in terms of buffer model, where it too stalls if the receiver's control token buffer is full. Data buffers and control token buffers are consumed at thread-end node at the receiver thread. Additionally, DT-nodes and CT-nodes are blocked if the receiver is occupied by other transactions or if the bus is occupied.

Above CT, DT, and DS nodes along with *thread-start* and *thread-end* nodes represent the software execution behavior that affect the states of MPSoC components including workloads of other threads. These nodes are designated as a new PTG-node class called *PTG-sync-nodes* that serve as anchor points for the simulator kernel to update MPSoC states such as busses and data buffers and, if needed, adjust its simulation clocks by inserting wait states on blocked events.

Consequently, these PTG-sync-nodes must not be removed during the PTG reduction process, as this will alter the sequence of system-level synchronization events. Therefore, we do not allow reductions to the PTG-SESE regions that contain PTG-sync-nodes.

20.4.3 MPSoC Workload Simulator Kernel

Our MPSoC trace-driven workload simulator kernel consists of two major components: thread workload simulator and global scheduler (Fig. 20.8). Thread workload simulator is responsible for updating the local processor clock by the method

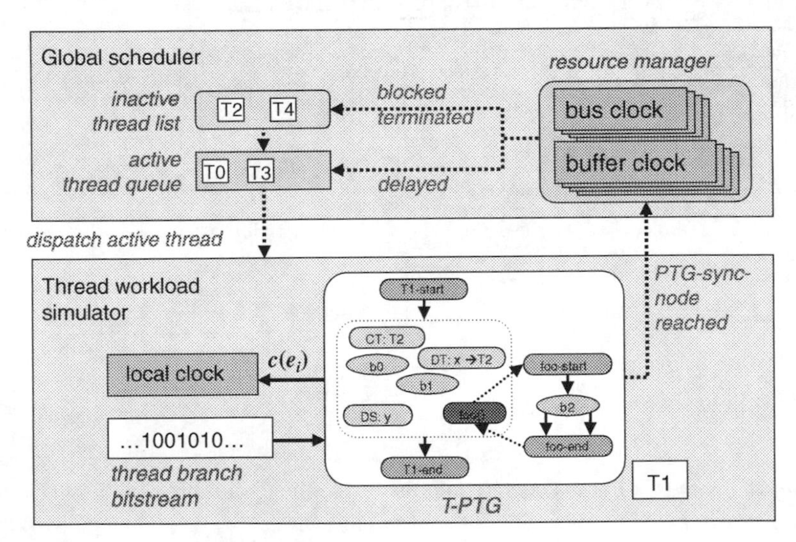

Fig. 20.8 MPSoC trace-driven workload simulator

described in the previous section (translating the thread branch bitstream into PTG-edge sequence and accumulating the edge cycles). Once activated, the thread workload simulator continues to increment its local processor clock until the generated PTG-edge sequence reaches a PTG-sync-node (CT, DT, DS, *thread-start*, and *thread-end* nodes) which requires a strict time-ordered simulation by the global scheduler. The global scheduler activates the thread trace simulator with the earliest local clock by maintaining a queue of active threads. Mutually exclusive resources (buses and data buffers) also maintain their own clocks to indicate their occupancy status during simulation.

20.5 Trace-Driven MPSoC Workload Simulation Framework

Figure 20.9 illustrates the overall flow of our MPSoC trace-driven workload simulation framework incorporating the methods explained in the previous sections. Below summarizes the basic steps for generating the MPSoC trace-driven workload model in conjunction with the TCT model.

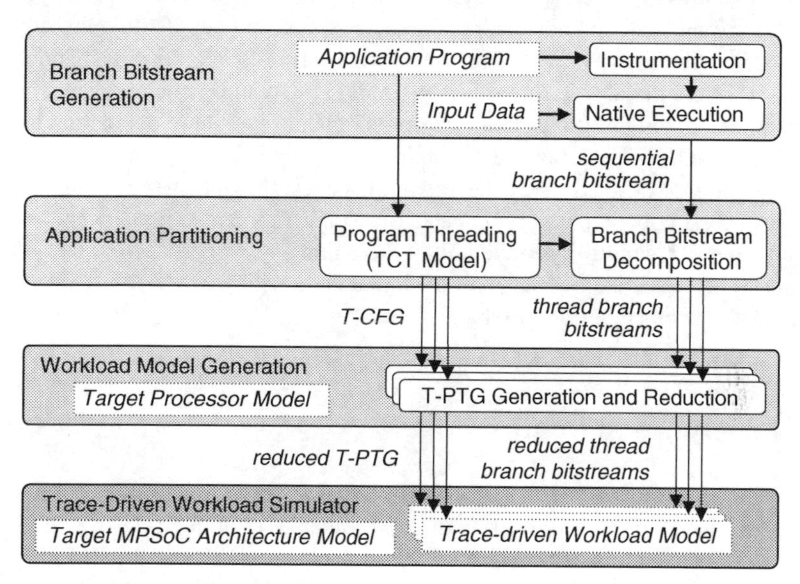

Fig. 20.9 MPSoC trace-driven workload simulation framework

1. Branch bitstream generation is performed on the original sequential code, as described previously, taking full advantage of the fast native code execution speed. In addition, this initial phase is decoupled with the application partitioning, which is very attractive in the sense that a single branch bitstream can drive multiple workload models derived from different application partitioning.
2. Application partitioning corresponds to the task of inserting the thread annotation to the original sequential code. TCT compiler partitions the thread-annotated

code with inter-processor communication instructions into separate CFGs for each thread (T-CFG). At the same time, the sequential branch bitstream is decomposed into separate thread branch bitstreams which is a straightforward task of scanning the sequential branch bitstream and redirecting it into different thread branch bitstream files according to the thread attribute on each branch node.

3. Workload model generation is performed separately on each T-CFG to obtain the T-PTG then undergoes the T-PTG reduction process, deriving the reduced T-PTG along with the reduced thread branch bitstreams.

MPSoC architecture model in our simulation framework is configurable on parameters explained below.

- *Processor type*: Currently supported processor model (TCT-PE) [18] is a simple four-stage RISC with 32 registers and separate memory ports for instruction and data.
- *Thread-to-processor mapping*: Each thread is statically allocated to one processor, where multiple threads can be allocated to each processor (this feature was not part of the prototype TCT-MPSoC, where TCT-PE could only handle one thread). The simulator kernel directly handles the context switching of threads within the processors using non-preemptive scheduling where the parameterized context switching overhead is also modeled.
- *Bus topology*: MPSoC interconnect is modeled as a parameterized multi-bus. At each processor, its output port connects to a single bus and its input port connects to all buses. This multi-bus model guarantees that any two processors have direct connectivity, which allows us to model a relatively wide variety of bus topologies from a single bus to a full crossbar.

20.6 Experimental Results

To validate the effectiveness of our proposed trace-driven workload simulation method, we have conducted a series of experiments on single-processor model as well as MPSoC models and compared them against the ISS for the TCT-PE. All simulations were executed on 3.4 GHz Intel Xeon machine with 3.25 GB memory.

The TCT framework includes a cycle-accurate instruction-set simulator (ISS) for the TCT-processor (TCT-PE). This ISS also has the capability to perform cycle-accurate multiprocessor simulations on an arbitrary number of processors that are connected by full-crossbar switch. For processor communication, data transfer setup time is 2–6 cycles, and transfer rate is 4 bytes/cycle, and the parallel-ISS accurately simulates all bus contentions due to simultaneous data transfer requests to the same destination processor. As this parallel-ISS was initially developed for the prototype TCT-MPSoC, this parallel-ISS cannot directly simulate the "multi-threaded" behavior on the processor.

20.6.1 Single-Processor Cycle Estimation

Table 20.1 shows the set of benchmarks used to evaluate the cycle estimation performance for the single-processor model. Here, "# inst" is the number of *TCT-PE instructions* in each program, "# cycles" is the total execution cycles, "ISS (s)" is the ISS simulation time, "ISS (cycles/s)" is the ISS simulation speed, and "native (s)" is the native execution time (all applications were compiled with Microsoft Visual C++ 2003 with full optimization).

Table 20.1 Benchmark programs for single-processor cycle estimation

	# inst	# Cycles	ISS (s)	ISS (cycles/s)	Native (s)
Prime500 k	100	629,940,586	13.328	47.26 M	0.083
String	273	900,232,697	31.266	28.79 M	0.172
Alphabet	113	716,650,704	25.328	28.29 M	0.193
JPEG	1,673	433,010,644	16.829	25.73 M	0.076

Table 20.2 shows our trace-driven workload simulator (TWS) performance without PTG reduction. Here, "# PTG-edges" is the total number of PTG-edges, "# branch bits" is the length of the branch bitstreams, "TWS (s)" is the TWS simulation time, "TWS (cycles/s)" is the effective TWS simulation speed, and "TWS speedup" is the speedup over ISS. Even without the PTG reduction, we can observe the enormous speed advantage of our TWS over ISS, where the effective TWS simulation speed ranges from 1.7 to 3.0 billion cycles/s, while achieving absolute cycle accuracy, that is, TWS cycle counts match *exactly* with the ISS cycle counts.

Table 20.2 Trace-driven workload simulator (TWS) performance (without PTG reduction)

	# PTG- edges	# Branch bits	TWS (s)	TWS (cycles/s)	TWS speedup
Prime500 k	29	53.99 M	0.218	2,889 M	61.14
String	52	136.18 M	0.515	1,748 M	60.71
Alphabet	21	101.04 M	0.375	1,911 M	67.54
JPEG	336	27.48 M	0.141	3,071 M	119.35

Figure 20.10 shows the processing times inside the TWS workflow that are normalized against the native execution time in Table 20.1. Here, "BB-gen" is the native execution time of the applications instrumented with branch bitstream generation, and "BB-gen+div" is the execution time with branch bitstream generation and integer division latency profiling. We can observe that the instrumentation overhead for the branch bitstream generation is quite small, ranging from 1.40 to 2.25, and even with the division latency profiling, the overhead increases only up to 2.75. "BB-scan" is the processing time of enumerating the PTG-edge occurrences from the generated branch bitstreams that is performed prior to the PTG reduction. As explained in Section 20.3, each of these applications in this single-processor case reduces to a single PTG-edge whose edge cycle match exactly with the ISS cycle count.

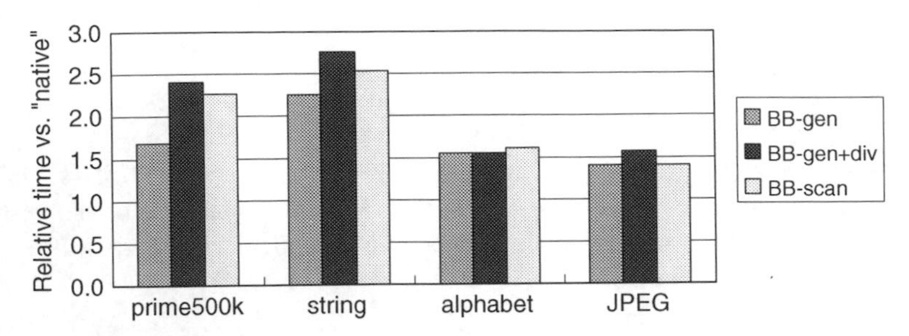

Fig. 20.10 Relative processing times against native execution

20.6.2 MPSoC Cycle Estimation

For evaluating the MPSoC cycle estimation performance, we have used the previous two applications, "string" with 9 threads and "JPEG" with 19 threads as shown in Table 20.3. Here, "# cycles" is the total execution cycles on the parallelized code, "para-speedup" is the speedup ratio from the sequential code, and "para-ISS (s)" is the parallel-ISS time for the parallelized code. As explained previously, current parallel-ISS can only handle one thread per processor, therefore it models a 9-PE MPSoC for "string" and a 19-PE MPSoC for "JPEG."

Table 20.3 Benchmark programs for MPSoC cycle estimation

	# Cycles	Para-speedup	Para-ISS (s)
String (9 threads)	118,581,455	7.59	51.281
JPEG (19 threads)	41,720,835	10.38	37.219

Table 20.4 shows the TWS simulation results for the target MPSoCs on the two applications, with and without PTG reduction. Here, the architecture parameters are set according to the prototype TCT-MPSoC. Unlike the single-processor cases, the MPSoC estimated cycles do not match exactly with the parallel-ISS cycles, although the estimation errors are well within 1%, primarily due to the simplistic communication model in our TWS which assumes a fixed setup time for all data transfers, whereas a precise communication hand-shaking is modeled in the parallel-ISS. The effects of the PTG reduction can be observed on the dramatic reductions in branch bitstream length, while the total cycles tend to decrease slightly since the coarse-grain workloads obtained by the PTG reduction have the effect of averaging out the temporal fluctuations of the fine-grain workloads before PTG reduction.

Table 20.4 TWS simulation results with and without PTG reduction

	# Branch bits	Est. # cycles	Est. error	TWS (s)	TWS speedup
String	136.18 M	118,893,200	0.263%	0.731	70.15
String (reduction)	41,946	118,562,675	0.016%	0.265	193.51
JPEG	27.48 M	41,745,317	0.059%	0.312	119.29
JPEG (reduction)	8,330	41,529,311	0.459%	0.156	238.58

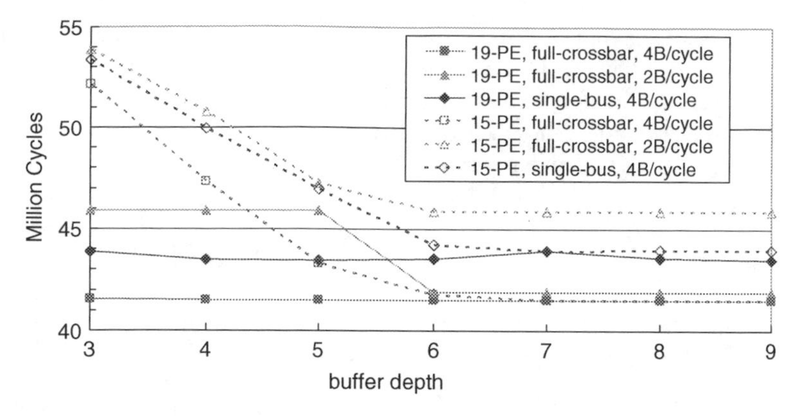

Fig. 20.11 TWS simulations on various MPSoC parameters (JPEG)

Figure 20.11 shows the simulation results on "JPEG" with PTG reduction enabled under different MPSoC parameters. A 19-thread to 15-PE mapping was derived by selecting six threads with low CPU utilization in the previous 19-PE simulation and merging them into two groups (three threads in each group). In addition to these PE configurations (19-PE and 15-PE), other MPSoC parameters such as bus topologies (full crossbar and single-bus), data transfer rates (4 bytes/cycle and 2 bytes/cycle), and buffer depth (3–9 data elements at each buffer) are altogether swept on our TWS framework from the same set of coarse-grain workload models and reduced branch bitstreams. The ability to quickly evaluate these MPSoC parameters (each only taking a fraction of a second) makes our TWS framework ideal for MPSoC design exploration.

20.7 Conclusion

In this chapter, we have introduced an approach based on trace-driven workload models for early software performance estimation of MPSoC architectures. The combined techniques of program trace encoding using branch bitstreams, program trace graph representation, and its reduction techniques for workload modeling of various granularities, as well as the workload coordination scheme for modeling processor communications can achieve considerable speedup in simulation time while sustaining high cycle accuracy. A single sequential branch bitstream generated through code instrumentation and fast native-code execution can drive a variety of workload models, each corresponding to different application partitioning and different MPSoC architectures, thus establishing a practical MPSoC design exploration framework. Due to its simplicity, this trace-driven workload model approach for software performance estimation can be fairly easily extended to other processor types (given an appropriate cycle estimation technique of the target instruction set) and MPSoC interconnect models. For more complex processor architectures with out-of-order executions and caches, their timing characteristics can change dynamically, thus making our timing model generation at compile time less reliable. Various performance estimation techniques for these complex processor cases have been addressed in Chapters 7–12. It will be interesting to see how these techniques can be incorporated into the trace-driven workload to address wider processor classes for a more effective solution.

References

1. Shukla, S.K., Pixley, C., Smith, G.: The true state of the art of ESL design. *IEEE Design Test Comput* **23**(5), 335–337, (2006).
2. Lazarescu, M.T., Bammi, J.R., Harcourt, E., Lavagno, L., Lajolo, M.: Compilation-based software performance estimation for system level design. In: *Proceedings of HLDVT'00*, pp. 167–172, (2000).
3. Schnerr, J., Bringmann, O., Viehl, A., Rosenstiel, W.: High-performance timing simulation of embedded software. In: *Proceedings of DAC 2008*, pp. 290–295, (2008).
4. Wang, Z., Herkersdorf, A.: An efficient approach for system-level timing simulation of compiler-optimized embedded software. In: *Proceedings of DAC*, (2009).
5. Kempf, T., Karuri, K., Wallentowitz, S., Ascheid, G., Leupers, R., Meyr, H.: A SW performance estimation framework for early system-level-design using fine-grained instrumentation. In: *Proceedings of DATE '06*, pp. 468–473, (2006).
6. Meyerowitz, T., Vincentelli, A.S., Sauermann, M., Langen, D.: Source-level timing annotation and simulation for a heterogeneous multiprocessor. In: *Proceedings of DATE'08*, pp. 276–279, (2008).
7. Chang, N.Y., Lee, K.B., Jen, C.W.: Trace-path analysis and performance estimation for multimedia application in embedded systems. In: *Proceedings of ISCAS '04*, vol. II, pp. 129–132, (2004).
8. Giorgi, R., Prete, C.A., Prina, G., Ricciardi, L.: A workload generation environment for trace-driven simulation of shared-bus multiprocessors. In: *Proceedings of 30th HICSS*, pp. 266–275, (1997).
9. Madsen, J., et al.: Network-on-chip modeling for system-level multiprocessor simulation. In: *Proceedings of 24th RTSS03*, pp. 82–92, (2003).
10. Thiele, L., Chakraborty, S., Gries, M., Kunzli, S.: A framework for evaluating design tradeoffs in packet processing architectures. In: *Proceedings of DAC 2002*, pp. 880–885, (2002).
11. Isshiki, T., Li, D., Kunieda, H., Isomura, T., Satou, K.: Trace-driven workload simulation method for multiprocessor system-on-chips. In: *Proceedings of DAC 2009*, (2009).
12. Johnson, R., Pearson, D., Pingali, K.: The program structure tree: Computing control regions in linear time. In: *Proceedings of ACM SIGPLAN '94*, pp. 171–185, (1994).
13. Urfianto, M.Z., Isshiki, T., Khan, A.U., Li, D., Kunieda, H.: Decomposition of task-level concurrency on C programs applied to the design of multiprocessor SoC. *IEICE Trans* **91-A**(7), 1748–1756, (2008).
14. Urfianto, M.Z., Isshiki, T., Khan, A.U., Li, D., Kunieda, H.: A multiprocessor SoC architecture with efficient communication infrastructure and advanced compiler support for easy application development. *IEICE Trans* **91-A**(4), 1185–1196, (2008).
15. Ceng, J., Castrillon, J., Sheng, W., Scharwächter, H., Leupers, R., Ascheid, G., Meyr, H., Isshiki, T., Kunieda, H.: MAPS: An integrated framework for MPSoC application parallelization. In: *Proceedings of DAC 2008*, pp. 754–759, (2008).

Index

R. Leupers, O. Temam (eds.), *Processor and System-on-Chip Simulation*,
DOI 10.1007/978-1-4419-6175-4,